# Modality and Propositional Attitudes

This book shows that the semantic analysis of modal notions of possibility and necessity can be used to enhance our understanding of the interpretation of reports of belief or emotional state. It introduces intuitive notation and terminology to express ideas in modern theories of modal interpretation that are normally represented in complex logical formulas, effectively updates the 1960s-era link between possible worlds and the semantics of propositional attitude ascriptions, and reconciles two disparate views of the role of events in semantic interpretation: that of Donald Davidson and that of David Lewis. It reduces a host of variable behaviors of propositional attitude ascription to an intuitive and precise distinction between ascriptions that merely express a commitment to propositional content versus ones that attribute a mental state to the holder of the propositional attitude. This leads to an explanation of the nature and effects of the language disorder of fluent aphasia.

MICHAEL HEGARTY was formerly an Associate Professor in the Linguistics Program at Louisiana State University, where he was the recipient of the LSU Alumni Faculty Excellence Award in 2013. He is currently a Test Specialist at the Law School Admission Council in Newtown, Pennsylvania.

# Modality and Propositional Attitudes

Michael Hegarty
*Law School Admission Council*

CAMBRIDGE
UNIVERSITY PRESS

**CAMBRIDGE**
**UNIVERSITY PRESS**

University Printing House, Cambridge CB2 8BS, United Kingdom

Cambridge University Press is part of the University of Cambridge.

It furthers the University's mission by disseminating knowledge in the pursuit of education, learning and research at the highest international levels of excellence.

www.cambridge.org
Information on this title: www.cambridge.org/9781107085763



First published 2016

*A catalogue record for this publication is available from the British Library*

*Library of Congress Cataloguing in Publication data*
Hegarty, Michael, 1959– author.
Modality and propositional attitudes / Michael Hegarty.
pages cm
Includes bibliographical references and index.
ISBN 978-1-107-08576-3 (hardback)
1. Modality (Linguistics) 2. Semantics–Psychological aspects. 3. Grammar, Comparative and general–Prepositional phrases. I. Title.
P325.5.P78H44 2016
415′.6–dc23 2015027884

ISBN 978-1-107-08576-3 Hardback

# Contents

# Acknowledgments

For comments on previous versions of parts of this work, and discussion of topics addressed here, I am grateful to audiences at various conference and workshop talks I have given dating back to 2006, when I first began working on the topics of this book. I particularly want to thank Nicholas Asher, Lucia Tovena, Pierre Larrivée, and Kai von Fintel. I am grateful to students who participated in my Louisiana State University seminar on topics related to this work, including Danielle Alfandre, Sara Monti, and Jerry Weltman. I particularly thank Kari Swingle and Jerry Weltman for giving detailed comments on the manuscript and for intensive discussion of some of its main points. I am grateful to Ann Mulkern also for detailed comments on the manuscript, and for preparing the index. Hugh Buckingham pointed me to his work and the work of others on jargonaphasia. I am grateful to him for this and for generously sharing his jargonaphasia data. I thank Cambridge University Press reviewers for helping to sharpen many points in the manuscript and for their efforts to help prevent such a large, long-unfolding project from falling too far behind the literature on topics addressed in the work. For financial support I am grateful to the State of Louisiana Board of Regents for a 2010–2011 Award to Louisiana Artists and Scholars, award number LEQSF(2010–11)-RD-ATL-03, and to the College of Humanities and Social Sciences at Louisiana State University (LSU – Baton Rouge) for a semester of paid leave in 2011. In Chapter 1, examples (44) and (45) as well as the figure in (46) are reproduced from Lassiter (2011), for which permission is gratefully acknowledged. The definitions associated with Lassiter's scalar theory are also drawn directly from his work. Some paragraphs of Chapter 5 and Chapter 6 (Section 6.2) are reproduced from my contribution, "Two Types of Belief Report," to the *Baltic International Yearbook of Cognition, Logic and Communication*, vol. 6. Permission to reproduce those paragraphs is also gratefully acknowledged.

# Introduction

A possibility is a way the world could be, including the way it actually is. One way to understand this is that a possibility is a way the world could be (or is) at a certain level of description, abstracting away from irrelevant particulars, so that if I oversleep tomorrow morning, all of the multitudinous ways in which that could happen count as one possibility. Another way to understand the notion that a possibility is a way the world could be (or is) is that it is a possible state of the world in all its particulars. When we say in everyday terms that the conditional statement in (1) addresses the possibility that Alex misses the train, our use of the singular definite description, *the possibility*, invokes the first notion of possibility.

(1) If Alex misses the train, Sam will pick him up.

But we cannot say this under the second conception of possibilities sketched earlier. In a typical context, (1) would not be used to state a contingency plan for a single, fully specified or fully conceived state of the world in which Alex misses the train – say, one in which Alex leaves his hotel fifteen seconds after 8:31 am with plenty of time, but is late because his taxi to the station gets stuck in traffic caused by an overturned pizza truck at the corner of Walnut Street and 18th Avenue, causing Alex to arrive at the station exactly ten minutes and thirty-six seconds after the train departed.[1] Rather, (1) would normally be used to state a contingency plan for the realization of any one of a vast number of possibilities, some of which could be considered variants of the one in which the taxi is stuck in traffic, and some of which would be very different, but all of which could be contextually reasonable ways in which Alex misses the train. The number of such ways the world could be is vast, even if we limit consideration to ones that are realistic and within the bounds of normal expectations. There are a multitude of basic patterns or templates of ways in which Alex could miss the train, each of which could be expressed as a reason he missed the train. "Alex's taxi to the station got held up in traffic" describes one such

[1] Of course, much more detail would have to be added – perhaps infinitely more – to truly describe a unique way the world could be.

pattern. But it could be realized in thousands of different ways, or millions of ways, depending on how finely we count them. Another would be, "Alex forgot his briefcase and had to back-track to his office," which could also be realized in innumerable different ways. Clauses of natural language can serve as descriptions of possibilities, but as such, they characterize vast sets of possibilities, not individual ones.

The present work pursues the consequences of the second view of possibilities in order to elucidate and defend the theory of modal interpretation developed by Angelika Kratzer (1981, 1991), and a kindred theory of the interpretation of ascriptions of beliefs and other propositionally contentful mental states, called propositional attitudes. Possibilities on this view are "first-order entities," that is, each possibility is a singular discrete entity in our conceptual and linguistic ontology, what is often called a possible world. Kratzer's theory involves quantifying over such possibilities based on their accord with a contextually determined set of propositions, which she called an ordering source. In recent years, Kratzer's theory has been argued to be faulty on a number of grounds. Lassiter (2011), in particular, alleges a multitude of logical and empirical faults stemming from its reliance on quantification over possible worlds, and argues for the superiority of a theory that interprets modal statements as expressing relationships involving degrees on a scale. And von Fintel and Iatridou (2008) argue that Kratzer's theory suffers from internal inconsistencies that arise in the case when there exists a set of possibilities that are all maximally in accord with the propositions in the ordering source. (The assumption that this case holds is called the Limit Assumption in modal interpretation.) The present work begins by arguing that, under the conception of possibilities sketched (the second one), most of Lassiter's critiques (which seem to implicitly assume the first one) fail to gain purchase, and the Limit Assumption fails to have the consequences that von Fintel and Iatridou attribute to it. These issues are taken up in Chapters 1 and 2.

Hintikka (1969) proposed that ascriptions of belief and other propositional attitudes are interpreted by quantifying over possibilities consistent with everything the experiencer believes, and a significant literature has adopted that proposal (see especially Heim [1992]). But this approach has a fatal flaw – it breaks down completely in implementation when the experiencer of the propositional attitude holds inconsistent beliefs. The view of possibilities advanced here permits an implementation of Hintikka's idea, using Kratzer's modal quantificational apparatus, which does not break down when beliefs are ascribed to someone with inconsistent beliefs. The approach developed handles the effects of presupposition accommodation in cases of belief ascription discussed by Heim (1992) without Heim's recourse to dynamic semantics for this purpose. These issues are taken up in Chapters 3 and 4, completing Part I of the book.

The interpretation of attitude ascriptions given in Part I does not involve positing any sort of event or state that would correspond to the mental or emotional condition of the experiencer of the attitude. This is fortunate since, while many ascriptions with attitude predicates such as *believe*, *think*, *suspect*, *be sure*, and *expect*, among others, do seem to characterize such an event or state, many don't; they simply describe the doxastic situation of the experiencer – in effect, where in logical space their beliefs or expectations reside. Katz (2000, 2003, 2008) argues that stative sentences, in general, do not characterize goings on or conditions of the world that we could package and reify as events or states. Chapter 5 contends with this claim, arguing that Katz has not convincingly ruled out associated states of the world for all stative attitude ascriptions. Toward developing this argument, Chapter 5 first summarizes a view of the role of events and states in semantics.

Contrary to Katz, Hacquard (2006, 2010) argues that attitude ascriptions are descriptions of localized goings on or conditions of the world, which we can reify as eventualities (events or states), and that the assumption of such an eventuality is a standard part of the conventional semantic interpretation of an ascription. Chapter 6 argues that there are, in fact, two types of attitude ascription. One type is a quantificational condition on possible worlds with no associated eventuality; the other posits an associated eventuality and uses the propositional content, expressed in terms of quantification over possible worlds, to characterize it, in much the way Hacquard's theory does. The rest of Part II is largely devoted to substantiating the claim that stative attitude ascriptions are of two types, with consequences that we can detect in a number of different domains. For example, (2) reports on an event of Alex saying something, and this event is characterized in part by the content of what is said, so that if the content were different, it would not be "the same" saying event.

(2) Alex said that Bill stole the ring.

(3) a. Alex thought that Bill stole the ring.
b. Alex long thought that Bill stole the ring.
c. Alex guiltily thought that Bill stole the ring.

But (3a) is essentially ambiguous. It can report on an event of Alex having a certain opinion over a period of time, with certain causal consequences for Alex's state of mind, and this is the interpretation involved when the main clause is modified by a temporal adverbial as in (3b), or the manner adverb in (3c). In this case, (3a), like (3b,c), reports on an eventuality going on in the mind of Alex. But (3a) has a more minimal interpretation, as one interpretive option, in which it is unlike (2) and (3b,c) in not characterizing an associated eventuality. Chapter 7 argues that that is the interpretation on which negation in (4) can be interpreted with the lower clause (called "Neg-Raising" in the literature).

(4) Alex doesn't think that Bill stole the file.

Chapter 8 explores what might differentiate two types of modal statement, epistemic ones based on what the speaker and addressee know, and root forms of modality based on desires or preferences, abilities or other circumstances of the world, or what the law or convention requires. A proposal by Hacquard (2006, 2010) for syntactic, semantic, and contextual differentiation of the two types of modal statement is examined. This approach comports to a large extent with the views that animate the present work, but with some departures in detail, which will be discussed in that chapter and at various earlier points throughout the book.

Building on the work by Diessel (2004), Chapter 9 argues that the more minimal interpretation of (3a) is acquired during child language development before the eventuality-characterizing interpretation, which requires (and reflects) the development of the child's Theory of Mind. In view of ideas put forth by Nespoulous et al. (1998), Chapter 9 argues that in the clinical condition of fluent jargonaphasia, the minimal interpretation of (3a) survives, but the eventuality-characterizing interpretation is lost, as part of a more general degradation of the capacity to refer consistently to objects, events, and states not present in the immediate environment.

# *Part I*

# Modality and propositional attitudes

Intensional statements are about ways the world could or might be, however it may be in actuality. This is exemplified by modal statements and reports of propositional attitude. Consider modal statements first.

If Kim and Jules walk by a student dorm, and Kim, noticing that Bill's window is lit, utters *Bill might be working in his room*, Kim is not saying that Bill *is* working in his room, nor that Bill *is not* working in his room. On a widely adopted view, she is saying something about possible ways the world could be, namely, that among those in which Bill's window is lit, and things are otherwise relevantly similar to what Kim and Jules know about the world, there are some (at least one) in which Bill is working in his room. If Kim, in the same circumstances, and mindful of Bill's work habits and aversion to wasting electricity, utters *Bill must be working in his room*, she is saying something different, and stronger, about all possible ways the world could be, namely, that Bill is working in his room in all of those possible worlds in which Bill has something like his actual habits, and which are in other relevant respects realistic enough to be worth considering.

On this analysis, the interpretation of modal statements invokes a set of possible ways the world could be, that is, a set of possibilities, consistent with what Kim and Jules know. But what is known to Kim and Jules significantly underdetermines what possibilities are in play, leaving too many of them. Assuming that Kim and Jules haven't seen Bill in a couple of days and have had no confirmation that he is still alive (not that they have any reason to suppose otherwise), they could not rule out the possibility that Bill is dead. So the set of possibilities consistent with what they know includes the ones in which Bill died the day before. These should be somehow excluded from play, if, in fact, Bill *didn't* die the day before, to prevent them from spuriously falsifying Kim's statement that Bill must be in his room when she has the typical sort of evidence for truthfully making that claim. In addition to making reference to the set of all possibilities consistent with what Kim and Jules know, it would be useful to order those possibilities according to their proto-typicality for, or relevance to, the truth of Kim's statements about Bill working in his room. Possibilities consistent with what Kim and Jules know, but in which Bill happened to die the day before, could ordinarily be ruled out of bounds since they involve something so beyond normal expectation that Kim's statement can reasonably be said not to have been about them.

Being indifferently educated, Kim and Jules know something about history, but not a lot. As a result, the set of possibilities consistent with what Kim and Jules know is truly vast. Suppose that all possibilities in which the history of the world unfolded as it actually did, and in which Bill's light is on, are possibilities in which he is working in his room. But suppose at the same time that there are other possibilities in which Bill's light is on, in which a battle during the Wars of the Roses unfolded rather differently than it did, while everything known to Kim and Jules are as they know them to be, but Bill is not working in his room. Suppose that in one of the alternate histories, there is a survivor of the battle who, in actuality, was killed in action before having children. Suppose further that, in this counterfactual world, a descendant of the survivor is responsible for Bill leaving his room with the light on just as Kim and Jules pass by. Then Kim's *must* statement is falsified by this possibility. It's too harsh to say that Kim must be counted wrong about the necessity of Bill working in his room if some such distant change in history would have sufficed to make it not so. And it works the other way as well: if Kim said that Bill might not be in his room, despite the light being on and Bill being frugal in habit, it would be too generous to Kim to say that she was right if the only possibilities in which Bill is not in his room are ones in which a battle during the Wars of the Roses, unbeknownst to Kim, unfolded slightly differently than it did (so that Kim was right only by coincidence, as it were).

For reasons such as this, Angelika Kratzer added a second component to modal interpretation.[1] In addition to delineating the set of possibilities consistent with what the discourse participants know (the modal base), there is a relation that serves to rank possibilities according to how suitable they are for the evaluation of the modal statement, one way or the other. Among possibilities consistent with what Kim and Jules know, those in which Bill is still alive, and the broader history of the world unfolded as it actually did, are "live" possibilities for the evaluation of Kim's statement. Others can be disregarded, so they do not impinge on the truth of Kim's statement in a way that could make it spuriously true or false.

Recognizing that the set of all possibilities consistent with what Kim and Jules know is generally too vast to serve as the correct basis for a modal statement, a ranking of possibilities with respect to their epistemic relevance is one

[1] Kratzer didn't put it this way. She was concerned to capture the notion of the accessibility of possible worlds, from the literature on modal logic, in a linguistically viable and natural way. But the concern is the same; we have merely chosen to enter the discussion here through some examples, rather than through the literature on modal logic.

way of taming that set. We would like to rank possibilities according to their suitability for the evaluation of the modal statement at hand. A natural idea at this point is that we could use the ranking to establish a certain threshold; then, disregarding possibilities that are outside that threshold (too far out to consider), we could evaluate *Bill might be working in his room* as true just in case the "prejacent," *Bill is working in his room*, is true in some possibility better than that threshold, and we could evaluate *Bill must be working in his room* as true just in case the prejacent is true in all possibilities better than the threshold.

But this idea cannot be implemented in such a simple way, since it isn't clear that we could always rank any two arbitrarily chosen possibilities according to whether one is better than the other for the evaluation of a given modal statement. Following Kratzer, define the epistemic modal base to be the set of propositions $\{p_i: i \in \text{I}\}$, characterizing what the speaker knows, for an index set I (which might be a subset of the natural numbers, if the modal base is finite, or at most countably infinite).[2] Each of these propositions is a set of possibilities, specifically, the set of possibilities in which the given proposition is true. So the modal generalizes over possibilities satisfying each of these propositions, which is the set of possibilities $\cap\{p_i: i \in \text{I}\}$. This set is called the set of epistemic alternatives for the epistemic modal statement. The ranking is derived from another set of propositions $\{q_j: j \in \text{J}\}$, called an ordering source, each one a set of possibilities, indexed by an index set J (which typically could also be a subset of the natural numbers). If a possibility $w_1$ satisfies ordering propositions $q_1$, $q_2$, and $q_3$, and another possibility $w_2$ satisfies just $q_1$ and $q_2$, then clearly $w_1$ is better than $w_2$ in providing conditions for the evaluation of the modal statement. Following Kratzer, when $w_1$ is at least as good as $w_2$ for evaluation of the modal, we write $w_1 \leq w_2$, putting the better-or-equal possibility on the "less than or equal to" side of the inequality symbol. In general, the principle is that if $w_1$ satisfies ordering propositions $\{q_j: j \in \text{J}_1\}$, for $\text{J}_1 \subseteq \text{J}$, and $w_2$ satisfies ordering propositions $\{q_j: j \in \text{J}_2\}$, for $\text{J}_2 \subseteq \text{J}$, then $w_1 \leq \text{w}_2$ just in case the set of ordering propositions for $w_1$ includes, as a subset, the set of ordering propositions for $w_2$, thus $\text{J}_2 \subseteq \text{J}_1$. But then if we pick any two possibilities, $w_1$, satisfying ordering propositions $\{q_j: j \in \text{J}_1\}$, and $w_2$, satisfying ordering propositions $\{q_j: j \in \text{J}_2\}$, there won't in general be any subset relation between the two sets of ordering propositions (e.g., $\{q_1, q_2, q_3\}$ and $\{q_3, q_4, q_5\}$). In that case, it is not true that either $w_1 \leq w_2$ or $w_2 \leq w_1$, so $w_1$ and $w_2$ are unranked with respect to one another. The relation $\leq$ defined in this way is not a complete ordering of any kind, since it is, in general, only defined on some pairs of possibilities.

And this means we can't adopt the idea mentioned earlier, where we simply take a candidate threshold of suitability, $w_1$, and say that *Bill must be working*

[2] For example, if $\text{I} = \{1, 2, \ldots, 27\}$, then the epistemic modal base is $\{p_1, p_2, \ldots, p_{27}\}$.

*in his room* is true just in case the prejacent expressed by *Bill is working in his room* is true in all possibilities in the epistemic alternative set that are ≤ $w_1$. For this wouldn't preclude the existence of another possibility in the alternative set, $w_2$, such that the prejacent is false on lots of possibilities ≤ $w_2$, however good they get for the evaluation of the modal statement. Our outcome for evaluating the truth of the modal statement would depend on our starting point within the set of epistemic alternatives. Instead, we have to step back and say that we get this sort of behavior for any candidate starting point $u$ that we pick, or if not at that point, then for some $v \leq u$. That is, if we pick candidate $u$ at random, then there is some point $v$ "inward of $u$" ($v \leq u$) that serves as a threshold in the simple sense we started with, namely, that the prejacent, *Bill is working in his room*, is uniformly true in possibilities ≤ $v$. This more elaborate approach doesn't assume that any two possibilities in the alternative set are ordered by ≤, but it does assert that when we get "far enough down on" ≤, and a possibility is ordered ≤ any candidate threshold, then the prejacent is true. In effect, the prejacent is true at progressively more possibilities as the rank of possibilities improves according to ≤.

But the more elaborate approach raises delicate questions of its own, which go to the heart of the question of how dense is the set of possibilities in the epistemic alternative set with respect to the order ≤. As we take possibilities that get progressively better in rank according to ≤, will they converge to a set of best possibilities, or will they continue to get progressively better, without end? The former supposition is called the Limit Assumption – that there is a set of best possibilities, whereby possibilities converge to an ideal set, members of which satisfy all propositions in the ordering source $\{q_j: j \in \mathrm{J}\}$. The Limit Assumption would permit a fairly simple formulation of the truth conditions of modal statements.

One of the distinguishing features of the present work is that this simplifying assumption is not made. For reasons given in the overall introduction, it seems to reflect a substantive claim about possibilities that we don't generally have reason to believe is true. The set of propositions in the ordering source does not have to be finite. It could be countably infinite, as we numerate conditions that a possibility must satisfy in order to be "in play" for the evaluation of a modal statement. In certain mathematical contexts, it is even imaginable that there are nondenumerably many propositions in the ordering source. In such a case, possibilities in the epistemic alternative set could get progressively better, with the set of possibilities that satisfy all propositions in the ordering source getting smaller and smaller, without ever reaching a stable set whose members meet all propositions in the ordering source.

The idea at play here, and the needs it is designed to meet, may seem rather abstruse – that we use an elaborate quantificational apparatus to interpret modal statements in order to obtain correct interpretations, even in the special case when the possibilities to consider keep getting progressively better for the evaluation of the modal statement. But this idea is the one we arrive at, following Kratzer, when we take seriously the idea that modal statements are really about all relevant possibilities. Furthermore, it turns out that this quantificational apparatus helps to meet several challenges that Lassiter (2011) and others pose to any theory of modal interpretation based on possible worlds. With these challenges met, we retain the apparatus of possible worlds to account for the sort of phenomena addressed by Kratzer, while also allowing, in a spirit of semantic pluralism, for Lassiter's preferred account of modal interpretation, based on mapping to scales and providing for a ready interface with numerical probability, to be retained for those purposes.

It also turns out that Kratzer's quantificational schemas, while accounting for what might seem to be abstruse problems in modal interpretation, play an even more critical role by rescuing the interpretation of propositional attitude ascriptions when they come up against real world conditions in which people have inconsistent beliefs, as will be previewed next.

It is common to analyze belief reports and other ascriptions of propositional attitudes as involving quantification over a set of possibilities. Consider the belief report, *Kim believes that Bill is working in his room.* On the view in question, this report makes a claim about all possibilities consistent with what Kim believes. Namely, it asserts that, in each of those possibilities, Bill is working in his room. The terminology used is parallel to that for modality. The set of all propositions characterizing what Kim believes is called Kim's doxastic base, and the set of all possibilities consistent with what Kim believes is called Kim's doxastic *context* or set of doxastic alternatives.

But just as in the evaluation of modal statements, we would want to restrict this quantification to accessible possibilities – those possibilities that are realistic enough to be considered "live alternatives," or to be considered "in play" for the evaluation of the ascription. It may be consistent with what Kim believes that aliens have abducted Bill so that Bill is no longer in his room (she may be somewhat skeptical of alien abduction, but not totally disbelieve in it), but ordinarily the existence of this far-fetched possibility wouldn't falsify the report that Kim believes Bill is working in his room. So the approach would naturally resort to an ordering source on doxastic alternatives to keep things in order.

The epistemic modal base is not the same as the doxastic base. Assuming that Kim is not infallible, she will have beliefs that are not among the things she knows. As a result, the epistemic modal base is a subset of the doxastic base: the propositions known are a subset of those believed. But given that the epistemic base is a subset of the doxastic base, there are fewer possibilities in the doxastic *context*, consistent with the greater number of propositions in the doxastic base, than there are possibilities in the epistemic alternative set, consistent with what Kim knows.[3] Think of it this way: the possibilities consistent with what Kim believes have to pass a first test, being consistent with what Kim actually knows, and then a second test, being consistent with what Kim merely believes. Assuming some possibilities get thrown out in the second stage of this process, the set of possibilities in the doxastic context, used in the interpretation of the belief report, is a proper subset of the set of possibilities in the epistemic alternative set. So a belief report is interpreted much like a modal statement, with a set of possibilities satisfying propositions in a base, and an ordering source on those possibilities, but the doxastic context is different from the set of epistemic alternatives.

So the standard account of belief ascriptions in possible worlds semantics is to quantify over possibilities in a doxastic alternative set. But there are several problems with this that will require us to formulate it more abstractly. First, in reality, a person can hold inconsistent beliefs. If the set of all of Kim's beliefs happens to be inconsistent, then Kim's doxastic context will be empty, and the interpretive procedure sketched earlier would not have anything to work with. We will take seriously the idea that the experiencer of belief may have inconsistent beliefs, in which case the interpretation of belief reports should not involve quantification over the doxastic context; if Kim's beliefs are inconsistent, that set will be empty, and the quantification would be vacuously true (if it is universal quantification, as it is usually taken to be). Or it would be automatically false if by chance it were existential quantification.

Next, consider how the ordering source works on this account in comparison with the parallel account of modal interpretation. Interpreting modal statements such as *Bill must be working in his room*, we used the ordering source to cut back on possibilities that would make this necessity modal statement spuriously false, for example, one in which some details of the Wars of the Roses diverge from actual history, and Bill is not in his room, when Bill *is* in his room in any of the possibilities in which the Wars of the Roses unfolded as they did in actuality. The ordering source pinned down possibilities to make them accord with the truth (actual world) enough to give us reliable results for modal evaluation.

Now consider a comparable situation in the interpretation of the belief report, *Kim believes that Bill is working in his room*. Consider the set of all possibilities that are consistent with everything that Kim believes. If Kim believes that Bill is working in his room, then Bill is working in his room in all of these possibilities, regardless of whether or not those details about the Wars of the Roses that are beyond Kim's ken happen to accord with reality. Belief ascriptions are fundamentally different from epistemic modal statements in their lack

[3] All talk of fewer and greater number of possibilities here is actually proxy for proper subset relations. We will formulate this more precisely in Chapter 3.

of required fidelity to the actual world. So there is no role for the epistemic ordering source in the interpretation of a belief ascription. When we interpret *Kim believes that Bill is working in his room*, the improvement we want to see is improvement in how much a possibility accords with Kim's beliefs. The report seems to be more faithfully interpreted as asserting that as possibilities accord more and more with what Kim believes, it should be the case that Bill is working in his room in progressively more of them. Of course, that ordering source is redundant if we are quantifying over the doxastic context, the possibilities that already satisfy all of Kim's beliefs. And the use of it is pointless since we would never get any improvement in the fit with Kim's beliefs if we are only quantifying over possibilities that are consistent with all of Kim's beliefs. This suggests that we might want to interpret the belief ascription by making reference to whether possibilities accord with Kim's beliefs, and how well they do so, but not by quantifying over, and thus "living on" the doxastic context, analogously to the way epistemic modal statements "live on" the epistemic alternative set.

Instead of restricting ourselves to possibilities consistent with what Kim believes, we interpret the belief ascription by quantifying over a much wider set of possibilities, in fact, the set of all possibilities, and interpret the belief report *Kim believes that Bill is working in his room* as asserting that the complement clause, *Bill is working in his room*, is progressively more true in possibilities as they accord with more and more of Kim's beliefs. Effectively, we use the set of propositions expressing what Kim believes as part of the ordering source, which orders possibilities according to how many of these propositions they satisfy. As with the epistemic ordering source discussed in Section 1.1, the resulting ranking of possibilities is not a complete order of any kind – given any two possibilities, it does not need to be the case that one accords better than the other with Kim's beliefs. Furthermore, even if we do have a sequence of possibilities that are all weakly ordered with respect to one another (in a transitive, reflexive, and complete relation) based on their accordance with Kim's beliefs, it need not converge on any possibilities that satisfy all of what Kim believes since there need be no such possibilities, and there won't be if Kim's beliefs are inconsistent.

Further issues arise in using the doxastic context for the interpretation of attitude ascriptions when we turn to other attitudes, such as the ones with *wish* or *want*. As we will discuss in Chapter 4, there is good reason to get the doxastic context involved somehow in the interpretation of these attitude ascriptions. For example, *Kim wants the spider off her neck* should plausibly involve possibilities that accord with Kim's belief (or knowledge, if that is what it is) that there is a spider on her neck. And even if there is no spider on Kim's neck (she has merely mistaken a small leaf for a spider, and she would have no concern to get the leaf off her neck), the ascription *Kim wishes that there wasn't a spider on her neck* should plausibly be interpreted through reference to possibilities in which there is a spider on Kim's neck, thus, with respect to what she believes. But if these interpretations just quantify over the set of possibilities consistent with what Kim believes, then the interpretation could not compare possibilities in which there is no spider on Kim's neck, with respect to their desirability, with ones in which there is. If Kim believes there is a spider on her neck, her doxastic context will not contain any of the more desirable possibilities in which (everything else being the same) there isn't a spider on her neck. Confining ourselves to possibilities that all accord with what Kim believes, it is hard to access contrary possibilities in order to compare the two sets according to Kim's bouletic criteria, expressing her wants, preferences, and desires.

Part I of the present work investigates the interpretation of intensional statements by invoking possibilities or possible worlds, and considers various challenges to that approach that have appeared in the literature. Each of the discussions begins with some review of ideas in the literature. However, these ideas have been put forward with many different formulations, in different semantic frameworks. A review of the literature as it exists is fortunately rendered unnecessary here by the excellent contemporary works that do that (and more), particularly Portner (2009). Instead, the relevant ideas in the literature will be reviewed and reconstituted (giving credit where credit is due to the maximum extent possible), while highlighting the tensions and conflicts that the literature has left unresolved. Part I aims mainly to restore ideas that have been seriously challenged, and rethink them to some extent from the ground up to see how they really fare in the face of the challenges. This will yield some emendations of the theories discussed, which will provide accounts of phenomena neglected in previous work (such as the possibility of an inconsistent doxastic base) or previously handled by more drastic steps than what is necessary with the restored theory (e.g., accommodating presupposition triggering and filtering in successive attitude ascriptions without resort to dynamic semantics, as in Heim [1992]).

Chapter 1 reviews the possible worlds semantics for modality developed by Kratzer (1981, 1991), sketched earlier, which uses complex quantificational schemas. The analysis is cast in some novel terminology and notation to facilitate the conceptualization and discussion of it. Lassiter's (2011) challenges for possible worlds semantics for epistemic modality are addressed and found to lend insight, but are found not to provide grounds for rejecting Kratzer's approach. The discussion of this chapter supports an understanding of epistemic modality arrived at in an extensive line of philosophical and semantic inquiry into modality (DeRose [1991], MacFarlane [2003], Hawthorne [2004], von Fintel and Gillies [2008a, 2008b]), which takes the epistemic modal base to reflect the knowledge of the speaker (and possibly other participants in the discourse). In this, it departs from some other recent semantic work, especially that of Hacquard (2006, 2010) and Anand and Hacquard (2008), which takes the epistemic modal base to be a doxastic or informational state of the speaker.[4]

Chapter 2 covers the same territory for root modals: Kratzer's possible worlds semantics, challenges from the literature, and where consideration of the two leaves us. The role of the Limit Assumption that von Fintel and Iatridou (2008) assume for weak modality will also be examined. Chapter 3 extends the approach to ascriptions of belief and other propositional attitudes, refining a possible worlds analysis originally proposed by Hintikka (1969), while showing that Kratzer's quantificational schemas for modality can serve to elucidate the semantics of propositional attitudes over and above what Hintikka accomplished. Lassiter's challenges to possible worlds semantics of attitude reports are also addressed. At this point, a novel twist will be introduced in the formulation of possible worlds semantics for attitudes by quantifying over all possible worlds and restricting the ordering source with the doxastic base. This will accommodate ascriptions of attitudes to people with inconsistent beliefs. And in Chapter 4, the use of such restricted ordering sources will be exploited further to provide an account of bouletic ascriptions and presupposition triggering and projection in successive doxastic–doxastic and doxastic–bouletic ascriptions, taking account of the rich observations made by Heim (1992), without the use of the rich formalism of dynamic semantics that Heim employed for this purpose.

Part II will examine how these intensional interpretations can be integrated with a semantics that otherwise interprets statements as being about the actual world, specifically, in the most widely accepted approaches, about eventualities (events and states) in the world.

[4] This point will be discussed further in Chapters 1 and 2. An anonymous reviewer claims that there is no departure, but Anand and Hacquard (2008: 42) explicitly state that their account departs from accounts of epistemic modal interpretation in terms of knowledge. Hacquard (2010: 92–93; footnotes 10 and 11) explicitly allows that her view of the speaker-dependence of a main clause epistemic modal may have to be refined in light of work by DeRose, von Fintel, and Gillies, and others mentioned earlier, which implicate a community to which the speaker belongs. In this case, the question is whether the informational state consists of propositions known to the speaker (or within a community to which the speaker belongs), or whether it is more akin to an information state in dynamic semantics and pragmatics, which would be a set of propositions presumptively shared between speaker and addressee at a given point in the discourse, but not necessarily known by speaker or addressee.

# 1 Epistemic modality

The present work, as a whole, argues that a theory of modal interpretation based on possible worlds, using the quantificational schemas of Kratzer (1981), provides a superior implementation of the long-standing idea that propositional attitude ascriptions be interpreted in terms of quantification over possible worlds. With further elucidation of the relationship between events and possible worlds, this approach can be used to mark a distinction between propositional attitude ascriptions that involve an eventuality that can be identified with an event or mental state, and ones that do not. The resulting account of that distinction illuminates Neg-Raising, the interactions of presuppositions with attitude ascriptions discussed by Heim (1992), and the nature of attitude ascriptions in the language production of fluent aphasics. To do all of this requires a particular understanding of the role of eventualities in semantic theory. The present chapter reviews Kratzer's theory of modal interpretation and develops a particular understanding and formulation of it. This possible worlds semantics of modality is then considered in light of critiques in the literature and responses by Kratzer (2012). Particular attention will be given to critiques leveled by Lassiter (2011), who argues in favor of a semantics of modality that involves interpreting modal statements on a scale that can be mapped to a probability space. Finally, the present chapter considers puzzles for the application of possible worlds semantics to epistemic modality, which have been pointed out and discussed in the literature.

The discussions in this chapter tend toward a position of semantic pluralism, by which interpretations of epistemic modal statements in possible worlds semantics stand alongside interpretations that involve mapping to scales realized in quantitative probabilistic terms, the former more appropriate for the interpretation of everyday modal statements that are not influenced by technical concepts of probability, and the latter to more technical or mathematical contexts, influenced by, or directly expressing, technical concepts of probability. This argument will be extended in Chapters 2 and 3, to assert the value of this sort of semantic pluralism for the interpretation of root modal statements and propositional attitude ascriptions. In addition, the present chapter will examine the conception of the epistemic modal base. It will develop a new argument concerning the way in which the epistemic modal base can inhere in a group of which the speaker is a member, supporting proposals by von Fintel and Gillies (2005, 2008a, 2008b) on this point. And it will provide arguments in favor of a traditional view of the epistemic modal base as what the speaker knows, as against a view put forth in some recent literature that the epistemic modal base can be identified with what the speaker believes, or what has been established as common ground within a discourse. The possible worlds interpretation of modality will underlie the approach to the interpretation of propositional attitude ascriptions developed in Chapters 3, 4, and 6.

Specialists in the semantics of modality are invited to skim through Section 1.1 and to note the novel terminology introduced in connection with examples (11), (16), and (19).

## 1.1 Possible worlds in the theory of epistemic modal interpretation

### *1.1.1 Kratzer's theory of epistemic modality*

Epistemic modal statements are statements about what must, might, should, or cannot be the case, based on available evidence, or what is known.[1] Along with other forms of modality, epistemic modality can be expressed by modal verbs such as *must*, *might*, *should*, adjectives such as *necessary*, *possible*, *likely*, *sure*, *certain*, and adverbs such as *certainly*, *possibly*, and *probably*. And it can be implicit in conditionals, as discussed by Kratzer (1986), whereby *If Bill's light is on, he is in his room* would typically mean that Bill must be in his room if his light is on.

Possible worlds accounts of modal interpretation in natural language follow the lead of possible worlds semantics developed for systems of modal logic under one or more different axiomatizations of these systems. These systems involve quantification over possible worlds, regulated by appropriate accessibility relations. To consider a model-theoretic interpretation of modal formulas in such a system, if $\varPhi$ is a well-formed formula of first-order logic, then the necessity modal $\Box\varPhi$ has a model-theoretic semantic value, $[[\Box\varPhi]]^{\mathbf{M},c,w}$, in model **M**, at world $w$, and at an indexical context $c = \langle$speaker, addressee, time, place$\rangle$, as universal quantification over possible worlds $v$ that bear a designated accessibility relation R to $w$. The possibility modal has a parallel semantic value with existential quantification.

(1) a. $[[\Box\varPhi]]^{\mathbf{M},c,w} = 1$ iff $(\forall v\text{: R}(v, w))\ [[\varPhi]]^{\mathbf{M},c,v} = 1$
b. $[[\Diamond\varPhi]]^{\mathbf{M},c,w} = 1$ iff $(\exists v\text{: R}(v, w))\ [[\varPhi]]^{\mathbf{M},c,v} = 1$

[1] For discussion of standard and traditional views of modality, and how they provide points of departure for a semantic theory of modality, see Palmer (1986) and Portner (2009).

For the task of modeling modality in natural language, modal logic has some serious limitations. It doesn't naturally capture changes in accessibility relations over time, such as those based on what a subject knows, for epistemic modality. We have a temporal variable in $c$ and the accessibility relation depends on $c$, but that just says that the accessibility relation changes over time, not why or how it changes. Furthermore, classical modal logic doesn't naturally model forces other than possibility and necessity, thus missing the apparent force of the modal auxiliary *should*, or the adjective *likely*. Along similar lines, it doesn't directly address graded and comparative modality, as in *There is a slight possibility we are lost*, or *It is more likely that the government will have to step in than that BP will clean up the mess itself.*

Kratzer (1981, 1991) adapted ideas from modal logic to fashion a model-theoretic account of modal interpretation in natural language that overcomes these limitations by evaluating modal statements relative to conversational backgrounds. One conversational background is a modal base, which contains relevant facts about the world – what is known, for epistemic modals, or the circumstances of a relevant situation for a root modal. A second conversational background, the ordering source, orders possible worlds in the modal base in order to limit the modal generalization to worlds that are pertinent to the type of modal statement being expressed, for example, how well they accord with rules, conventions, or laws operative in the context of a deontic modal statement, or with the goals of the speaker or subject referent in the context of a teleological modal, or with the preferences and desires of the speaker or subject referent in the context of a bouletic modal statement. This second conversational background consists of propositions that effectively limit the modal generalization to worlds that are genuinely relevant in the context of the natural language utterance. This naturalizes the interpretation, and the implementation produces a ranked ordering of worlds that provides resources for modeling graded and comparative modality.

For epistemic modal statements, the first part of the conversational background is generally taken to be what the speaker knows (this formulation will be examined critically in Section 1.3), realized as a set of propositions $\{p_i\}$ expressing what the speaker knows. This set is usually assumed to be finite or denumerable, so that the indices $i$ can be drawn from the set of natural numbers, $\{1, 2, 3, \ldots\}$, but the index set will remain unspecified here, and simply labeled Ind.[2] Let $c$ be the indexical context of an utterance, specifying its speaker, addressee(s), time, and place. Then the epistemic modal base for a modal utterance in context $c$, at world of evaluation $w$, will be written as in (2).

[2] We can formulate nondenumerably infinite sets of propositions, but it is not clear that an epistemic modal base could realistically be nondenumerable. We circumvent the need to settle this philosophical issue by choosing the present notation, which is neutral on this matter.

(2) $E(w) = \{p_i$: the speaker in $c$ knows $p_i$ in $w$, $i \in \text{Ind}\}$.

But each $p_i$ is a set of possible worlds (the worlds in which $p_i$ is true). So, interpreting the epistemic modal will involve generalizing over the set of worlds in (3), called the epistemic alternative set.

(3) $\cap E(w) = \cap\{p_i$: the speaker in $c$ knows $p_i$ in $w$, $i \in \text{Ind}\}$
= set of worlds in which everything the speaker in $c$ knows holds true.

Suppose now that Jill observes Max, apparently sleeping on a couch. On the basis of her observation that a car alarm is blaring rather noisily nearby, and her knowledge that Max had a full night of sleep and just finished a full cup of caffeinated coffee, Jill utters (4) to express the idea that we (Jill and her interlocutors) can be sure that Max is awake.

(4) Max must be awake.

To develop Kratzer's account of the interpretation of (4) as quantification over worlds, assume that the modal auxiliary takes scope over the rest of the sentence, *Max is awake*, called the prejacent of the modal sentence; how this is derived compositionally will be discussed in Chapter 8. Let $\mathbf{S_1}$ be the phrase marker of the prejacent, and let **must** $\mathbf{S_1}$ be the LF phrase marker of (4), encoding and determining the scope of the modal auxiliary. As discussed in Chapter 8, following proposals in the literature, the epistemic modal base will be supplied by the context, not taken as an argument of the modal operator; thus, the modal auxiliary will have the simple type ⟨t, t⟩. The interpretation of (4) can be represented as restricted quantification over worlds in $\cap E(w)$, as given in (5), a preliminary formulation that remains close to (1a).[3]

(5) $[[\textbf{must } \mathbf{S_1}]]^{c,w} = 1$ iff $(\forall v: v \in \cap E(w))\ [[\mathbf{S_1}]]^{c,v} = 1$

On this approach, to evaluate **must** $\mathbf{S_1}$, **might** $\mathbf{S_1}$, or the LF phrase marker of any other epistemic modal statement, we look at worlds in $\cap E(w)$ and see in which of them $\mathbf{S_1}$ is true. The epistemic modal base is determined by the context; how it is determined by context is a matter of substantive investigation, which will be taken up in Section 1.3 of this chapter, as well as in Chapter 8.

This interpretation treats all worlds in $\cap E(w)$ as accessible. But, in fact, some worlds in $\cap E(w)$ might be more relevant than others. We might want to order the worlds in $\cap E(w)$ and only take the best of them, or only take those that rank better than a designated accessibility floor. To see why, consider (6), uttered by Jill, based on normally trustworthy indicators of market behavior.

[3] The context will sometimes be suppressed when the speaker and addressee(s), and time and place of utterance, are either clear or not referenced in the discussion.

(6) The stock market must be crashing.

Jill's statement will evaluate as false if $\cap E(w)$ contains far-fetched worlds in which the market is not crashing but aliens lurking on the far side of the moon are manipulating computer readings to produce false signs of a crash, things that cannot be ruled out based on Jill's knowledge (for all she knows, there are aliens who could do that), but which would not ordinarily impinge on the evaluation of (6). To address this, among other issues, Kratzer introduced a second conversational background, a set of propositions $\Omega(w) = \{q_j: j \in J\}$, for J an index set (typically, an initial segment of the natural numbers), defined with respect to the world of evaluation $w$, where each $q_j$ specifies a condition on more stereotypical or realistically relevant worlds for the modal generalization being made. Kratzer calls this set of propositions the ordering source. The members of $\Omega(w)$ are propositions, each of which corresponds to a set of worlds, the worlds in which that proposition is true. Now, the power set of $\Omega(w)$, $\wp(\Omega(w))$, is ordered by set inclusion, $\subseteq$. But each subset of $\Omega(w)$ corresponds to a set of worlds, the worlds in which every proposition in that subset is true. This yields a derivative ordering of any two worlds, according to the set inclusion order of the members of $\wp(\Omega(w))$ that are true of those two worlds. For example, if $u$ is in the set A of worlds in which $\{q_j, q_k, q_l\}$ are true, and $v$ is in the set B of worlds in which $\{q_j, q_k\}$ are true, then $u$ is a more stereotypical or relevant world for the modal generalization than $v$. Define $u \leq_{\Omega(w)} v$ derivatively from $\cap A \subseteq \cap B$ (which is true iff $B \subseteq A$), as in (7).

(7) For worlds $u$ and $v \in \mathbf{W}$, $u \leq_{\Omega(w)} v$ just in case all propositions in $\Omega(w)$ that are true in $v$ are true in $u$.

This effectively orders worlds according to how relevant versus how far-fetched they are for the evaluation of the modal statement at hand, relative to $w$, so that $u \leq_{\Omega(w)} v$ just in case $u$ is better than $v$ for evaluating the type of modality at hand. We use the notation as Kratzer does, so that $u$ is higher ranked than $v$, even though it is on the "less than" side of the inequality, since this corresponds to satisfying more of the conditions in the ordering source, and thus being a "closer," more accessible world to an ideal for the evaluation of the modal generalization.

There is, however, a wrinkle, based on the nature of the ordering relation defined in (7). To review relevant concepts first, a relation R on a domain D is reflexive if, for any element $x$ of D, $x$R$x$. R is transitive if $x$R$y$ and $y$R$z$ jointly entail $x$R$z$, for any $x$, $y$, and $z$ in D. R is antisymmetric if $x$R$y$ and $y$R$x$ jointly entail that $x$ is identical to $y$ in D. R is a **complete order** if, for any elements $x$ and $y$ of D, either $x$R$y$ or $y$R$x$. A **preorder** is reflexive and transitive. A **weak order** is reflexive, transitive, and complete. A **total** (or **simple**) order is a weak order that is antisymmetric. There is reason to assume that the order $\leq_{\Omega(w)}$ on worlds in the epistemic modal base is not antisymmetric, reflecting the view that two distinct worlds (representing conceptually distinct possibilities) might be equally in accord with everything the speaker knows, and equally plausible or relevant in the context of the modal utterance.

Now, the ordering of elements of $\wp(\Omega(w))$ by $\subseteq$, and thus the ordering of worlds by $\leq_{\Omega(w)}$, is a preorder, reflexive, and transitive, but not necessarily antisymmetric or complete. So if we want to look at a sequence of worlds that get successively better (strictly speaking, successively better-than-or-equal-to) for the evaluation of a modal statement, those on the "less than" end of $\leq_{\Omega(w)}$, we cannot generally do this for all worlds in **W**, but only for sequences (or "strands") of worlds among which we have a local weak order (reflexive, transitive, and complete) under $\leq_{\Omega(w)}$.

Here is how these ideas apply to the interpretation of (6) in the world of evaluation $w$, where the prejacent, *The stock market is crashing*, is $\mathbf{S_1}$. The epistemic modal base $\cap E(w)$ is the set of worlds consistent with everything the speaker knows in $w$. Instead of asserting that $\mathbf{S_1}$ is true in all of these worlds, we want to use the ordering source $\Omega(w)$ to screen out the irrelevant ones. If $\leq_{\Omega(w)}$ were a weak order, we could interpret **must** $\mathbf{S_1}$ as asserting the existence of a "threshold" world $v$ such that $\mathbf{S_1}$ is true for all worlds "within" the threshold $v$, that is, at all worlds $z$ such that $z \leq_{\Omega(w)} v$. This would say that, starting at $v$, any way of tracing progressively better (or "better than or equal to" worlds) "inward" on $\leq_{\Omega(w)}$ yields a sequence of worlds in which $\mathbf{S_1}$ is uniformly true.

But $\leq_{\Omega(w)}$ is not a complete order. As a result, even if there is a world $v$ such that $\mathbf{S_1}$ is uniformly true on all worlds $z$ such that $z \leq_{\Omega(w)} v$, that doesn't prevent there being another world $v'$, unordered by $\leq_{\Omega(w)}$ with respect to $v$, but satisfying just as many propositions in the ordering source, such that $\mathbf{S_1}$ is not uniformly true on all worlds $z$ such that $z \leq_{\Omega(w)} v'$. That is, even if all sequences of worlds going "inward" starting at $v$ are ones in which $\mathbf{S_1}$ is uniformly true, this isn't enough to make **must** $\mathbf{S_1}$ true. If **must** $\mathbf{S_1}$ is true, we should get this result for any starting point that is good enough to be relevant for the evaluation of the modal statement. At some point as we go inward, it must be the case that all weakly ordered sequences of worlds are ones in which $\mathbf{S_1}$ is uniformly true.

Now, if $\Omega(w)$ is consistent, then there could be one or more worlds that meet all the conditions in $\Omega(w)$, and thus realize the ideal for the evaluation of the modal generalization. Then $\cap\Omega(w)$ would be the set of all such worlds, which we could write as $\mathrm{Max}(\Omega(w))$, and we could interpret modals by quantifying over $\mathrm{Max}(\Omega(w))$. But even if $\Omega(w)$ is consistent, we can imagine that **W** has no lower bound (or "Maximum") with respect to $\leq_{\Omega(w)}$. If $E(w)$ and $\Omega(w)$ are vast enough, we can imagine that there are sequences of worlds weakly ordered by $\leq_{\Omega(w)}$, which get better and better for the evaluation of a modal statement, but don't converge on any best world or worlds. In this case, the best we could do is to look at what happens on the "tails" of (potentially) nonconvergent sequences of weakly ordered worlds.

Kratzer (2012: 39) characterizes the ordering source as a "stereotypical conversational background," which "can be used to rank worlds according to how close they come to the normal course of events in the world of evaluation, given a suitable normalcy standard." On this conception, the ordering source is potentially nonveridical – it could contain propositions that characterize the normal course of events, but which happen to be false in the instance at hand, and $\Omega(w)$ itself, can, correspondingly, be inconsistent. At this point, we face a fateful choice between two theoretical suppositions. We can assume that $\Omega(w)$ is never inconsistent in practice, for example, by framing its normalcy standards in entirely generic terms, thus insulating $\Omega(w)$ from inconsistency. (This theoretical maneuver creates more work when the propositions of the ordering source are invoked in a particular example.) Alternatively, we can take the propositions of $\Omega(w)$ to be nongeneric propositions that simply state the normal course of affairs, which can be false or inconsistent as applied to the context of the utterance being interpreted. We choose the latter, in part to avoid problems with the interpretation of generic statements when they are applied to a particular context. But the real payoff for making this choice will come in later chapters, where we see that the semantics of modality based on this choice is directly applicable to the interpretation of propositional attitude reports in a way that circumvents various empirical difficulties noted by Heim (1992), and some taxing theoretical maneuvers that have been made to address those difficulties.

In fact, it is easy to construct examples in which $\Omega(w)$ is not consistent, especially if we assume a certain degree of fallibility on the part of the speaker whose utterance is being interpreted. Suppose that X and Y are two economic measures that are alleged indices of the New York stock market. Let $p$ be the proposition that X tracks the market, and Y does not, and let $q$ be the proposition that Y tracks the market, and X does not. Suppose that Jill and her addressees used one of these two indicators in the past, and found it to be a reliable indicator of market behavior, but nobody remembers which it was. On a particular morning, Jill notes that both X and Y indicate an imminent market crash, and to her associates, utters, *The market must be crashing*. This could evaluate as true by virtue of two different weakly ordered sequences of worlds, in one of which worlds satisfy $p$ and get progressively better according to other propositions in $\Omega(w)$, and in the other of which worlds satisfy $q$ and get progressively better according to other propositions in $\Omega(w)$.

In such a case, there are no worlds that satisfy all of the propositions in $\Omega(w)$, so any sequence of worlds weakly ordered by $\leq_{\Omega(w)}$ would fail to converge to any world that satisfies all of the ordering propositions in $\Omega(w)$. In this case, we can't quantify over best ranked worlds in order to ensure the relevance of worlds quantified over.[4] But we can revise the analysis by imposing an accessibility floor for each weakly ordered sequence of worlds, asserting the existence of a world $v$ for each such sequence that exemplifies a minimum degree of accessibility for the interpretation of (6), and thus for quantification in (5), for that sequence.

In the present work, we will make no assumptions about the consistency of $\Omega(w)$, or the convergence of sequences of worlds with respect to $\leq_{\Omega(w)}$. And since $\leq_{\Omega(w)}$ is only a preorder, we cannot simply modify (5) by introducing an accessibility limit $v$ and requiring that all worlds quantified over in (5) be $\leq_{\Omega(w)}$ $v$. We must, instead, do this for each sequence of weakly ordered worlds, where worlds across sequences may happen not to be ordered by $\leq_{\Omega(w)}$. The result is in (8), formulated as Kratzer states it (modulo superficial notational changes).

(8) $[[\textbf{must } \mathbf{S_1}]]^{c,w} = 1$ iff
$\forall u \in \cap E(w)$: $(\exists v \in \cap E(w)$: $v \leq_{\Omega(w)} u)$ $(\forall z \in \cap E(w)$: $z \leq_{\Omega(w)} v)$
$[[\mathbf{S_1}]]^{c,z} = 1$

This says that anywhere we start in $\cap E(w)$, we can find a local accessibility floor $v$ inward from that starting point, at which $\mathbf{S_1}$ is uniformly true on $z \leq_{\Omega(w)}$ $v$. That is, for every world $u$ on a completely – and therefore weakly – ordered sequence of worlds (this quantifier effectively quantifies over weakly ordered sequences of worlds), there is an accessibility limit $v$ such that $\mathbf{S_1}$ is true in every world $z$ within the limit set by $v$ (with respect to $\leq_{\Omega(w)}$). More simply, if we call the worlds $z$ such that $z \leq_{\Omega(w)} v$ is a tail of the sequence, (8) states that every sequence of worlds has a tail on which $\mathbf{S_1}$ is uniformly true. Finally, for simplicity, we will say in this case that every weakly ordered sequence of worlds has an $\mathbf{S_1}$–uniform tail.

As Kratzer (1991) notes, the truth conditions in (8) relativize the inference pattern in (9) in a way that renders it invalid.

(9) If $\mathbf{S_1}$ is necessarily true, then $\mathbf{S_1}$ is true.

(8) asserts that $\mathbf{S_1}$ is true as we trace weakly ordered sequences of worlds "inward" as worlds satisfy progressively more of the members of $\Omega(w)$. Even if $w \in \cap E(w)$, we aren't guaranteed that $w$ is on any of the $\mathbf{S_1}$–uniform tails of sequences of worlds weakly ordered by $\Omega(w)$ that are guaranteed to exist by (8). However, if $w \in \cap E(w)$, and $w$ is as much in accord with $\Omega(w)$ as any other world, then assuming the antecedent of (9), we can choose $z = v = w$ in (8) to establish the consequent, in this case, that $\mathbf{S_1}$ is true in $w$. In connection with this, Kratzer (1991) observes that **must** $\mathbf{S_1}$ is both weaker and stronger than $\mathbf{S_1}$,

[4] In Section 1.2.1 below, we return to the question of whether there is a set of "best ranked" worlds under the ordering source – called the Limit Assumption in the literature – and implications it has been claimed to have.

stronger in that it asserts the truth of $\mathbf{S_1}$ in many worlds, but weaker in that it relativizes the truth of $\mathbf{S_1}$ to epistemic conditions.

If we interpret **might** $\mathbf{S_1}$ as **not** (**must** (**not**-$\mathbf{S_1}$)), so that *might* and *must* are duals of one another, then the truth conditions for **might** $\mathbf{S_1}$ are as given in (10).

(10) $[[\textbf{might } \mathbf{S_1}]]^{c,w} = 1$ iff $[[\textbf{not (must (not-}\mathbf{S_1}\textbf{))}]]^{c,w} = 1$ iff $\exists u \in \cap E(w)$: $(\forall v \in \cap E(w)$: $v \leq_{\Omega(w)} u)$ $(\exists z \in \cap E(w)$: $z \leq_{\Omega(w)} v)$ $([[\textbf{not-}\mathbf{S_1}]]^{c,z} = 0)$
iff $\exists u \in \cap E(w)$: $(\forall v \in \cap E(w)$: $v \leq_{\Omega(w)} u)$ $(\exists z \in \cap E(w)$: $z \leq_{\Omega(w)} v)$ $[[\mathbf{S_1}]]^{c,z} = 1$

We will say that $\mathbf{S_1}$–worlds persist on a sequence of worlds weakly ordered by $\leq_{\Omega(w)}$ just in case $(\forall v \in \cap E(w))$ in the sequence, $(\exists z \in \cap E(w)$: $z \leq_{\Omega(w)} v)$ in the sequence such that $\mathbf{S_1}$ is true in $z$. Then (10) can be paraphrased as asserting that **might** $\mathbf{S_1}$ is true iff there is a weakly ordered sequence of worlds in $\cap E(w)$ on whose tail $\mathbf{S_1}$–worlds persist. That is appropriate: it means that however good worlds get on the ordering $\leq_{\Omega(w)}$ according to the ordering source, we can always find a world where $\mathbf{S_1}$ is true. We will say that such a sequence has an $\mathbf{S_1}$–persistent tail. Using the new terminology, the interpretations in (8) and (10) are summarized in (11).

(11) a. $[[\textbf{must } \mathbf{S_1}]]^{c,w} = 1$ iff every sequence of worlds in $\cap E(w)$ that is weakly ordered by $\Omega(w)$ has an $\mathbf{S_1}$–uniform tail.
b. $[[\textbf{might } \mathbf{S_1}]]^{c,w} = 1$ iff there is a sequence of worlds in $\cap E(w)$ that is weakly ordered by $\Omega(w)$, with an $\mathbf{S_1}$–persistent tail.

Kratzer (1981, 1991) presented an analysis of comparative modality, leading up to truth conditions for *should*, which will be formulated in this section. Section 1.2.2 examines objections brought forth by Yalcin (2010) and Lassiter (2010), elaborated in Lassiter (2011) as a serious challenge to possible worlds semantic accounts of modality. That challenge will be adjudicated through examination of a revised analysis developed by Kratzer (2012) in response to such objections, and objections to the revised analysis spelled out by Lassiter (2011). The issue will be resolved in favor of a new possible worlds analysis of comparative possibility, to be developed in Section 1.2.2, which nevertheless leads to the same truth conditions for *should* originally given by Kratzer.

Following Kratzer (1991), $\mathbf{S_1}$ is defined to be a good possibility in a world of evaluation $w$ if there is a sequence of worlds on which $\mathbf{S_1}$ is a "local necessity" – a necessity on the tail of that sequence.

(12) $\mathbf{S_1}$ is a good possibility in $w$ iff
$\exists u \in \cap E(w)$: $(\forall v \in \cap E(w)$: $v \leq_{\Omega(w)} u)$ $[[\mathbf{S_1}]]^{c,v} = 1$

The comparative notions, "$\mathbf{S_1}$ is at least as good a possibility as $\mathbf{S_2}$" and "$\mathbf{S_1}$ is a better possibility than $\mathbf{S_2}$," are useful not only for directly modeling comparative notions of modality in natural language, but also because they can be used to define weak necessity, a plausible candidate interpretation of *should*. "$\mathbf{S_1}$ is at least as good a possibility as $\mathbf{S_2}$" is true just in case, for any world in which $\mathbf{S_2}$ is true, there is an equal or higher ranked world in which $\mathbf{S_1}$ is true.

(13) $\mathbf{S_1}$ is at least as good a possibility as $\mathbf{S_2}$ in $w$ iff
$(\forall u \in \cap E(w): [[\mathbf{S_2}]]^{c,u} = 1)\ (\exists v \in \cap E(w): v \leq_{\Omega(w)} u)\ [[\mathbf{S_1}]]^{c,v} = 1$

And "$\mathbf{S_1}$ is a better possibility than $\mathbf{S_2}$" is true just in case $\mathbf{S_1}$ is at least as good a possibility as $\mathbf{S_2}$, but it is not the case that $\mathbf{S_2}$ is at least as good a possibility as $\mathbf{S_1}$. This is written out in (14), coordinating the condition in (13) with the negation of the result of swapping $\mathbf{S_1}$ and $\mathbf{S_2}$ in (13); for convenience, we write $[[\mathbf{not}\ (\mathbf{S_2})]]^{c,v} = 1$ in (14b) in place of $\sim([[\mathbf{S_2}]]^{c,v} = 1)$, taking advantage of their logical equivalence under the bivalence of truth.

(14) $\mathbf{S_1}$ is a better possibility than $\mathbf{S_2}$ in $w$ iff
a. $(\forall u \in \cap E(w): [[\mathbf{S_2}]]^{c,u} = 1)\ (\exists v \in \cap E(w): v \leq_{\Omega(w)} u)\ [[\mathbf{S_1}]]^{c,v} = 1\ \wedge$
b. $(\exists u \in \cap E(w): [[\mathbf{S_1}]]^{c,u} = 1)\ (\forall v \in \cap E(w): v \leq_{\Omega(w)} u)\ [[\mathbf{not}\ (\mathbf{S_2})]]^{c,v} = 1$

Kratzer's notion of the weak necessity of $\mathbf{S_1}$ is that $\mathbf{S_1}$ is a better possibility than **not**-$\mathbf{S_1}$, as spelled out in (15), which we use as the interpretation of **should** $\mathbf{S_1}$. This is obtained by replacing $\mathbf{S_2}$ by **not**-$\mathbf{S_1}$ in (14).

(15) $[[\mathbf{should}\ \mathbf{S_1}]]^{c,w} = 1$ iff $\mathbf{S_1}$ is a weak necessity in $w$ iff
$\mathbf{S_1}$ is a better possibility than **not**-$\mathbf{S_1}$ iff
a. $(\forall u \in \cap E(w): [[\mathbf{not}\text{-}\mathbf{S_1}]]^{c,u} = 1)\ (\exists v \in \cap E(w): v \leq_{\Omega(w)} u)\ [[\mathbf{S_1}]]^{c,v} = 1\ \wedge$
b. $(\exists u \in \cap E(w): [[\mathbf{S_1}]]^{c,u} = 1)\ (\forall v \in \cap E(w): v \leq_{\Omega(w)} u)\ [[\mathbf{S_1}]]^{c,v} = 1$

By the first condition, in (15a), every sequence with a **not**-$\mathbf{S_1}$–world has an $\mathbf{S_1}$–persistent tail (because, it asserts that wherever you find a **not**-$\mathbf{S_1}$–world, you can find a better $\mathbf{S_1}$–world, and if a sequence has no **not**-$\mathbf{S_1}$–worlds, then that sequence consists uniformly of $\mathbf{S_1}$–worlds, and so, ipso facto, has an $\mathbf{S_1}$–persistent tail). So (15a) states that every sequence has an $\mathbf{S_1}$–persistent tail. The condition in (15b) states that there is a sequence with an $\mathbf{S_1}$ –uniform tail. We record the resulting paraphrase of (15) in (16).

(16) $[[\mathbf{should}\ \mathbf{S_1}]]^{c,w} = 1$ iff
a. every sequence weakly ordered by $\Omega(w)$ has an $\mathbf{S_1}$–persistent tail, and
b. there is a sequence weakly ordered by $\Omega(w)$ with an $\mathbf{S_1}$–uniform tail.

Using (15), we can define a dual of **should**, notated $\mathbf{should}^{\perp}$, to satisfy (17).

(17) $[[\textbf{should}^{\perp}\ (\textbf{not-S}_1)]]^{c,w} = 1$ iff $[[\textbf{not}\ (\textbf{should}\ \textbf{S}_1)]]^{c,w} = 1$ iff $[[\textbf{should}\ \textbf{S}_1]]^{c,w} = 0$ iff

a. $(\exists u \in \cap E(w)\text{: } [[\textbf{not-S}_1]]^{c,u} = 1)\ (\forall v \in \cap E(w)\text{: } v \leq_{\Omega(w)} u)\ [[\textbf{not-S}_1]]^{c,v} = 1 \lor$

b. $(\forall u \in \cap E(w)\text{: } [[\textbf{S}_1]]^{c,u} = 1)\ (\exists v \in \cap E(w)\text{: } v \leq_{\Omega(w)} u)\ [[\textbf{not-S}_1]]^{c,v} = 1$

That is, some sequence has a tail uniform in **not-S**$_1$–worlds, **or** every sequence has a tail persistent in **not-S**$_1$–worlds. Replacing **S**$_1$ with **not-S**$_1$ in (17), that is, expanding (17) for **not-S**$_1$ in place of **S**$_1$, we obtain the truth conditions of **should**$^{\perp}$ **S**$_1$ in (18).

(18) $[[\textbf{should}^{\perp}\ \textbf{S}_1]]^{c,w} = 1$ iff

a. $(\exists u \in \cap E(w)\text{: } [[\textbf{S}_1]]^{c,u} = 1)\ (\forall v \in \cap E(w)\text{: } v \leq_{\Omega(w)} u)\ [[\textbf{S}_1]]^{c,v} = 1 \lor$

b. $(\forall u \in \cap E(w)\text{: } [[\textbf{not-S}_1]]^{c,u} = 1)\ (\exists v \in \cap E(w)\text{: } v \leq_{\Omega(w)} u)\ [[\textbf{S}_1]]^{c,v} = 1$

We record this by paraphrase in (19), where the two conditions are reversed in order to facilitate comparison with (16).

(19) $[[\textbf{should}^{\perp}\ \textbf{S}_1]]^{c,w} = 1$ iff

a. every sequence weakly ordered by $\Omega(w)$ has an **S**$_1$–persistent tail, or

b. there is a sequence weakly ordered by $\Omega(w)$ with an **S**$_1$–uniform tail.

We can see that (16) is identical to (19), except for the coordination of conditions (a) and (b) with "and" for **should S**$_1$, versus "or" for **should**$^{\perp}$ **S**$_1$. In this respect, the interpretation of **should**$^{\perp}$ is just slightly weaker than the interpretation of **should**. This matches the behavior of the dual of **should** in a simpler theory using generalized quantifiers, behavior that would be matched in a probabilistic account (such as that of Yalcin [2005, 2007]). To see this, consider the interpretation of **not (should S**$_1$**)**, which is equivalent to **should**$^{\perp}$ **(not-S**$_1$**)**, in a numerically based theory. Suppose that "**S**$_1$ should be the case," means that in > 50% of worlds, **S**$_1$ is the case. Then, "it is false that **S**$_1$ should be the case" means that in ≥ 50% of worlds, **not-S**$_1$ is the case. Thus, if **should** means > 50% likely, then its dual means ≥ 50% likely. Likewise, if "**S**$_1$ should be the case" means that **S**$_1$ is the case in > 60% of worlds, then "it is false that **S**$_1$ should be the case" means that in ≥ 40% of worlds, **not-S**$_1$ is the case. The stronger the interpretation of **should**, the weaker its dual is. In all of these interpretations where **should** is interpreted near the midpoint of its scale, as probability greater than 50% or greater than some threshold slightly above 50%, its dual is slightly weaker than it. And specifically, when **should S**$_1$ is interpreted as weak necessity, that **S**$_1$ is a better possibility than **not-S**$_1$, then **should** is as close to the midpoint of its scale as it could be, and so the dual of **should** is minimally weaker than it. This slight weakening under dualization will be critical in discussion of the LIN in Chapter 7. Comparison with probabilistic accounts will be addressed in Section 1.2 of this chapter.

### *1.1.2 Kratzer's theory: problems, reformulations, revisions*

*1.1.2.1 The limit assumption and strong and weak necessity modals* If the ordering source $\Omega(w)$ is a consistent set of propositions, then there are possible worlds that meet all conditions in $\Omega(w)$, and that are therefore all equally good for the evaluation of the epistemic modal statement. But even if $\Omega(w)$ is inconsistent, there could be a set of propositions Max($\Omega(w)$) that are "best possible worlds" under the ordering source, in that there are no worlds that are better than them in satisfying members of $\Omega(w)$, although, when $\Omega(w)$ is inconsistent, these "best worlds" cannot meet all conditions in $\Omega(w)$. They might all satisfy a particular subset of the conditions in $\Omega(w)$, or they might all satisfy equally many conditions in $\Omega(w)$.

The Limit Assumption stipulates that the ordering of worlds under $\Omega(w)$ converges on a set of "best worlds" Max($\Omega(w)$), such that all worlds in Max($\Omega(w)$) are equally good for the evaluation of the epistemic modal statement, whether $\Omega(w)$ is consistent or inconsistent. Under the Limit Assumption, von Fintel and Iatridou (2008) claim that Kratzer's truth conditions for weak necessity collapse into those for strong necessity. They offer an analysis of the interpretation of weak necessity *ought* under which, in contrast, this does not happen. The focus here will be on their claims regarding Kratzer's analysis, but we note their own analysis here as a benchmark to which we return at the end of this section. Applied to an epistemic use of the statement *Morris ought to be in his office*, von Fintel and Iatridou (2008: 119) "propose that epistemic *ought* differs from epistemic *must/have to* in being sensitive not just to the hard and fast evidence available in a situation but also to a set of propositions which describe what is *normally* the case." The middling force of *ought* then comes not from a mid-scalar quantificational strength of a quantifier over possibilities compatible with what the speaker knows ("hard and fast evidence available"), but from a secondary modal base, as it were, of worlds in accord with what is normally the case, the commitment to which is quite soft and can be surrendered more readily.

Let's now examine their claim that Kratzer's analysis gives strong and weak necessity identical truth conditions under the Limit Assumption. To do this, we go back to the beginning, recalling the definition of the order $\leq_{\Omega(w)}$, given in (7) and repeated here.

(7) For worlds $u$ and $v \in \mathbf{W}$, $u \leq_{\Omega(w)} v$ just in case all propositions in $\Omega(w)$ that are true in $v$ are true in $u$.

If the Limit Assumption means that all worlds in Max ($\Omega(w)$) are equal with regard to $\leq_{\Omega(w)}$, and if Max ($\Omega(w)$) is finite, and if all sequences have tails that are immersed in Max ($\Omega(w)$), then, since Max ($\Omega(w)$) is finite, the tail of every sequence will cover all the worlds in Max ($\Omega(w)$). In that case, if one sequence has an $\mathbf{S_1}$–uniform tail, then all do. So **should-$\mathbf{S_1}$** entails **must-$\mathbf{S_1}$** in such a case, and we get that truth conditions for the two do indeed collapse into one another.

But this train of reasoning doesn't just trade on the Limit Assumption. It also ignores the potential density of possible worlds outside Max ($\Omega(w)$) on tails of sequences weakly ordered by $\leq_{\Omega(w)}$. This is easy to do if we tend to think of the set of possible worlds as a small, discrete set of relevant alternatives.

If the set of possible worlds is dense, the situation is quite different. When $\Omega(w)$ is consistent and infinite, there are ways for the Limit Assumption to fail. Possible worlds on the tail of a sequence could steadily improve with respect to the ordering source, satisfying progressively more members of $\Omega(w)$, and yet never converge on a maximum. If there are enough worlds, worlds could keep "getting better" (or at least maintaining their place in the finite case) without getting optimally good. The Limit Assumption is premised on a view that possible worlds are basically not very plentiful. Even when $\Omega(w)$ is consistent and finite (presumably the most favorable case for the Limit Assumption), possible worlds on the tails of some sequences weakly ordered by $\Omega(w)$ in the truth conditions for **should-$\mathbf{S_1}$** and **must-$\mathbf{S_1}$** could simply maintain their position, satisfying (say) half the members of $\Omega(w)$ and not improving, as possible worlds on other tails improve beyond that.

When $\Omega(w)$ is inconsistent, there are other problems for the Limit Assumption. It is not clear that Max ($\Omega(w)$), the set of all worlds that are "as good as can be" with regard to satisfying members of $\Omega(w)$, is even defined in this case. And even if it is, and all sequences have tails immersed in Max ($\Omega(w)$), it does not follow that all worlds in Max ($\Omega(w)$) are essentially equal with regard to $\leq_{\Omega(w)}$.

The main idea behind this response is that ordering source propositions can be stringent, individually or collectively, and possibilities can be multitudinous. As a result, tails of sequences weakly ordered by $\Omega(w)$ are not necessarily immersed in Max ($\Omega(w)$) – they can sustain their place under $\leq_{\Omega(w)}$, and when $\Omega(w)$ is infinite, even keep improving, without "reaching" Max ($\Omega(w)$). So the Limit Assumption may hold, but Max ($\Omega(w)$) will not play the role that is required of it in von Fintel and Iatridou's argument if possibilities are multitudinous and dense enough to maintain their place under $\leq_{\Omega(w)}$, or to keep improving, without "reaching" Max ($\Omega(w)$). With such possibilities, Kratzer's truth conditions for **should-$\mathbf{S_1}$** do not reduce to those for **must-$\mathbf{S_1}$**.

This is the case for Kratzer's theory of epistemic modality. Recall what the ordering source is for an epistemic modal. The modal base is the set of worlds in accord with everything that the speaker knows. But the knowledge of any person is limited; the set of worlds compatible with what a speaker knows is far too vast and liberal a set to use to evaluate a typical epistemic modal statement. The ordering source channels the interpretation to worlds that are more and more relevant. The point is that this is not achieved with a few propositions cutting back a small set of worlds to an even smaller set. It is in general the successive pruning of a vast torrent of worlds. The interpretation has to take account of all the ways the world could be, consistent with what the speaker knows, but also controlling for things the speaker does not know. This is the situation if we are fairly realistic about possible worlds, in the ontological sense that we take them to be all the possibilities that exist, not just the ones that speaker and addressees have overtly in mind as salient alternatives. That is the main point of difference between the potential of Kratzer's analysis that the present work seeks to give full credit to and to develop, on the one hand, and the arguments of von Fintel and Iatridou on the other, which seem to cast the issue of modal interpretation in terms of processing a limited number of possible worlds instantiating well-defined salient alternatives.

What about the example *Morris ought to be in his office*, and the analysis that von Fintel and Iatridou give for it? Their claim is that the *ought* statement is not a weaker quantificational condition on worlds; it is responsive to a set of conditions reflecting what is normally the case, and these are more readily given up. Thus, it has a dual base: a "hard and fast" one and the weaker "normality" one. This analysis of *ought* is intuitively plausible and seems merited. However, it should probably be considered not as a replacement for Kratzer's analysis showing that the latter is flawed, but as a different and special case. I think we can do "hard-nosed" cases of "S is likely" according to strict epistemic criteria where Kratzer's analysis is in force. *Ought* has a more specialized possible dimension for epistemic interpretation, the one highlighted by von Fintel and Iatridou.

In sum, the claim by von Fintel and Iatridou that Kratzer's truth conditions for weak modality reduce to those for strong modality under the Limit Assumption does not hold generally, even in the most favorable case for that claim when $\Omega(w)$ is consistent and finite. This can be said while acknowledging that they have identified a very common interpretation of *ought*-statements that does not proceed along the lines of Kratzer's theory of modality.[5]

[5] See Rubinstein (2012) for another view of weak necessity, by which it differs from strong necessity not in quantificational strength but in conditions on commitments to propositions in discourse.

*1.1.2.2 Lassiter's objections to possible worlds semantics for comparative modality* Lassiter (2011: 58) shows that Kratzer's definition of "better possibility" or "more likely" in (14) , which is the basis of the truth conditions for *should* in (16), makes valid the Disjunctive Inference schema given in (20), which is, in fact, clearly and intuitively invalid.[6] (This inference schema was noted by Halpern [1997, 2003] as a problem for various theories of modal meaning.)

(20) a. $p$ is at least as likely as $q$.
b. $p$ is at least as likely as $r$.
c. $\therefore$ $p$ is at least as likely as $q \vee r$.

This inference fails in a lot of cases. In particular, it fails when $r = {\sim}q$, and neither $q$ nor ${\sim}q$ is more likely than the other; in this case, $p$ could be only slightly or moderately more likely than ${\sim}p$ for (20a,b) to be satisfied, and yet (20c) would state that $p$ is at least as likely as the tautology $q \vee {\sim}q$, and therefore necessary in this case. In more recent work, in response to such objections from Yalcin (2010) and others, Kratzer (2012: 41) revises her approach by replacing her definition of "at least as good a possibility" in (13), on which (14) is based, with that in (21).

(21) $p$ is at least as good a possibility as $q$ iff $(\forall u \in q - p)$ $\exists v \in p - q$: $\sim (u <_{\Omega(w)} v)$

But Kratzer (2012: 41) finds a problem with this new definition when there is a world $w_1$ that is a "best world" in **W**, according to $\leq_{\Omega(w)}$. Then (21) counts the propositions **W** and $\{w_1\}$ equipossible. In particular, all of **W** is at least as good a possibility as $\{w_1\}$, vacuously, because $\{w_1\} - \mathbf{W} = \varnothing$ $(q - p = \varnothing)$. Nevertheless, Kratzer's new definition in (21) has the virtue that it doesn't render the faulty Disjunctive Inference schema valid. If $p$ is at least as likely as $q$, we have (21). If $p$ is at least as likely as $r$, we have (22a), and for the special case of $r = {\sim}q$, (22b).

(22) a. $(\forall u \in r - p)$ $\exists v \in p - r$: $\sim (u <_{\Omega(w)} v)$
b. $(\forall u \in {\sim}q - p)$ $\exists v \in p - ({\sim}q)$: $\sim (u <_{\Omega(w)} v)$

Then (20c) has the truth conditions given in (23), or in (24) for the special case of $r = {\sim}q$.

(23) $p$ is at least as likely as $q \vee r$ iff
$(\forall u \in (q \vee r) - p)$ $\exists v \in p - (q \vee r)$: $\sim (u <_{\Omega(w)} v)$

[6] For the purposes of this discussion, following the sources to be discussed, we express modal relations in a logical meta-language that includes propositions $p$, $q$, and $r$, regarded as sets of possible worlds. The results of this discussion apply directly to the proposals for the semantics of statements of modal relations in natural language examined and used in the present work.

(24) $p$ is at least as likely as $q \vee {\sim}q$ *iff*
$(\forall u \in (q \vee {\sim}q) - p)\ \exists v \in p - (q \vee {\sim}q)\text{: } {\sim}\,(u <_{\Omega(w)} v)$
iff $(\forall u \in {\sim}\,p)\ \exists v \in \varnothing\text{: } {\sim}\,(u <_{\Omega(w)} v)$

And this is clearly false. So the Disjunctive Inference Schema is not rendered valid in general by (21). More generally in (21) and (22a), the existence of $v_1$ in $p - q$, subject to ${\sim}\,(u <_{\Omega(w)} v_1)$, and $v_2$ in $p - r$, subject to ${\sim}\,(u <_{\Omega(w)} v_2)$, doesn't guarantee the existence of $v_3$, subject to a similar condition, in the (generally, assuming $q \neq r$) smaller set $p - (q \vee r)$, which is smaller than both $p - q$ and $p - r$ since it is obtained by subtracting the larger set $(q \vee r)$ from $p$.

Unfortunately, Lassiter (2011: 84) shows that Kratzer's new definition in (21) validates a special case of the Disjunctive Inference schema, just as invalid, formulated when $p$, $q$, and $r$ are all mutually exclusive. (His example: $p$ = the Blue Jays will win the World Series, $q$ = the Red Sox will win, $r$ = the Braves will win.).

(25) a. $(p \wedge {\sim}q \wedge {\sim}r)$ is at least as likely as $({\sim}p \wedge q \wedge {\sim}r)$.
b. $(p \wedge {\sim}q \wedge {\sim}r)$ is at least as likely as $({\sim}p \wedge {\sim}q \wedge r)$.
c. $\therefore$ $(p \wedge {\sim}q \wedge {\sim}r)$ is at least as likely as $({\sim}p \wedge q \wedge {\sim}r) \vee ({\sim}p \wedge {\sim}q \wedge r)$.

Lassiter (2011: 84) presents a proof. The idea is that the exclusivity among alternatives in (25) ensures that the worlds $v_1$ and $v_2$, obtained by applying (21) to (25a) and (25b), each suffice to make the "$\exists v$" condition in application of (21) to (25c) true. So the inference is valid. In effect, the exclusivity among these defined propositions prevents us from getting any mileage out of the sets-of-worlds subtractions in Kratzer's (21). Applying (21) to the baseball example, (25a) states that, for any world in which the Red Sox alone win, there is an equally good or better world in which the Blue Jays alone win; (25b) states that, for any world in which the Braves alone win, there is an equally good or better world in which the Blue Jays alone win. Exclusivity then ensures that, for any world in which the Red Sox alone win or the Braves alone win, there is an equally good or better world in which the Blue Jays alone win, so the conclusion of the inference schema follows.

This, for Lassiter, is the death knell for a Kratzer-type possible worlds semantics for modality. Furthermore, Lassiter (2011: 90) rejects the prospect of developing a hybrid between Kratzer's theory of natural language modality and one based on quantitative probability theory on the grounds that such a hybrid would fail to validate the condition in (26), Lassiter's example (3.92).

(26) If must ($p$) is true, and $q$ is at least as likely as $p$, then must ($q$) is true.

This condition is intuitively sound, and satisfying it is a reasonable additional criterion for a theory of comparative modality.

But Kratzer (2012) herself noted that (21) is just one possible formulation of comparative modality. Although the Disjunctive Inference Schema in (25) may spell the end of the particular proposal given by Kratzer (2012: 41), it isn't a problem for all formulations of a Kratzer-style possible worlds semantics, as discussed next.

Consider an analysis of "*p* is at least as likely as *q*," based on the idea that *p* should be at least the equal of *q* in persistence and uniformity on sequences weakly ordered by the ordering source, as stated in (27), and more simply in (28), where we adopt the shorthand of saying that a sequence is *p*-persistent just in case it has a tail that is persistent in *p*, and *p*-uniform just in case it has a tail uniform in *p*.

(27) *p* is at least as good a possibility as *q* iff
  a. every sequence weakly ordered by $\Omega(w)$ that has a tail persistent in *q* has a tail persistent in *p*, and
  b. every sequence weakly ordered by $\Omega(w)$ that has a tail that is uniform in *q* has a tail that is uniform in *p*.

(28) *p* is at least as good a possibility as *q* iff
  a. every weakly ordered sequence that is *q*-persistent is *p*-persistent, and
  b. every weakly ordered sequence that is *q*-uniform is *p*-uniform.

For this to meet the concerns posed by Lassiter, it must render the condition in (26) valid, but not make the inference schema in (25) valid. Taking the latter concern first, supposing that (25a,b) are true, consider whether (25c) follows using (27). For convenience, the proposition $(p \wedge \sim q \wedge \sim r)$ will be called *p*-alone, and likewise the propositions at the end of (25a,b) will be called *q*-alone and *r*-alone. To check whether (25c) is true, we need to check both the persistence and uniformity provisions of (27). Any sequence persistent in $(\sim p \wedge q \wedge \sim r) \vee (\sim p \wedge \sim q \wedge r)$, which is to say, persistent in (*q*-alone) ∨ (*r*-alone), is either persistent in *q*-alone or persistent in *r*-alone, or persistent in both. In any of those three cases, (25a,b) guarantee that it is persistent in *p*-alone. So the application of the condition in (27a) counts toward (25c) being true. To check the application of (27b), suppose that a sequence is uniform in (*q*-alone) ∨ (*r*-alone). But this could be because *q*-alone and *r*-alone essentially alternate on the sequence, as worlds improve according to the ordering source, so that tails of the sequence, while uniform in (*q*-alone) ∨ (*r*-alone), are not uniform in (*q*-alone) or uniform in (*r*-alone). Thus, the condition in (27b) blocks the inference: (25c) does not follow from (25a,b).

Applying (27) using the terminology in (28), it's easy to check that the condition in (26) is satisfied. If $q$ is at least as likely as $p$, then, by (28b), every sequence weakly ordered by $\Omega(w)$ that is $p$-uniform is $q$-uniform. If must ($p$) is also true, that is, every weakly ordered sequence is $p$-uniform, then it follows that every weakly ordered sequence is $q$-uniform, so must ($q$) is true.

The truth conditions in (27) can serve as the foundation for developing truth conditions for should ($p$) that match those given by Kratzer. First, the truth conditions for "$p$ is a better possibility than $q$" are given in (29) and elaborated in (30).

(29) $p$ is a better possibility than $q$ iff
($p$ is at least as good a possibility as $q$) ∧ ~($q$ is at least as good a possibility as $p$).

(30) $p$ is a better possibility than $q$ iff
a. (i) every $q$-persistent sequence is $p$-persistent ∧
(ii) every $q$-uniform sequence is $p$-uniform
and
b. (i) ∃ a $p$-persistent sequence that is not $q$-persistent ∨
(ii) ∃ a $p$-uniform sequence that is not $q$-uniform.

Substituting ~$p$ for $q$ in (30), we obtain (31), which we will reduce to the truth conditions for should ($p$).

(31) $p$ is a better possibility than ~$p$ iff
a. (i) every ~$p$-persistent sequence is $p$-persistent ∧
(ii) every ~$p$-uniform sequence is $p$-uniform
and
b. (i) ∃ a $p$-persistent sequence that is not ~$p$-persistent ∨
(ii) ∃ a $p$-uniform sequence that is not ~$p$-uniform.

Now, any sequence that is not persistent in $p$ is persistent (in fact, uniform) in ~$p$. So every sequence must be persistent in $p$ or ~$p$ (or both). If it is persistent in ~$p$, then, by (31a–i), it is persistent in $p$. Thus, (31a–i) reduces to the condition that every sequence is $p$-persistent. The condition in (31a–ii) is vacuously quantifying, so is dropped: the conjunction of the condition in (31a–i) with one that is vacuously true reduces to just the condition in (31a–i). Thus, (31a) reduces altogether to the condition that every weakly ordered sequence is $p$-persistent. The condition in (31b–i) asserts that there is a $p$-persistent sequence that is $p$-uniform (if it is **not** ~$p$-persistent, then it is $p$-uniform), and (31b–ii) asserts that there is a $p$-uniform sequence that is, ipso facto, not ~$p$–uniform. Thus, (31b) reduces altogether to the condition that there is a $p$-uniform sequence. Putting these together, we get the truth conditions for should ($p$) in (32).

(32) should ($p$) iff $p$ is a better possibility than ~$p$ iff
  a. every weakly ordered sequence has a $p$-persistent tail
  b. there is a weakly ordered sequence with a $p$-uniform tail.

In natural language semantics, this recapitulates the results of Kratzer's analysis of weak necessity modals given in (16).

In summary, we skirt the problems posed by the Disjunctive Inference Schema for Kratzer's original analysis, and the revised analysis of Kratzer (2012: 41) in (21), by defining modal notions on weakly ordered sequences of worlds that do not converge to an ideal or have a maximum defined on them with respect to the ordering source. In doing so, we recover Kratzer's truth conditions for the weak necessity modal *should*, which we express as truth conditions in (16), and in the logical metalanguage as in (32). Pursuing this, we arrive at the same dual for *should*, given in (19) in Section 1.1.1.

## 1.2 A scalar approach to modality: Lassiter (2011)

Lassiter (2010, 2011) proposes a scalar theory of modality, and develops it in conjunction with a battery of critiques of Kratzer's possible worlds theory of modal interpretation. Lassiter's scalar approach to epistemic modal interpretation is sketched in Section 1.2.1. Lassiter's main critiques of Kratzer's theory of epistemic modality are presented in Section 1.2.2. Section 1.2.3 argues that, to handle iterated modal statements, Lassiter's theory must be more fully articulated in a way that brings it closer in line with possible worlds semantics, thus at least partly counterbalancing Lassiter's critique of the possible worlds approach to the semantics of modality. Section 1.2.4 argues for a reconceptualization of the problem, in which we admit the Semantic Pluralism of actual modal interpretation, and thereby give up the tit-for-tat comparison of contending theories, seeking instead to identify the different data structures involved in modal interpretation and to articulate the best formulations of each, without assuming that one must be a comprehensive theory. This supports a view of natural language semantics not as a fixed formal semantic system in which composition rules operate on specified lexical representations, but instead as a formal scaffolding of compositional rules that operate on a mix of lexically specified conventional semantic meanings and opportunistically constructed data structures that are used in appropriate contexts.

### *1.2.1 Modal scales and probability: Lassiter's theory*

Following Kennedy (1999b, 2007), a gradable adjective is a function of type ⟨e, d⟩, from individuals to degrees on a scale. A scale, on this approach, is just an ordered set of degrees, which can be bounded (or closed) at the bottom end, the top end, both, or neither. For example, *tall* maps an individual to the degree of height of the individual, a degree on a lower-bounded scale, $S_{tall}$, of height: $[[\textbf{tall}]]^{c,w} = \lambda x$ [height $(x)$]. The adjective *full*, applied to a glass to say how full it is of liquid, maps an individual to a degree on a scale of fullness, $S_{full}$, which is bounded at both ends. This permits a direct interpretation of comparatives applied to scalar adjectives, resulting in comparative expressions such as *taller* or *less full*. An unadorned use of a scalar adjective with no upper bound on the scale, as in *Max is tall*, and with no overtly expressed degree modification or comparison, may be understood as relative to an implicit threshold on the associated scale of the adjective, determined by that adjective, its scale, and the context. Kennedy (2007) proposes that this threshold is introduced compositionally by a silent morpheme, **pos**, with the denotation in (33), where $A_{\langle e,d\rangle}$ is the denotation of the adjective; then the phrase marker **Max is pos tall** has the truth conditions given in (34).

(33) $[[\textbf{pos}]]^{c,w} = \lambda A_{\langle e,d\rangle}\, \lambda x_e\, [A(x) \geq \theta_A]$

(34) $[[\textbf{Max is pos tall}]]^{c,w} = 1$ iff height $(\textbf{m}) \geq \theta_A$

For a maximum standard adjective, such as *full*, the threshold $\theta_A$ in (33) is understood to be the top point of the associated scale, $S_{full}$, so in this case, "≥" reduces to "=", and consequently **The glass is pos full** is true just in case the degree of fullness of the glass is equal to $\max(S_{full})$.

The denotation of a scalar adjective, and the associated scale invoked, do not have to be taken as primitives of the theory. As Lassiter (2011: 28–33) notes, they can be derived as the real-number images under a homomorphism from a more basic scale obtained from an order relation $\succcurlyeq_P$ defined on elements of a set X according to how much they exemplify a property P associated with the adjective. Assuming that a weak order (transitive, reflexive, complete on X) can be defined on elements of X, reflecting their relative size or loudness, for example, a corresponding simple order (transitive, reflexive, complete, and antisymmetric) can be defined on equivalence classes of members of X under $\succcurlyeq_P$. The structure $\langle X, \succcurlyeq_P\rangle$ can be mapped by a homomorphism $\mu$ to the real-number numerical structure $\langle R, \geq\rangle$ as in (35).

(35) For any $x$, $y$ in X,
  a. $\mu(x) \in R$ and $\mu(y) \in R$, and
  b. if $x \succcurlyeq_P y$, then $\mu(x) \geq \mu(y)$.

Lassiter (2011: 66) argues that *possible* is a minimum-standard adjective, like *bent*, which asserts attainment of a minimum threshold corresponding to the bottom of the associated scale. The scale for possibility, $S_{possible}$, is defined in terms of the set of propositions $\Phi$ to which *possible* can apply, a preorder or weak order $\succcurlyeq_{possible}$, expressing comparative possibility, and a bottom element, $\perp_{possible}$.

(36) $S_{possible} = \langle \Phi, \succcurlyeq_{possible}, \bot_{possible} \rangle$, where
a. $\Phi$ is a set of propositions;
b. $\succcurlyeq_{possible}$ is a preorder or weak order on $\Phi$; and
c. $\forall p \in \Phi, p \succcurlyeq_{possible} \bot_{possible}$.

With respect to this scale, Lassiter expresses the interpretation of the possibility modal as in (37), where iff $p \succ_{possible} \bot_{possible}$ means $p$ is strictly greater than $\bot_{possible}$ on $S_{possible}$. The corresponding condition in real numbers under the homomorphism $\mu_{possible}$ is given in (38).

(37) $[[p \text{ is } \textbf{pos} \text{ possible}]]^{c,w} = 1$ iff $p \succ_{possible} \bot_{possible}$
(38) $[[p \text{ is } \textbf{pos} \text{ possible}]]^{c,w} = 1$ iff $\mu_{possible}(p) > \mu_{possible}(\bot_{possible})$

The interpretation of *likely* or *probable* is similar, asserting a position on the associated scale at or above an appropriate threshold, $\theta_{likely}$, as in (39).

(39) $[[p \text{ is } \textbf{pos} \text{ likely/probable}]]^{c,w} = 1$ iff $p \succcurlyeq_{likely} \theta_{likely}$

However, Lassiter proposes a more richly structured scale for *likely* since it must accommodate proportional modification (e.g., *That the economy will grow this year is 80% likely*), which is commonplace with *likely* and *probable*. The structure he proposes is a ratio scale, in which $\succcurlyeq_{likely}$ is a weak order that respects combinations of disjoint propositions under a concatenation operation $\circ$, by obeying a set of axioms stating additivity and multiplier effects on those combinations.[7] These constraints make $\succcurlyeq_{likely}$ tractable for expressing proportional statements. The resulting scale for *likely* is as given in (40), from Lassiter (2011: 67).

(40) $S_{likely}$ is a ratio scale $\langle \Phi, \succcurlyeq_{likely}, \circ \rangle$, where
a. $\Phi$ is a set of propositions;
b. $\succcurlyeq_{likely}$ is a weak order on $\Phi$; and
c. $\circ$ is a partial binary operation on disjoint members of $\Phi$ that is associative, and whose interactions with $\succcurlyeq_{likely}$ satisfy a set of properties (monotonic, Archimedean, positive, regular) that make proportional statements well-behaved under the homomorphism $\mu_{likely}: \Phi \rightarrow \mathrm{R}$.[8]

[7] The definition is in Lassiter (2011: 44–45), developed from standard assumptions distilled from Krantz et al. (1971), which Lassiter (2011: 39–45) presents, adapts, and revises.

[8] We record here what Lassiter assumes, although the exact definition of a ratio scale will not concern us here. $\langle \Phi, \succcurlyeq, \circ \rangle$ is monotonic: $\forall x, y, z \in \Phi, x \succcurlyeq y$ iff: $(x \circ z) \succcurlyeq (y \circ z)$ iff $(z \circ x) \succcurlyeq (z \circ y)$. $\langle \Phi, \succcurlyeq, \circ \rangle$ is positive: $\forall x, y \in \Phi$, if $x \circ y$ is defined, then $x \circ y \succ x$. $\langle \Phi, \succcurlyeq, \circ \rangle$ is regular: $\forall x, y \in \Phi$, if $x \succ y$, then $\exists z$: $x \succcurlyeq (y \circ z)$. $\langle \Phi, \succcurlyeq, \circ \rangle$ is Archimedean: every bounded sequence of equally spaced elements of $\Phi$ under $\succcurlyeq$ is finite; that is, there are no infinite sequences of elements of $\Phi$, monotonically increasing or decreasing under $\succcurlyeq$, which are bounded. The Archimedean condition ensures that successive elements of the domain do not become arbitrarily small, or infinitesimal, thus assuring that proportional statements are well defined (e.g., do not effect division by zero).

With these properties, we can reliably interpret the statement that $p$ is 80% likely, for example, in terms of $\mu(p)/\mu(p \circ {\sim}p)$, for any admissible measure $\mu$ on $\Phi$.

Finally, Lassiter (2011: 68) argues that *certain* has the scalar properties of the adjective *full*, which admits proportional statements (half full, 80% full), a bottom element characterized by the antonym *empty*, and a top element that can be realized (*the glass is full*) or approximated (*almost full*, *nearly full*). Lassiter concludes that the scale for *certain* is a fully closed ratio scale, with a top element, $\top_{certain}$, as well as a bottom element, $\bot_{certain}$. The scale is given in (41), and the scalar and measure-theoretic interpretations (which are equivalent) in (42).

(41) $S_{certain}$ is a fully closed ratio scale $\langle \Phi, \succcurlyeq_{certain}, \circ, \bot_{certain}, \top_{certain} \rangle$ where

a. $\Phi$ is a set of propositions;

b. $\succcurlyeq_{certain}$ is a weak order on $\Phi$;

c. $\forall p \in \Phi$, $\top_{certain} \succcurlyeq_{certain} p \succcurlyeq_{certain} \bot_{certain}$; and

d. $\circ$ is a partial binary operation on disjoint members of $\Phi$ that is associative, and whose interactions with $\succcurlyeq_{certain}$ satisfy a set of properties (monotonic, Archimedean, positive, regular) that make proportional statements well-behaved under the homomorphism $\mu_{certain}: \Phi \rightarrow \mathbb{R}$.

(42) a. $[[p \text{ is } \mathbf{pos} \text{ certain}]]^{c,w} = 1$ iff $p \approx_{certain} \top_{certain}$

b. $[[p \text{ is } \mathbf{pos} \text{ certain}]]^{c,w} = 1$ iff $\mu_{certain}(p) = \mu_{certain}(\top_{certain})$

Each of the three epistemic scales, $S_{possible}$, $S_{likely}$, and $S_{certain}$, was defined by Lassiter to have only the minimum structure required to accommodate the clearest and most uncontroversial properties of its associated natural language modal operators. The question at this point is whether each range of modal operators has its own scale, or whether they all access a single, quite richly structured epistemic scale, which has all the structural properties that any one of them would need, but in such a way that each uses only those properties of the scale that pertain to it. Lassiter acknowledges that this question is not easy to settle without subtle argumentation. For example, for the adjective *tall*, the infelicity of the expression *seventy percent tall* clearly implicates a minimum-bounded ratio scale such as that for *likely* in (40), not a maximum-bounded scale with a top element such as (41). In such a case, a clear strategy for explaining the infelicity is to assume that interpretive properties of the adjective are conferred by the scale on which the adjective is interpreted and then invoke a scale that lacks the features required to support the missing interpretive property evidenced by the infelicity. Unfortunately, as Lassiter (2011: 70) notes, the adjectives *likely*, *probable*, and *possible*, which we might expect to follow suit with *tall* if they are interpreted on the scales in

(36) and (40), actually occur in proportional statements with percents (*percent likely*, *percent possible*) much more often than *percent tall*, so we cannot so readily infer that their scales lack a top element. And since *possible* is attested quite often in proportional statements, we cannot readily conclude that it is interpreted on a minimum-bounded scale such as (36), rather than a ratio scale such as in (40). Lassiter (2011: 68–75) discusses this at some length, and although he acknowledges some semantic differences (e.g., #*slightly probable* is infelicitous, contrasting with *slightly possible*), he takes the ubiquity of degree modification across all three classes of epistemic modals to count as preponderant evidence that they are all interpreted on a single scale, a fully closed ratio scale, $S_{epistemic} = \langle \Phi, \succcurlyeq_{epistemic}, \circ, \bot_{epistemic}, \top_{epistemic} \rangle$ with the structural properties the same as those in (41). Hereafter, reference to Lassiter's scalar theory of epistemic modality will be neutral between the single-scale and multiple-scales views, but will be presented notationally using the single scale, $S_{epistemic}$.

Lassiter (2011: 109–119) proceeds to argue that modal adjective interpretation is sensitive to alternatives, which are reflected in information-theoretic focus/ground distinctions, including alternatives formulated under Roberts's (1996) notion of a "Question under Discussion" in a discourse. Lassiter credits the restriction to alternatives provided by pragmatics or information structure for many of the asymmetries in modification possibilities exhibited among *possible*, *likely*, and *certain*, attributing such asymmetries to this factor rather than to any difference in the underlying scales.

Lassiter argues that the scalar theory of modal interpretation successfully accounts for semantic properties of modal statements, while skirting the objections and pitfalls that he alleges for possible worlds semantics of modal interpretation, including the Disjunctive Inference Schema discussed in Section 1.1.2. The claims regarding his scalar semantics will be granted here without further discussion. The next three sections (1.2.2–1.2.4) discuss whether these successes entitle the scalar theory to supplant possible worlds semantics in the theory of modal interpretation.

### *1.2.2 Lassiter's case against Kratzer's theory of epistemic modal interpretation*

Lassiter's case against possible worlds semantics for epistemic modality, presented at length in Lassiter (2011: 51–101), finds three major faults with Kratzer's (1991) theory. The first is that it wrongly predicts that the Disjunctive Inference Schemas presented in (19) and (24) should be valid. As discussed in Section 1.1.2, this is a problem for Kratzer's formulation of "*p* is at least as likely as *q*," but not for the alternative given in (27) in that section, repeated here in (43).

(43) $p$ is at least as good a possibility as $q$ iff
a. every sequence weakly ordered by $\Omega(w)$ that has a tail persistent in $q$ has a tail persistent in $p$, and
b. every sequence weakly ordered by $\Omega(w)$ that has a tail that is uniform in $q$ has a tail that is uniform in $p$.

This formulation is in the spirit of Kratzer's approach, and it was shown in Section 1.1.2 that this formulation leads to Kratzer's truth conditions for the weak necessity expressed by *likely* or *should*. The novel terminology used here may confer some advantage in lending facility with the notions in Kratzer's semantics for modality in a way that favors finding more intuitive semantic analyses, and ones that, as with (43) vis-à-vis the Disjunctive Inference Schema, are more successful. Lassiter's first objection is thus countered on grounds that stay well within the parameters of Kratzer's approach to epistemic modality.

Lassiter's second objection is that Kratzer's theory of modal interpretation does not provide the resources to interpret proportional modal statements, such as that $p$ is 90% certain, or that $p$ is twice as likely as $q$. In contrast, Lassiter's theory of modal interpretation integrates interpretation on an epistemic scale with the representational theory of measurement, capturing proportional modal statements in a way that is exactly analogous to proportional statements with other relative adjectives, and therefore, from the point of view of semantic theory generally, at the cost of no special theoretical machinery.

Starting with Kratzer's possible worlds theory of epistemic modal interpretation, it might be possible to address proportional modal statements by defining associated scales, and mapping them to a probability space. But this would simply recapitulate Lassiter's scalar approach to epistemic modality, and in the resulting theory, the scalar part would clearly be doing all the work in accommodating proportional epistemic modal statements. Alternatively, one might seek to map sets or sequences of possible worlds in possible worlds semantics directly to a probability space. It's not immediately clear how this should be done, and if it were done, it might be claimed that this really vindicates the scales and probability approach after all, since it seeks to reproduce half of that approach in the theoretical context of Kratzer's possible worlds semantics. I will take it that Lassiter's scalar theory has a clear advantage over possible worlds semantics in accounting for proportional modal statements. But Section 1.2.4 will argue that this does not confer on Lassiter's theory the status of an exclusively correct, comprehensive theory.

Lassiter's third objection is that Kratzer's theory predicts that many pairs of propositions (construed as sets of possible worlds) are not comparable, in principle, according to epistemic modal relations such as the statement that $p$ is more likely than $q$, including cases where they are, in fact, intuitively comparable. In particular, Kratzer's theory renders incomparable many pairs of propositions that would be considered readily comparable when we employ any kind of quantitative accounting of likelihood or probability.

Lassiter's (2011: 62) example of Bill coming home can serve as a focal point for the discussion. The ordering source $\Omega(w)$ is the set of expectations E1, E2, and E3 given in (44), and the relevant worlds are given in (45), accompanied by a specification of which members of $\Omega(w)$ each world satisfies.

(44) E1. Bill is at home by 6 PM.
E2. Bill drives his car.
E3. Bill has macaroni for dinner.

(45)
| | |
|---|---|
| $w_1$: | All of E1, E2, E3 are satisfied. |
| $w_2$, $w_3$: | Only E2, E3 are satisfied. |
| $w_4$: | Only E1, E3 are satisfied. |
| $w_5$, $w_6$: | Only E1, E2 are satisfied. |
| $w_7$: | Only E3 is satisfied. |
| $w_8$: | Only E2 is satisfied. |
| $w_9$, $w_{10}$: | Only E1 is satisfied. |
| $w_{11}$, $w_{12}$: | None is satisfied. |

The ordering $\leq_{\Omega(w)}$ is diagrammed in (46), where $[w_i]$ is the equivalence class of worlds that satisfy the exact same constraints as $w_i$.

(46)

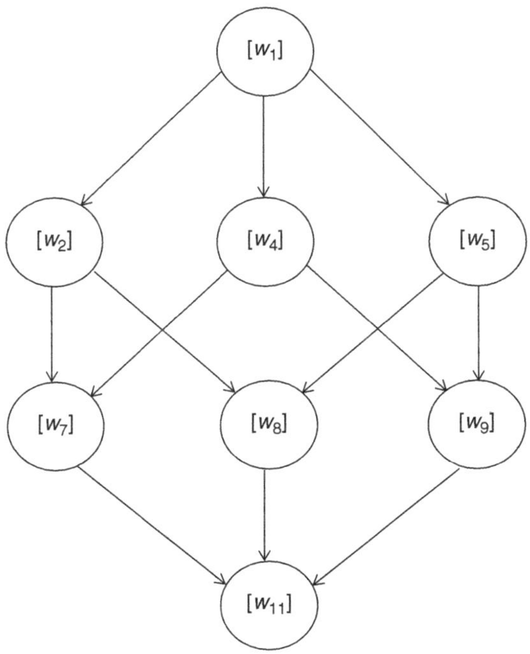

With this backdrop, Lassiter invites us to compare the likelihood or plausibility of two propositions, $p = \{w_5, w_6, w_8, w_9, w_{10}, w_{11}\}$, and $q = \{w_7, w_{11}, w_{12}\}$, according to Kratzer's theory.

Kratzer's (1991) truth conditions for "*p* is at least as likely as *q*," given initially in (12), are repeated in (47) in notation suitable for the purposes of the present discussion.

(47) $p$ is at least as likely as $q$ in $w$ iff $(\forall u \in q)\ (\exists v \in p)$: $v \leq_{\Omega(w)} u$

As Lassiter notes, with this definition, *p* and *q* are incomparable with regard to likelihood – neither *p* nor *q* is at least as likely as the other. The revised definition proposed here, given in (43), has a mirror image situation: *p* and *q*, while distinct, are each at least as likely as the other. But the intuition, Lassiter claims, is that *p* should be more likely than *q* since it contains more worlds that are plausible, at least in terms of satisfying the constraints E1, E2, and E3.

But we can note that this intuition holds only on the supposition that the probabilities of E1, E2, and E3 are roughly equal. If, in fact, E3 were very strongly held, considered much more likely to be satisfied than the others, then *q* would contain more plausible worlds than *p*, and might be deemed more likely overall. How we really feel about the relative likelihood of *p* versus *q* depends on the relative strengths with which we hold the constraints. To capture this dimension of our intuitions, we would have to perform a sophisticated calculation using probability theory. This suggests that Lassiter's diagram is an idealization towards mathematical formulation of the problem. And this is a perfectly legitimate thing to do in a context that calls for it. The import of the example, then, is that it highlights where we make a transition from a possible worlds conception of modal notions to a more precise, mathematical one. I am willing to grant that Lassiter has provided a fully adequate formalization of the latter. But that does not, by itself, show that the former is useless.

The present claim is that if we have quantitative measures of the probabilities of the three constraints in this example, then we can get a reliable quantitative reckoning of the relative likelihood of *p* versus *q*, and this will provide a basis for interpreting comparative modal statements whose truth conditions can be mathematically sustained. But if we don't have any precise measures of the relative strengths or likelihood of the three constraints, then it might be entirely appropriate to say that, for all we know, each of *p* and *q* is as likely as the other in this case, reserving the statement that *p* is at least as likely as *q*, and not vice versa, for situations where every *q*-persistent sequence is *p*-persistent, and every *q*-uniform sequence is *p*-uniform, but not vice versa. Our intuitions demand different levels of granularity and precision depending on our purposes. Lassiter's critique, in effect, assumes that when we get serious, our purposes always favor quantitative precision. That assumption is true in certain scientific and professional contexts, and may be true in casual contexts in which the discourse is modeled after professional quantitative contexts. But most causal contexts don't demand quantitative precision, and more plausibly track merely the persistence and uniformity of worlds that satisfy epistemic conditions.

To summarize the argument, the case when we can simply look at a diagram and count worlds is a special, mid-range case. More typically, either E1, E2, and E3 have different probabilities, and then comparability of the likelihood of *p* and *q* is a nonintuitive probability calculation, or we have lots of infinite sequences of worlds, and a proposition *p* with nonzero probability typically involves at least one infinite sequence that is at least persistent in ~*p*, in which case, probability calculations are probably off the table, and we are more likely to rely on intuitions of persistence and uniformity on weakly ordered sequences of worlds.

### 1.2.3 *Iterated modal statements: scales and possible worlds*

In this section it is argued that there are iterated modal statements that have natural interpretations in a Kratzer-style possible worlds semantics for modality, because this interpretation can be readily iterated. Lassiter's scale-based approach can handle these only if it is more fully articulated in a way that brings it closer to traditional possible worlds semantics for modality. Consider first the modal statement in (48).

(48) No matter what, it is possible that *p*.

This statement can be used to emphasize that, no matter how wild things get, you can always find a *p*-world "eventually," when worlds get realistic enough under the ordering source to really count for the modal generalization. Or it can mean that, no matter in how many cases *p* is eliminated as a possibility as we get closer to the way the world of evaluation is assumed to be, we never totally eliminate *p* as a possibility. This second meaning critically appeals to the persistence of *p* on sequences of worlds. Or it can mean (49).

(49) It is necessarily possible that *p*.

The modal statement in (49) is readily interpretable in Kratzer's theory with the truth conditions in (50), which we can derive compositionally.

(50) every sequence of worlds in the epistemic modal base that is weakly ordered by $\leq_{\Omega(w)}$ has a tail uniform in the existence, from each world on that tail, of a sequence that is persistent in *p*, $=_{\mathrm{df}}$
$\forall u \in \cap\mathrm{E}(w), (\exists v \in \cap\mathrm{E}(w)\text{: } v \leq_{\Omega(w)} u)\ (\forall z \in \cap\mathrm{E}(w)\text{: } z \leq_{\Omega(w)} v)$
$\exists u' \in \cap\mathrm{E}(z)\text{: } (\forall v' \in \cap\mathrm{E}(z)\text{: } v' \leq_{\Omega(z)} u')\ (\exists z' \in \cap\mathrm{E}(z)\text{: } z' \leq_{\Omega(z)} v')$
$[[p]]^{c,z'} = 1.$

Furthermore, we can interpret (49) using the same epistemic modal base and ordering source, or we can switch them in midstream. For example, we might switch them if we are talking about the possibility that the stock market will hold firm (not drop precipitously) today under a variety of external conditions. A speaker might conclude, based on the usual internal indicators, that it is possible that the market will hold its value. An interlocutor might then bring up external events – the possibility of war in the Middle East, or recent events affecting the stability of the euro – and after running through how the internal factors fare under each scenario constructed on the basis of various external factors, the speaker might conclude that in all scenarios based on external factors, it is possible that the market will hold firm. In (50), $w$ is the world of evaluation of the entire iterated modal utterance, $z$ is arrived at through consideration of the external factors, and for each such $z$, the possibility of $p$ based on internal factors is computed.

In possible world semantics of modality, we can interpret (49) because a world that is arrived at in the interpretation of the outer modal operator can serve as the starting point for the interpretation of the inner modal operator. On a compositional interpretation, the truth conditions of *(it is) possible that p* are computed first, using the semantics of the possibility modal, yielding as interpretation the second row of quantified conditions in (50), expressed as truth conditions of $p$ at $c$, $z'$. Applying the interpretation of the necessity modal to this, compositionally, we get the first row of quantified conditions in (50).

For a measure-theoretic approach, Lassiter (2011: 75) defines a scale for epistemic modality, $S_{epistemic} = \langle \Phi, \succcurlyeq_{epistemic}, \circ, \bot_{epistemic}, \top_{epistemic} \rangle$, with the same properties as the scale for *certain*, given in (41), and argues that statements of epistemic necessity, likelihood, and possibility are all interpreted on this scale. Points on the scale are mapped by a measure $\mu$ to a Finitely Additive Probability Space, defined on the real number interval [0, 1]. The truth conditions for *p is possible*, or *It is possible that p*, are given in (51a,b), repeated from (37) and (38); in these representations, **pos** is a silent operator, adopted from Kennedy (2007), that establishes the inequality $\succcurlyeq_{epistemic}$ with respect to a contextually determined threshold, which, for possibility, is the bottom element of the scale, $\bot_{epistemic}$, and for certainty, is the top element of the scale, $\top_{epistemic}$. This is matched by an inequality in $\mu_{epistemic}(p)$.

(51) a. [[$p$ is **pos** possible]]$^{c,w}$ = 1 iff $p \succ_{epistemic} \bot_{epistemic}$
b. [[$p$ is **pos** possible]]$^{c,w}$ = 1 iff $\mu_{epistemic}(p) > \mu_{epistemic}(\bot_{epistemic})$

Consider (49) from this perspective. A compositional interpretation would first compute the truth conditions of *(it is) possible that p*, yielding the results in (51). The interpretation of *necessarily* in (49) is that, for some proposition (set of worlds) $q$, $q$ is coincident with the top element of the scale, and therefore has a measure equal to the measure of the top element.

(52) a. $[[q \text{ is } \mathbf{pos} \text{ necessary}]]^{c,w} = 1$ iff $q \approx_{epistemic} \mathrm{T}_{epistemic}$
b. $[[q \text{ is } \mathbf{pos} \text{ necessary}]]^{c,w} = 1$ iff $\mu_{epistemic}(q) = \mu_{epistemic}(\mathrm{T}_{epistemic})$

How do we compose the two? It would be straightforward to define a composite operator necessarily possible, with its own scale. But that seems to miss the point.

To obtain a proposition $q$ – a set of worlds – in applying (52) to the interpretation of (49), we must interpret ***pos** be possible that p* as the set of worlds $v$ at which ***pos** be possible that p* would be true, using an epistemic scale and measure appropriate for each world. That is, we must first interpret ***pos** be possible that p* at world $v$, with the epistemic scale and its components dependent on $v$, as in (53), where the dependence of the epistemic scale and its components on the world $v$ is recorded in the subscript. The measure $\mu$ would depend on the scale, and therefore also on $v$.

(53) a. $[[p \text{ is } \mathbf{pos} \text{ possible}]]^{c,v} = 1$ iff $p \succ_{epistemic,v} \perp_{epistemic,v}$
b. $[[p \text{ is } \mathbf{pos} \text{ possible}]]^{c,v} = 1$ iff $\mu_{epistemic,v}(p) > \mu_{epistemic,v}(\perp_{epistemic,v})$

The argument of the higher predicate (*necessarily*) is obtained as the λ-abstract on $v$, on the right-hand side of (53).

To interpret (49), we must build compositionally from the interpretation of ***pos** be possible that p* in (53), by applying (52), with

(54) a. $q = \lambda v\, [p \succ_{epistemic,v} \perp_{epistemic,v}]$, or
b. $q = \lambda v\, [\mu_{epistemic,v}(p) > \mu_{epistemic,v}(\perp_{epistemic,v})]$.

On this approach, the scale $S_{epistemic,v}$ is relative to the world from which the modal relation (certainty or possibility) is being computed. The result of abstracting on $v$ on the right-hand side of (53) is the property of a world $v$ that holds of $v$ just in case $p$ is possible **at** $v$.

Once we realize this, it sheds new light on the interpretations of unembedded epistemic modal statements. For the most vivid illustration, consider Lassiter's interpretation of *p is **pos** likely*, given in (39), which, in view of relativization of the scale to the world of evaluation, is elaborated in (55).

(55) a. $[[p \text{ is } \mathbf{pos} \text{ likely}]]^{c,w} = 1$ iff $p \succcurlyeq_{likely,w} \theta_{likely,w}$
b. $[[p \text{ is } \mathbf{pos} \text{ likely}]]^{c,w} = 1$ iff $\mu_{epistemic,w}(p) \geq \mu_{epistemic,w}(\theta_{likely,w})$

Suppose that many of the worlds in $p$ are, in fact, incompatible with $w$ (or inaccessible from $w$, in older parlance). How do we then place $p$ on the scale $S_{epistemic,w}$ to determine whether it is above the threshold $\theta_{likely,w}$, and thus to determine whether the modal statement is true? Only the accessible worlds count for this placement. If $R$ is the accessibility relation, we are in effect not placing all of $p$ on the scale $S_{epistemic,w}$, but rather only the part $\{z \in p: wRz\}$.

Likewise, in interpreting (49) using (52) and (54), the property that defines $q = \lambda v\ [p \succ_{epistemic,v} \bot_{epistemic,v}]$ needs, for each world $v$ to which it applies, to be sensitive to how much of $p$ is accessible from $v$. The property is more properly defined as in (56).

(56) $\lambda v\ [\{z \in p: vRz\} \succ_{epistemic,v} \bot_{epistemic,v}]$

It seems that Lassiter's scalar account implicitly incorporates some elements of traditional possible worlds semantics for modality. Furthermore, if we sought to remove the dependence on $v$ from (56) in order to recover a unitary scale $S_{epistemic}$, the order $\succcurlyeq_{epistemic}$ on that scale, in view of inaccessibilities, would not be a complete order, therefore not a weak order on $\Phi$, compromising the status of $S_{epistemic}$ as a ratio scale that can be mapped to a probability space.

On the quantificational theory, the relativization to possible worlds is built in since the worlds in the nested condition are constrained to be members of weakly ordered sequences with respect to the ordering source, as specified by the semantics of the modal operators.

### *1.2.4 Semantic pluralism in epistemic modal interpretation*

The example from Lassiter in (44)–(46) was intended to show that Kratzer's possible worlds semantics makes a demonstrably false prediction for this case, that the propositions $p$ and $q$ are incomparable. The response offered in Section 1.2.2 was that this is essentially question-begging. If we want mathematical precision in this example, then we have to specify the strengths or likelihoods of each of E1, E2, and E3 and do a probability calculation. If we aren't after mathematical precision, and especially if the strengths or relative likelihoods of E1, E2, and E3 are not known, then we might well be prepared to say that $p$ and $q$ are incomparable in this example. The discussion of Section 1.2.3 suggested that Kratzer's theory of epistemic modal interpretation may be more directly suited to interpret examples of modal iteration such as (49). It seems that the lesson to be drawn from these discussions transcends tit-for-tat comparisons of semantic theories. Lassiter's scalar theory illuminates the interpretation of epistemic modal statements in some contexts, especially ones calling for the mathematical precision of probability measures. Kratzer's theory may illuminate epistemic modal interpretation better in other contexts. For example, in stressing the impossibility of a particular proposition $p$, a speaker might choose to say that there is "no way" $p$ could be true, "no matter what" – that is, no matter how favorable the circumstances (expressed in a delineated set of background propositions) may be, thus stressing the persistence and uniformity of $\sim p$ on weakly ordered sequences of worlds. Or the same speaker could choose to emphasize that the probability of $p$ being true is absolutely zero. This suggests that language users choose from a range of data structures that they might employ in producing and interpreting these utterances. To the extent that this is correct, semantic theory will not find unique lexical semantic meanings for these predicates, but rather will elucidate the different data structures that are employed, and how each is interpreted in their employment on a particular occasion of use.

It is plausible that some adjectives have associated scales as part of their conventional lexical semantic meaning. And at least for mathematically literate speakers, a measure, homomorphically mapping the scale to the real numbers, may also be part of the conventional semantic meaning. But it is clear that in other cases, scales and measures are imposed by speakers. A speaker might claim to be 90% in love with a song, to indicate the strength of their emotional response to it, or their appreciation of the combination of musical cleverness and restraint that went into composing it. The predicate *in-love-with* would not have to be endowed with a suitable scale as part of its lexical semantic meaning for this communication to work; the speaker and addressees construct suitable structures and choose to "see" the interpretation of *in-love-with* in scalar terms. We do not have to assume that *in-love* is polysemous, having divergent lexical semantic representations, to get these interpretations. We can, instead, take a stance of Semantic Pluralism – that various data structures and modes of interpretation are available to language users, some of them quite culturally specific (such as mathematical or technical notions), and users can choose one or another of these quite freely, to structure the conceptualization of a domain as they choose, to communicate a particular idea, or to cast the subject matter in a particular light.

The present proposal is that epistemic modal interpretation exhibits such Semantic Pluralism – that we employ at least two quite distinct systems of semantic interpretation of epistemic modal statements. One, captured by Lassiter, originates in quantitative domains, and deals with fully mathematized notions of possibility, likelihood, and necessity. Another originates in our intuitions of possible ways the world could be, a notion that can be readily displaced from actuality by having its center moved to a nonactual possible world. The quantitative version can be employed in everyday life, but it need not be, and is not necessarily implicit in more casual modal notions, and is not necessarily continuous with them. In fact, as Lassiter is well aware and acknowledges, probability theory itself often thwarts "common sense" intuitions of probability.

## 1.3 What is the epistemic modal base?

Having considered the forms of interpretation of modal statements in possible worlds semantics, let us turn to a consideration of the epistemic modal base, which is part of the foundation of those interpretations.

The epistemic modal base is often characterized as what is known to the speaker, or evidence available to the speaker. That for which a speaker has evidence could, at least on a casual understanding, be quite different from that which is known to the speaker, since knowledge entails veridicality – what is known is true – whereas that for which the speaker has evidence need not be true. But in some special cases, these two conceptions of the epistemic modal base can be reconciled with one another. If a man is approaching Max and Max thinks to himself, "That man might be Fred," the evidence available to Max on the basis of which he makes this judgment would be the appearance with which he is presented, together with his memories of Fred, and these can all be quite veridical: Max might have accurate memories of Fred, and Max might be fully correct in evaluating the appearances presented to his senses, on the basis of which he makes the epistemic modal judgment. The approaching man might have Fred's gait and hairstyle, and be identical to Fred in any other respect that Max can judge from a distance, and yet not be Fred. The evidence available in such a solipsistic case as this can be construed quite narrowly as appearances about which the speaker has incontestable knowledge. We can characterize this as knowledge at the phenomenal level – knowledge of what is presented to the senses.

But, as well as it works for this special case, the conflation of available evidence with knowledge of appearances may not be sustainable in general. If the basis for a modal judgment involves expected interactions among people and things external to the speaker, then reduction to appearances presented to the speaker would be essentially sophistical, since it would provide no basis for how relevant properties cohere as they enter into interactions with one another, as a basis for the modal judgment. In such a case, the evidence available to the speaker might seem to be a more promising conception of the epistemic modal base than phenomenal-level knowledge. The theory of modal interpretation therefore has a substantive issue to settle. Does the epistemic modal base reflect evidence available to the speaker, or does it reflect knowledge of the speaker?

The theory of modal interpretation has another substantive issue to settle: whether the evidence or knowledge at issue is invariably that of the speaker, or whether it could be the evidence or knowledge of another party to the discourse in which the modal statement is made, or somebody external to the discourse. This has been extensively discussed in the literature.

We pursue these two issues, reviewing relevant proposals in the literature and making some supplementary observations and proposals as we go. We will adopt views that have been developed in the literature on these points, reviewed in Section 1.3.1, but informed by the observations and arguments contributed by the present work, given in Section 1.3.2. Examples cited from the literature will be reworded or slightly reworked as needed for expository convenience.

### *1.3.1 The epistemic modal base according to von Fintel and Gillies*

This section presents and discusses examples and ideas from the literature, in particular, recapitulating the argument of von Fintel and Gillies (2008b) in some detail since their conclusion, stated at the end of the section, will play an essential role later in the present work. Section 1.3.2 presents some new examples and observations, elaborating the views of von Fintel and Gillies regarding the nature of the epistemic modal base for purposes of the present work.

Consider the example of contrary modal judgments in (57), from Kratzer (1986). A man is approaching Jill and Max, but he is partially obscured by foliage. Jill speaks first, straining to see through the leaves of a low-lying tree branch, although it is mutually manifest to Jill and Max that Max is better placed to see the approaching man.

(57) Jill: That guy [ostending the approaching man] might be Fred.
Max: That guy cannot be Fred.

If we evaluate each statement with respect to evidence or information available to the speaker, then Jill's utterance is true, based on her evidence, and Max's utterance is also true, based on his evidence. This captures the intuition that each of Jill and Max were justified in making their respective utterances, but, of course, that doesn't necessarily mean that both are true. But if the epistemic modal bases in (57) are taken to be information available to the speaker for each utterance, or knowledge at the phenomenal level, Kratzer's theory of modal interpretation predicts that each of the two modal statements in (57) actually has truth conditions that make it true in the context of the utterance. So far, so good. But this doesn't capture the respect in which the utterances of Jill and Max seem to be in tension with one another, or even contradictory to one another.

And it doesn't account for why we can have outright contradiction, as in the constructed discourse in (58), adapted here from versions given by McFarlane (2003) and Hawthorne (2004), which were also discussed by von Fintel and Gillies (2008b).

(58) [Monday. Mary, Sally, and George at their office in New York City]
Mary: Where's Joe this week?
Sally: He might be in Boston. The draft duty roster for this week listed him there.
George: No, he can't be in Boston. I just saw him in the hallway five minutes ago.

If Sally's statement is to be evaluated with respect to her evidence, then George has no grounds on which to contradict Sally since her statement that Joe might be in Boston is true when evaluated with respect to Sally's evidence. And on this view, George cannot contradict Sally based on his own evidence since Sally's utterance was not made based on George's evidence. So George seems to have no basis on which to contradict Sally.

These examples are also a problem for the view that the modal statement is evaluated with respect to evidence shared by the speaker and addressees in the discourse. To favor this explanation, we could assume that evidence that accompanies the modal statement can count as shared for the establishment of the modal base. That way, Sally's utterance establishes the fact of the duty roster as shared for the evaluation of her modal statement, and George's utterance establishes the fact that he saw Joe in the hallway five minutes ago as shared for the evaluation of his modal statement. However, there is still a problem with (58) in the fact that George contradicts Sally. At the moment of Sally's utterance, George's information about seeing Joe in the hallway hadn't yet been introduced, and therefore could not have been shared. Thus, it seems that Sally's modal statement should still count as true, given the context in which she uttered it. This is exacerbated by the observation of MacFarlane (2003) that (59a) is an appropriate follow-up for Sally, but not (59b).

(59) Sally:
  a. Oh, then I guess I was wrong.
  b. # Oh, then I guess he can't be in Boston. But I nevertheless uttered a true statement when I said "He might be in Boston."

Although George establishes a shared context in which Joe cannot be in Boston, this context should not be operative for the evaluation of Sally's original modal statement, if the context of evaluation is Sally's knowledge, or shared knowledge, at the moment she makes her statement. Likewise, in (57), the intuition is that Jill's modal statement must be regarded as false in light of Max's statement from his better vantage point. But what Max observes from his vantage point is not shared evidence at the moment when Jill makes her statement, so Jill's modal statement is not clearly false; in fact, if Jill's informationally more limited observations count as the shared minimum between the two of them, as seems more reasonable, and modal statements are to be evaluated with respect to shared evidence or shared knowledge, then Jill's utterance would even count as true, and Max's as false.

A more promising approach to (57), (58), and (59) is to suppose that epistemic modal statements are evaluated with respect to evidence distributed through the group consisting of speaker and addressee(s). On the simplest formulation of this view, a *might* modal is true if it is consistent with evidence available to one or more members of the group. In (58), the evidence available to the group consisting of Mary, Sally, and George includes the fact about the duty roster and the fact that George saw Joe in the hallway. Clearly, the fact that George saw Joe in the hallway nullifies the force of Sally's fact about the duty roster, so George's modal statement is true, and Sally's false, which Sally can acknowledge as in (59a).

DeRose (1991) gives some examples that militate powerfully for a group-based analysis of epistemic *might* statements, and which also lead to refinements of the view.[9] Two of DeRose's examples are given in (61a,b); they have a common background, given in (60).

(60) Background: John has some symptoms indicative of cancer. John's doctor decides to run a "filtering" test, which has two possible results: If the results are "negative," then cancer is conclusively ruled out; if the results are "positive," then John might, but also might not, have cancer, and further tests will have to be run.

(61) DeRose cancer example 1: Two independent but simultaneous dialogues.

a. Bill: I have heard John has cancer. Is it true?
Jane, John's wife: It's possible that John has cancer. He has some of the symptoms. But it's by no means certain that he's got it. They've run a test on him that may rule cancer out, but the lab work for the test takes many hours, and John and I won't be informed of the results until tomorrow.

b. John's doctor, to a colleague: The lab work is done. It's impossible that John has cancer, so we should start planning tests for other diseases.

These two dialogues favor quite solipsistic epistemic bases: John's wife gives her modal judgment relative to the evidence available to her, with no regard for what the doctors may have already found out. And her modal statement is one that we can comfortably judge to be true. The doctor gives his contrary modal statement relative to evidence available to John's doctors, after the tests, and we can comfortably count his modal statement as true. This pair provides a strong case for the evaluation of a modal statement relative to the context of the speaker: the modal statement that it's possible that John has cancer is true relative to Jane's evidence, and false relative to the doctor's evidence.

[9] For these examples, the evidence available to the speaker need not be construed as more than knowledge at the phenomenal level, so in discussing them, we put aside the question of whether the epistemic modal base should be defined as the set of possible worlds consistent with evidence available, or what is known.

Now consider the alternate example in (62), presented by DeRose (1991) with the same background.

(62) DeRose's cancer example 2:
Bill: I've heard that John may have cancer. Is that possible?
Jane: I don't know whether it is possible that John has cancer; only the doctors know. They've run a test on him that may rule cancer out, but the lab work for the test takes many hours, and John and I won't be informed of the results until tomorrow.

In (62), Jane denies knowing whether or not it is possible that John has cancer. She could have replied as in (61a), using her own evidence as the epistemic modal base. But we understand that in (62), she is taking the results of the test, about which she does not yet have any information, as part of the evidence on which the claim that John might have cancer should be evaluated. This follows if she is taking the evidence for the modal statement to be distributed across members of a relevant group, including the doctors who have run the test on John. Since the doctors are in possession of overriding evidence, and Jill doesn't know what it is, Jill cannot know whether the possibility modal statement is true.

These examples show that the epistemic modal base is variable, in this case, as determined by the choice of the speaker. In (61a), Jane's modal statement can be comfortably interpreted as based on what she knows; in (62), Jane's statement is best interpreted as based on evidence that resides in a group of which she is a member.

What it means for the epistemic modal base to reside in a group will be considered at greater length in Section 1.3.2. In that section, we will also address the question of whether the modal base should be construed as knowledge (whether at the phenomenal level, or more broadly), or whether it can or should be construed as fallible evidence available to the speaker or the group.

But for now, we review some further examples. The example in (63), from MacFarlane (2004), is designed to show that a modal statement can be evaluated with respect to the knowledge or evidence of an eavesdropper.[10]

(63) Bond has infiltrated SPECTRE headquarters and planted evidence indicating that he might be in Zürich, and has hidden a bug in the conference room. He has then escaped to London, where he and Leiter monitor transmissions from the bug. Blofeld convenes a meeting in the conference room and announces, "Bond

[10] See also discussion in Egan (2005), Egan, Hawthorne, and Weatherson (2005), and von Fintel and Gillies (2008b).

> might be in Zürich." Eavesdropping on this, Leiter says, "Ha! Blofeld is wrong again."

Leiter could make allowances for Blofeld. "Well, he's right, for all he knows." And that would implicate a solipsistic modal base for the evaluation of Blofeld's modal statement, which seems to be one available interpretation. But Leiter's reaction in (63) is more straightforward, and we can comfortably credit as true his judgment that Blofeld is wrong, thus accessing an interpretation on which Blofeld's modal statement is false. MacFarlane offers this as evidence that the truth value of Blofeld's modal statement is relative to an assessor, which must be specified for modal interpretation. If we choose Leiter as the assessor, Blofeld's modal statement is false.

This is a result that seems quite generally available for the evaluation of modal statements intercepted by eavesdropping, especially when the eavesdropper has more information than the speaker. Following von Fintel and Gillies (2008b), we can assimilate it to cases in which information for modal evaluation is distributed among members of a group centered at the speaker, provided we count an eavesdropper as a peripheral member of the group. This idea is not as ad hoc as it might sound at first; an eavesdropper is a participant in the overheard discourse, although the eavesdropper's participation in the discourse is one-sided. Furthermore, if an eavesdropper or other assessor is far removed from the speaker and the speaker's immediate addressees, the choice of such an assessor becomes increasingly strained the greater the degree of removal, lending credence to the idea that closer, felicitous assessors are felicitous because they can count as peripheral participants in the discourse. The example in (64), from von Fintel and Gillies (2008b: 86), illustrates this.

> (64) Detective Parker has been going over some old transcripts from Al Capone's court case in the 1920s. Capone is being asked whether his underlings might, without his knowledge, have stashed money from a bank heist in a particular safe.
> Capone [on tape]: The loot might be in the safe.
> Parker [eighty-five years later]: ??Capone was wrong. /??What Capone said is false. The safe was cracked in the 1980s, and there was nothing there.

Detective Parker can certainly say that Capone (or the 1920s prosecutor) was barking up the wrong tree since the money was definitely not in the safe. We could even endorse Parker's response in (64), that Capone was wrong, if we choose to interpret Capone's modal statement as uncharitably as possible. But this strict standard for the evaluation of Capone's statement doesn't seem to be an isolated possibility; it seems to be an extreme point on a range of available options. The judgments of von Fintel and Gillies recorded in (64) reflect the choice of a viable and more natural context for the evaluation of Capone's statement, the choice to grant Capone the context of his era, without interference with what we know, since Capone's era is so far removed from our own. In short, we are not in Capone's group.

As von Fintel and Gillies (2005, 2008b) show, even inanimate repositories of information can function as points of assessment for a modal statement, if they are linked to the speaker or other participants in a discourse. This is illustrated in the example in (65), from DeRose (1991), based on similar examples offered by Ian Hacking.

(65) DeRose cancer example 3

a. In this case, the test has been run, but not even the doctor knows the results of the test. A computer has calculated the results and printed them. A hospital employee has taken the printout and, without reading it, placed it in a sealed envelope. The policy of the hospital is that the patient should be the first to learn the results. Jane has made an appointment to pick up the results tomorrow. She knows that the printout of the results has been generated and that nobody knows what the results are.

b. Jane to Bill: "I don't yet know whether it's possible that John has cancer. I'm going to find that out tomorrow when the results of the test are revealed."

No information known to any participant in Jane's circle (John, her doctors, Bill, or others) would lead her to disclaim knowledge of whether it is possible that John has cancer. That she does so indicates deference to the results of the test, recorded and sealed in an envelope. We can assimilate this to the previous discussion by assuming that the sealed test results count as a member of the relevant group, as von Fintel and Gillies propose. If this seems like unnecessary personification, the next best alternative is to say that the information sealed in the envelope counts for modal evaluation since it is in some sense "available" to Jane and other human members of the group (itself a modal notion), or by appealing, as DeRose (1991) does, not just to what members of a relevant community know or have evidence for, but relevant ways they have of coming to know whether the prejacent of an epistemic modal statement is true. But von Fintel and Gillies show that merely being available is not enough; they note that if I have been inside all day with curtains drawn and have not looked out, but remember a forecast of rain on the morning news, it is more natural for me to say "It's possible that it is raining" than to say "I don't know whether it's possible that it is raining." And this holds despite the accessibility of information – a readily available way of coming to know – to settle the matter (by just pulling aside the curtain). So the availability of the information or a way of coming to know what is needed is not what is critical in (65); it is the physical instantiation of the information in the sealed printout of the test results. This physical instantiation of the information seems to be critical, giving credence to the idea that the sealed printout is functioning as an entity in the discourse in (65), with the standing, as it were, of an honorary participant.

A peculiar wrinkle is presented by another example cited by von Fintel and Gillies (2009b: 83), in (66), a case in which a modal statement seems to be evaluated relative to the addressee's knowledge or evidence.

(66) Mordecai and Pascal are playing Mastermind, a board game with colored pegs that are inserted into holes in a board. During each turn, one player, the codemaker, places a row of pegs behind a screen, and the other player, the codebreaker, tries to figure out the sequence of colors in the hidden row, using feedback from the codemaker. Pascal, playing codebreaker, hesitates over his next move to guess the sequence that has been set by Mordecai, and the following dialogue ensues.
Pascal: There might be two reds.
Mordecai: That's right. There might be two reds.

In this context, Mordecai knows that either there can't be two red pegs, or that there are (at least) two red pegs in the sequence he has created. So based on his knowledge, his own reiteration of Pascal's conjecture is either false, or a blatant violation of Grice's Quantity maxim. In fact, Mordecai's utterance is an affirmation of Pascal's modal statement, evaluated relative to Pascal's knowledge, and thus, an affirmation of Pascal's thought processes. In cases such as these, as well as in a number of pedagogical cases, the addressee's context seems to be a viable context for the evaluation of the speaker's modal statement.

Given this range of possible contexts of evaluation – the knowledge or evidence of the speaker, of the addressee, of another member of a group involved in the discourse, or of a peripheral member of the group such as an eavesdropper – von Fintel and Gillies (2005, 2008b) propose that the epistemic modal base can be chosen from a fairly broad set of alternatives. When the epistemic modal base is overtly specified, as in (67), this resolves the question of choice of assessor.

(67) a. Jill: From what I can see, the approaching man might be Fred.
b. Sally: From what I know, Joe might be in Boston.

c. Sally: As far as we can tell from the duty roster, Joe might be in Boston.
d. Jane: Based on what I know, John might have cancer.

Otherwise, there is a set of contexts that are available, with respect to which it would be possible to evaluate the modal. They refer to this as a cloud of contexts, and we will use that expression as shorthand for this view.

In the next section, we will explore some fine points concerning the ways in which knowledge can inhere in a group. And we will finally address the question of whether the modal base consists of knowledge, however phenomenal, or whether it can contain evidence that is available, but not known.

### *1.3.2 Further discussion of the epistemic modal base*

In this section, we look at some further examples that support the proposals of von Fintel and Gillies (2005, 2008b) reviewed in Section 1.3.1, and address some contrary views of what the epistemic modal base might be.

Let $\{a_i: i = 1, 2, 3, \ldots\}$ be the set of individuals in a discourse context, and for those individuals capable of knowledge, let $K_i(w)$ – short for K-subscript-$a_i$ at $w$ – be a set of propositions characterizing what $a_i$ knows in a world of evaluation $w$. This is the epistemic base of $a_i$ in $w$. Then the epistemic alternative set, $\cap K_i(w)$, is the set of possible worlds consistent with what $a_i$ knows. The solipsistic epistemic modal base for speaker $a_1$ is then just $K_1(w)$. In a group $G = \{a_i: i = 1, \ldots, n\}$ of two or more people, the one who knows the most has the largest epistemic base $K_i(w)$, and the smallest epistemic alternative set $\cap K_i(w)$.

Now, reconsider the example in (62), where Jane disclaims knowledge of whether John has cancer, in deference to the expert knowledge of John's doctor. We could say that Jane is simply taking all modal claims about John's cancer to be evaluable relative to a specified expert, John's doctor, whom she has in mind as someone to whom such claims must be deferred. Or we could construct group knowledge in such a way that evaluation of modal statements relative to group knowledge will automatically yield deference to the member with the most expertise about the matter at issue. These two choices correspond to two different views of epistemic modal statements. One view is that they must be interpreted according to an intention of the speaker to defer to a particular expert or experts. The other view is that there is a group interpretation by which expertise is automatically accessed for modal interpretation, wherever it resides in a group, and regardless of whether the speaker knows where it resides. These two views are empirically distinguishable, as we discuss next. In fact, we will see that the speaker does not have to know where expertise resides in a group; in fact, the expertise does not have to reside in a single member, and those who contribute to the expertise do not have to know how much of the relevant expertise they contribute.

One possible construction of the knowledge of a group G is $\cap\{\cap K_x(w)$: $x \in G\}$, the intersection of the epistemic alternative sets $\cap K_i(w)$ for members of G. If $a_d$ is John's doctor, possibilities in $\cap K_d(w)$ will be retained in this intersection, while possibilities in $\cap K_i(w)$, for $a_i \neq a_d$, will not be retained if they are incompatible with the greater expertise of the doctor. Thus, the epistemic alternative set $\cap\{\cap K_x(w)$: $x \in G\}$ effects deference to expertise within the group, even if there is no intention to specify the expert. This works well not just for (62), but for examples where the speaker may not know who in the group knows the most, or has relevant expertise. For example, in (58), Sally doesn't know that one of her addressees actually happens to know more than she does about Joe's whereabouts. This explains why we can readily evaluate Sally's modal statement as false, and she would likely concur when she finds out, as in (59a).

The set $\cap\{\cap K_x(w)$: $x \in G\}$ is a model of knowledge as distributed through the group. But it is possible for a modal statement to be true relative to knowledge in a group, when no single member of the group was in possession of that knowledge at the moment a modal statement is made. The constructed discourse in (68) takes place among members of a Wall Street investment group.

(68) Tom: Look at the numbers for the past hour! It's possible that the market is crashing.

George: No, I talked to all the traders this morning, and they made it clear that there are no major sell-offs in the works today. That's just a temporary dip.

Molly: Actually, somebody changed their mind at the last minute: 2G stock is being sold off today. That's what you're seeing. But here's a list of indexings, and you can see that 2G stock is not indexed to any others, so this is clearly an isolated event. Nothing to worry about.

Sam: Oh, no! There's a mistake in that list. 2G stock is indexed to Intel, Google, GE, and other heavyweights. So Tom was right: it *is* possible that the market is crashing.

George doesn't know that there are no major stock sell-offs, and particularly no major sell-off of 2G stock today. He can't know that because it's false. But he knows that the traders told him that there would be no major sell-offs, and that they are generally reliable (or have always been reliable in the past). But in the actual world, and ones like it, the traders are wrong this time: there is a sell-off of 2G stock in the works. Among these worlds, there are many in which 2G stock is not indexed to any others, because the list of indexings says so and it has always been accurate in the past. But in the actual world, and ones close to it, the indexing list is wrong this time, since 2G stock is actually indexed to several major stocks. So there is a set of worlds compatible with what George and Molly know (and with what Tom and Sam know) in which the traders reported that there were no major sell-offs, but they were wrong this once; in which the indexing list said that 2G was not indexed with others, but it was wrong this once; in which 2G stock is actually selling off and is indexed to several major stocks. On the basis of the existence of such possibilities in $\cap\{\cap K_x(w): x \in G\}$, Tom was right: it is indeed possible that the market is crashing.

But this set, $\cap\{\cap K_x(w): x \in G\}$ for G containing the individuals Tom, George, Molly, and Sam, is not equal to $\cap K_i(w)$ for any of these individuals at the moment of Tom's modal utterance, since each of these individuals contributed something to revealing the knowledge of the group, but each was also ignorant of, or wrong about, something else.

Thus, the set $\cap\{\cap K_x(w): x \in G\}$ as an epistemic alternative set has the virtue of picking out expertise in the group wherever it resides, and this seems to be the operative set of possibilities in the evaluation of modal statements relative to a group. In line with the "cloud of context" view of von Fintel and Gillies, this does not prevent us from evaluating a modal statement with respect to the speaker's knowledge only, or even with respect to an addressee's knowledge. It is also worth noting that the union $\cup\{\cap K_x(w): x \in G\}$ is not an adequate construction of the epistemic alternative set of a group. It would permit a statement of possibility to be true by virtue of worlds consistent with what the least informed member of the group knows, which could be something quite misleading. This can be seen in (69).

(69) Tom: Look at the numbers for the past hour! It's possible that the market is crashing.

Molly: No. I talked to all the traders this morning, and they said that the only sell-off in the works is a sell-off of 2G stock, which they planned to do around this time of day. It's not a large portfolio, and 2G stock is not indexed to anything else. So this is just a temporary dip.

Tom does not know about the day's scheduled trades, which explain the numbers he is looking at as merely a reflex of the scheduled sell-off of one minor stock. So there are possibilities consistent with his knowledge in which the market is crashing, and these possibilities will be present in the union $\cup\{\cap K_x(w): x \in G\}$, making Tom's modal statement true. Since Molly is present as a group member with knowledge that explains why Tom's knowledge is misleading, we can comfortably judge Tom's modal statement as false relative to the knowledge that resides in the group. We would judge it as true only on an unrealistically strict view of knowledge, which would disavow knowledge of what is really going on by anyone in the group since it is logically possible that they are deceived and that truly exceptional events are unfolding.

So far, we have adopted the view that the epistemic modal base consists of propositions consistent with what is known, by the speaker or relevant group. Some researchers have proposed that the epistemic modal base is more akin to the informational context of the speaker, or the common ground of the participants in a discourse, reflecting that which is justifiably, but not infallibly, believed, rather than that which is known.[11] The discussion of example (68) earlier gives us some reason to doubt this view. If we replace the knowledge sets $\cap K_x(w)$ by doxastic contexts, their intersection across members $x \in G$ would be empty – since the participants in this discourse believe incompatible things, the set of all possibilities compatible with what they all believe is empty. And yet the interpretation of (68) seems to rest on the intersection of the epistemic alternative sets of the discourse participants. We further consider the idea that the epistemic alternative set is a doxastic or informational context with some new examples mentioned in the following pages.

Consider the medieval anatomist Galen, who knew much about the human body, but well short of what modern anatomists know. Anything that Galen knew about the body, he believed, but not vice versa. That is, the set of propositions about the body that Galen knew is a proper subset of the set of propositions that he believed. Thus, any possible world consistent with the set of propositions that Galen believed is also consistent with the set of propositions that he knew. For convenience, this is recorded in (70), where $\{p_i^K: i \in I\}$ is the set of propositions Galen knew, and $\{p_j^B: j \in J\}$ is the set of propositions Galen believed (for index sets I and J that can probably be taken to be initial segments of the natural numbers, assuming these sets are finite).

(70) a. $\{p_i^K: i \in I\} \subset \{p_j^B: j \in J\}$
b. $\cap\{p_j^B: j \in J\} \subset \cap\{p_i^K: i \in I\}$

Any possibility consistent with what Galen believed is consistent with what he knew, but not vice versa. Now, suppose Galen had good grounds to believe that the primary function of the pancreas is different from the primary function of the liver. Nevertheless, suppose he recognized that he was not sure of this, and consequently uttered the possibility modal in (71).

(71) Galen: It is possible that the primary function of the liver is identical to the primary function of the pancreas.

[11] Tancredi (2007a) suggests this, as do Hacquard (2006, 2010) and Anand and Hacquard (2008). So did earlier versions of the present work, presented between 2006 and 2010.

We evaluate Galen's possibility modal in (71) as true since there exist possible worlds (let $v_1$ be one of them) that are consistent with everything that Galen knows, in which the primary function of the liver is identical to the primary function of the pancreas. So in this case, letting "1" be the index for Galen, the epistemic alternative set is $\cap K_1(w)$, the set of worlds consistent with what Galen knew.

Could it be otherwise? To see, suppose that Galen utters a modal statement such as that in (72), based on what he justifiably believes (meaning he has evidence that makes his belief justified, but not infallible), not based on what he knows.

(72) Galen: It is not possible that the primary function of the liver is the same as the primary function of the pancreas.

The intuition is that Galen is wrong in (72). His modal statement is too strong. Even if he has some justification for believing that the functions differ (as, in fact, they do), the negated possibility modal in (72) is false. This result is not obtained by the theory if the epistemic alternative set could be just the set of possible worlds compatible with what Galen believes. With just this set, his statement in (72) would be true. The correct result is obtained if the epistemic alternative set is the set of possible worlds consistent with what Galen knew, since then the negated modal in (72) is falsified by worlds like $v_1$.

In conclusion, the epistemic alternative set is the set given in (73), for G as a group containing the speaker $a_1$. In the limiting case where $G = \{a_1\}$, it reduces to the knowledge of the speaker.

(73) Epistemic alternative set = $\cap\{\cap K_x(w): x \in G\}$.

The epistemic alternative set will never be smaller than this for the reasons given earlier. And the epistemic alternative set will never contain worlds that are inconsistent with what the speaker knows.

In some branches of the literature on modality, it is claimed that there is a basic dichotomy (a basic distinction within the typology of the theory) between subjective modals, which are evaluated with respect to the evidence of the speaker, and nonsubjective modals evaluated with respect to assumed-shared evidence.[12] On the basis of the foregoing discussion in this section, following von Fintel and Gillies (2008b) and others, this dichotomy can be derived from variability in how the epistemic modal base inheres in a community to which the speaker belongs. To see this, however, it is critical to examine data that are alleged to demonstrate the special status of subjective modals – that they don't embed.[13] In (74a,b,c), modal statements appear, respectively, in the protasis

[12] For development of this view, see Lyons (1977), Nuyts (2001b), and Papafragou (2006).
[13] See Lyons (1977) and Papafragou (2006). For discussion, see Portner (2009).

(antecedent) of the conditional, in the complement to the attitude adjective *surprising*, and in the asserted attitude complement to the verb *tell*.[14] These examples are all infelicitous when the modal claim is evaluated with respect to the knowledge of the speaker. Imagine (74a) being uttered directly after the speaker concludes that Max must/may be lonely, based on evidence the speaker has just collected or observed. Likewise, (74b,c) are infelicitous when the evidence for the modal claim is the speaker's evidence.

(74) a. ? If Max must/may be lonely, his wife will be worried.
b. ? It is surprising that Superman must be jealous of Lois.
c. ? Superman told me that Superman must be jealous of Lois.

Each of these means, "according to my evidence." Objective counterparts such as those in (75a,b,c), respectively, are intended to be interpreted relative to evidence available to everyone in the discourse, and, indeed, anyone outside the discourse who might be attending to the issues at hand.

(75) a. If Paul may get drunk, I am not coming to the party.
b. It is surprising that the victim must have known the killer.
c. The police told reporters that the victim must have known the killer.

This contrast is not surprising. On the intended subjective interpretation, (74a) means, in effect, that if I must or can conclude that Max is lonely, then his wife will be worried. There is a gap here between what is known to the speaker, and whatever basis Max's wife might have to be worried. In (74b), it might be surprising that there is any evidence that Superman is jealous of Lois, since the very idea of Superman being jealous of Lois might be surprising. But (74b) says more than this, such that it is felicitous only if what is surprising is the existence of conclusive evidence for Superman's jealousy. In the right context, and with stress on *must*, (74b) can felicitously be interpreted to mean that, but it is odd if there is no justification for the necessity modal in the context. And (74c) means, in effect, that Superman told me that I have conclusive evidence that he is jealous of Lois, which is itself odd. With these meanings, it is not mysterious why these are odd in normal circumstances. In contrast, (75a) means, in effect, that if generally available evidence renders it possible that Paul will get drunk, then I am not coming to the party. (75b) means that it is surprising that, according to generally available but conclusive evidence, the victim knew the killer. And (75c) means that the police told reporters that the available evidence conclusively indicates that the victim knew the killer. This contrast does not require a special typological class of subjective modal

[14] The examples are from Papafragou (2006: 1690), reproduced and discussed by Portner (2009: 145–146).

statements. The behavior falls out from the choice of a solipsistic modal base in (74), versus the choice of more widely based group knowledge as the modal base in (75).

## 1.4 Conclusion

On the formulation of Kratzer's analysis of modality adopted here, we do not assume that the ordering source $\Omega(w)$ is a consistent set of propositions. Thus, a sequence of possibilities in $\cap E(w)$ that happens to be weakly ordered under the ordering $\leq_{\Omega(w)}$ derived from $\Omega(w)$ does not necessarily converge to a "maximum" that satisfies all propositions in $\Omega(w)$. A sequence of possibilities in $\cap E(w)$ could keep getting progressively better for the evaluation of the modal statement, with successive possibilities (each $\leq_{\Omega(w)}$ the other) satisfying progressively more propositions in $\Omega(w)$, but never satisfying all of them. In this case, the best we can do is to evaluate a modal statement in terms of the truth of the prejacent on the "tails" of sequences of possibilities in $\cap E(w)$, which are weakly ordered by $\leq_{\Omega(w)}$. This yields the truth conditions for necessity, possibility, weak necessity and its dual, repeated in (76)–(79).

(76) [[**must** $\mathbf{S_1}$]] $^{c,w} = 1$ iff every sequence of possibilities in $\cap E(w)$ that is weakly ordered by $\Omega(w)$ has an $\mathbf{S_1}$–uniform tail.

(77) [[**might** $\mathbf{S_1}$]] $^{c,w} = 1$ iff some sequence of possibilities in $\cap E(w)$ that is weakly ordered by $\Omega(w)$ has an $\mathbf{S_1}$–persistent tail.

(78) [[**should** $\mathbf{S_1}$]] $^{c,w} = 1$ iff for sequences of possibilities in $\cap E(w)$ that are weakly ordered by $\Omega(w)$,
   a. every such sequence has an $\mathbf{S_1}$–persistent tail, and
   b. there is such a sequence with an $\mathbf{S_1}$–uniform tail.

(79) [[$\mathbf{should^{\perp}}$ $\mathbf{S_1}$]] $^{c,w} = 1$ iff
   a. every sequence weakly ordered by $\Omega(w)$ has an $\mathbf{S_1}$–persistent tail, or
   b. there is a sequence weakly ordered by $\Omega(w)$ with an $\mathbf{S_1}$–uniform tail.

With this theory of modal interpretation, what is lexically specified for the modals is their quantificational machinery. For an epistemic modal, the modal base seems to be the set $\{K_x(w): x \in G\}$, for a group G to which the speaker belongs, with the alternative set $\cap\{\cap K_x(w): x \in G\}$. When G consists of just the speaker, this yields the solipsistic epistemic modal, as one option for interpretation, at one extreme. Or G can be chosen to be a larger group, including expertise beyond that of the speaker, or physically instantiated information, or even knowledge of an eavesdropper, as other options for interpretation. Each of these can be instantiated in the group G. Each such epistemic modal base would be accompanied by an ordering source appropriate to it.

Following the "cloud of contexts" view of the determination of the epistemic modal base from von Fintel and Gillies (2005, 2008b), it is clear that the exact determination of the epistemic modal base is contextual, if it is not overtly specified. This much of the nature of the epistemic modal base, then, does not need to be lexically specified, as properties of modal lexical items.

This leaves open whether anything about the modal base (and associated ordering source) is lexically specified, or whether it is entirely determined on a combination of structural syntactic and contextual grounds. In Chapters 3 and 4, we will demonstrate that the quantificational machinery of the modal operators is evident in the interpretation of ascriptions of belief and other propositional attitudes, but with a different sort of set entirely playing the role of the modal base. This suggests that modal-like quantificational structure is available more generally in natural language semantics, with characteristic modal strengths, beyond the scope of modality proper. These quantificational structures will be denoted "MUST," "MIGHT," "SHOULD," and "SHOULD$^{\perp}$." These are metalanguage operators that combine with the phrase marker of a clause, given a base and ordering source determined through some combination of lexical, structural, and contextual factors, to yield the truth conditions in (76)–(79), which are themselves an abbreviation for the truth conditions in (8), (10), (15), and (18). In particular, the right-hand side of (76) will be abbreviated as MUST ($\mathbf{S}_1$ | E($w$) || $\Omega(w)$), as recorded in (80).

(80) [[**must** $\mathbf{S}_1$]]$^{c,w}$ = 1 iff every sequence of possibilities in $\cap$E($w$) that is weakly ordered by $\Omega(w)$ has an $\mathbf{S}_1$–uniform tail
$=_{def}$ MUST ($\mathbf{S}_1$ | E($w$) || $\Omega(w)$)

These abbreviations will be used in later chapters.

For some purposes, epistemic modal statements alternate with other constructions for the expression of evidential qualification, as in the different answers to Max's question in (81), where Sally's different responses convey evidential qualification using different means, including modality, first person attitude ascriptions, a first person perception report, and intonation.

(81) Max: Why isn't Bill at this meeting?
Sally: a. He might/should be in Boston.
b. I guess/think he's in Boston.
c. I've heard he's in Boston.
d. He's in Boston. [emphatically]
e. He's in Boston (?). [tentatively]

The kinship among these, and the extent to which they are identical or just similar in effect, will be among the issues investigated in subsequent chapters.

# 2 Root modality

## 2.1 Introduction

Root modal statements are statements about what must, may, should, or cannot be the case according to some set of circumstances in the world, rather than what is known by the speaker or a community to which the speaker belongs. For deontic statements, those circumstances are obligations or expectations imposed by laws, regulations, or convention. For bouletic or teleological statements, the circumstances are preferences, desires, or goals of the referent of the subject, or of the speaker and/or addressee(s). For ability modals, the circumstances are abilities of the referent of the subject of the clause. Still others, whose circumstances don't fit any of these descriptions, are properly called circumstantial modal statements. Examples are given in (1), with the most salient types of interpretation.

(1) a. Sam must appear in court at 8 am tomorrow. (deontic)
b. Sam must have a cup of coffee now! (bouletic or deontic or teleological)
c. With this traffic, we should get to the airport in time for our flight. (circumstantial)
d. Max can run a mile in four minutes. (ability)
e. Max could play the villain in our production. (bouletic or ability)
f. I can walk to the train station in five minutes. (circumstantial or ability)

Many modal statements can express more than one variety of root modality. The modal statement in (1b), for example, can express a necessity based on the preferences or desires of Sam (bouletic), or based on what is specified in his contract (deontic), or what is required by his doctor to maintain Sam's health, or by the speaker and addressee(s) to wake Sam up (teleological). The possibility modal in (1e) can be interpreted as expressing a possibility that accords with the preferences of the speaker and addressee(s) or the abilities of Max. The possibility modal in (1f) can be interpreted as a statement about the speaker's location with respect to the train station (circumstantial), or about the speaker's ability to walk fast (ability). Ability modals do not seem to be expressible with a necessity modal, and there is considerable variation in the forms and strengths of root modality, which can be expressed by modal auxiliaries versus modal adjectives and adverbs – for example, in Modern English, *With the traffic so light, we are sure to get to the airport in time for our flight*, can be uttered as a circumstantial modal statement with the force of necessity, but the counterpart with modal *must* is not naturally interpreted as circumstantial. For discussion of the quite specialized forms and modes of expression of root modality, across syntactic constructions, types and forces of modality, and languages, see Johannesson (1976), Palmer (1979), Coates (1983), Abraham (2001a, 2001b, 2012), Condoravdi (2002), Abusch (2012), and Kratzer (2012).

Although root modality differs in some ways from epistemic modality in morpho-syntactic expression (as mentioned earlier and explored further in Chapter 8), Kratzer (1981, 2012) argues for formally parallel employment of conversational backgrounds, with the identification of the backgrounds serving to semantically distinguish epistemic from root modality, and different forms of root modality from one another. Kratzer (2012) proposes that circumstantial and ability modals have a realistic modal base of facts about the world, confining attention to worlds that are in accord with those facts or circumstances. Bouletic modal bases, on the contrary, need not be realistic, as will be explored in this section. The appropriate ordering source for different types of root modality will also be considered. The modal base and ordering source are employed within the quantificational schemas to yield different strengths of modal assertion, as in Chapter 1.

In the interpretation of epistemic modals, we quantify over an epistemic context consisting of worlds consistent with all the propositions in the epistemic modal base, expressing everything that the speaker knows. But this admits a lot of irrelevant worlds (in which history unfolded differently from the way it actually did, unbeknownst to the speaker, for example). With an epistemic modal, the role of the ordering source is to effectively confine things to successively more relevant or realistic worlds.

For some root modals, this is not good enough. Consider the ability modal statement in (1d).[1] Alternate physical properties of the world or alternate histories of the world could pose a problem here. If we use a modal base of realistic accessible worlds, **Acc**, to screen out unrealistic worlds, we should probably specify **Acc** to screen out worlds in which (say) gravity is stronger than in the actual worlds, or Max's ancestry was different from the one in the actual world in ways that would enhance or diminish his mile-running ability. (1d) is a

[1] I am indebted to Jerry Weltman for discussion of this example.

statement about Max as he is – worlds in which he had different ancestry, with results for his mile-running ability, are simply not in play. We don't just want to impose an ordering source to screen them out more and more as we get further in accord with the ordering source; we want to throw them out entirely, from the beginning. (1d) is effectively essentialist about Max's mile-running ability. This is one rationale for the realistic modal base proposed by Kratzer for this case.

The ordering source for the interpretation of (1d) should reflect what is fixed, which is the realization of Max's four-minute mile potential under different circumstances. If in the actual world, Max does not compete in a given race because he has to attend a wedding that day, but he would have been able to run a four-minute mile if he had competed, then that world should not falsify (1d). A race in which Max competes with a touch of flu and fails to run a four-minute mile should not falsify (1d) as a generic statement about Max. Instead, as a generic statement about Max, (1d) asserts that sequences of worlds that reflect progressively more and more what Max is capable of in the running department are persistent in worlds in which Max runs a four-minute mile.

Bouletic modal bases, as mentioned earlier, need not be realistic. Referencing statements of mathematical necessity or possibility, Kratzer (2012: 55) acknowledges that the realistic modal base can, in limiting cases, be empty, leaving recourse to quantification over an ordering source. This is an approach that we effectively have to take with some bouletic ascriptions, as well. A bouletic ascription makes a statement about what happens on sequences of worlds that are weakly ordered by the bouletic ordering source – a set of propositions expressing the preferences and desires of the referent of the prejacent subject, or of the speaker and/or addressee(s). The example in (1b) can be used to assert that such a sequence, from some point onward, is uniform in the truth of the prejacent, that Sam has a cup of coffee, and this can be asserted even if all coffee has disappeared from the face of the earth. In the case of such a bouletic ascription expressing counter-to-fact desires or preferences, the statement quantifies over possibilities that are known to be counterfactual, and one approach to modeling it is to quantify over all possibilities; the ordering source does the work of delineating the grounds of the assertion, even if no realistic worlds, compatible with the facts, are in play. In such a case, we can throw the interpretation open to the set **W** of all possible worlds. The modal statement pertains to worlds in **W** that constitute sequences weakly ordered by the bouletic ordering source, $\Omega(w)$. The necessity modal asserts that every such sequence of worlds in **W** has a tail that is uniform in the truth of the prejacent. A range of bouletic root modals could then be interpreted as in (2).

(2) a. [[**must** $\mathbf{S_1}$]] $^{c,w}$ = 1 iff every sequence of worlds in **W** that is weakly ordered by $\Omega(w)$ has an $\mathbf{S_1}$–uniform tail.

b. [[**might/can** $\mathbf{S_1}$]] $^{c,w}$ = 1 iff there is a sequence of worlds in **W** that is weakly ordered by $\Omega(w)$, with an $\mathbf{S_1}$–persistent tail.

c. [[**should** $\mathbf{S_1}$]] $^{c, w}$ = 1 iff
every sequence of worlds in **W** that is weakly ordered by $\Omega(w)$ has an $\mathbf{S_1}$–persistent tail, and
there is a sequence of worlds in **W** which is weakly ordered by $\Omega(w)$ that has an $\mathbf{S_1}$–uniform tail.

At this point, it is tempting to conjecture that all root modal statements could be interpreted as in (2), including circumstantial and ability modals. On this approach, (1f) would be interpreted by looking at all possibilities, and it would assert that, as possibilities become progressively more in accord with the speaker's abilities, we can always find worlds in which the speaker walks to the station in five minutes. The problem with this is that realism is imposed, on this interpretation, only by the accord with the speaker subject's abilities. Nothing prevents the worlds that make (1f) true from being ones in which, say, gravity is weaker than in the actual world, facilitating the speaker's progress in those worlds. This is counter to the intent of typical uses of circumstantial modals, which normally invoke a realistic modal base. Likewise, ability modals are normally useful only when they talk about actual abilities (or lack thereof) in realistic contexts. In contrast, bouletic modal statements have force even if the prejacent is true only on rather fanciful, counterfactual worlds. Deontic statements may also have force by virtue of the truth of the prejacent in worlds that cannot be realized, when used to state more idealized obligations (e.g., Sam is due in court at 8 am, even if he is physically unable to make it), although a failure to hold in realistic worlds may undermine the practical import of a deontic statement. In practice, deontic modal statements tend to have a realistic modal base, confining consideration to worlds in which the relevant actor can abide by the obligations that apply in the context at hand. If those obligations are $\{O_1, \ldots O_n\}$, the ordering source is the set of propositions $p_i$ (i = 1, …, n) such that $p_i$ is true in all and only realistic worlds in which the relevant agent abides by $O_i$. Teleological modals are typically interpreted with respect to a realistic base – there is usually no point in saying what someone could or must do to attain a particular outcome under wildly counterfactual conditions (such as the possession of superhuman powers), although some counterfactuality can be tolerated to express the hopelessness of a situation, or the bounds of possibility. (*To get out of this quarry, he has to scale that sheer wall.*) If the relevant actions that pertain to the achievement of an end are $\{A_1, \ldots A_k\}$, the ordering source for a teleological modal is the set of propositions $q_i$ (i = 1, …, k) such that $q_i$ is true in all and only realistically accessible worlds in which $A_i$ is performed by the relevant actor(s).[2]

[2] This is not to reject the proposals of von Fintel and Iatridou (2008) that weak necessity root modals involve quantification over worlds that are optimal by a measure over and above that of the basic ordering source for the type of modal, as discussed in the following pages.

So there is variation in base among root modals. Circumstantial, ability, and teleological modals favor a realistic modal base much more than bouletic modals, and deontics vary across the scale, from more idealized to more realistic uses. With **W** containing all possible worlds, the interpretations in (2) suit bouletic modality, and more rigid or idealistic deontic modals. Realistic ability, circumstantial, and teleological modals would favor parallel truth conditions to those in (2) with **W** replaced by a set **Acc** of more accessible realistic worlds. The ordering source in each case is pertinent to the sort of modality expressed. For ability modals, the ordering source would contain propositions that sort realistic worlds according to their suitability for the realization of the ability under discussion. To evaluate *Max can run a mile in four minutes* as a general statement, we don't want to count days when he is seriously ill. For teleological modals, the ordering source orders worlds according to the extent to which actions favorable to a given end are performed. And in deontic modals, the ordering source orders worlds according to the extent to which relevant obligations are observed in them. The bouletic ordering source, in contrast, orders worlds directly according to outcomes, with more desirable worlds ranked higher.

Bouletic modal statements also readily provide examples that show that a root modal assertion can be interpreted with an inconsistent ordering source. Suppose that Max would like to spend the afternoon at the zoo with his young children, but would also like to play golf, and no realistic world would permit him to do both. Then Max's bouletic ordering source is inconsistent. But both *Max would like to go to the zoo with his children this afternoon*, and *Max would like to play golf this afternoon* could be interpreted as bouletic modal statements, and both even evaluate as true, by virtue of, on the one hand, weakly ordered sequences whose worlds satisfy progressively more members of the bouletic ordering source, but never Max's desire to play golf, which are persistent in worlds in which he goes to the zoo with his children, and, on the other hand, weakly ordered sequences whose worlds also satisfy progressively more members of the ordering source, but never Max's desire to spend time with his children, which are persistent in worlds in which he goes to play golf.

As they did with epistemic modality, von Fintel and Iatridou (2008) propose that weak necessity modals are not interpreted using the quantificational schema for SHOULD. The teleological modal in (3a) can be interpreted as making a claim about all worlds in which you get to Ashfield, but (3b), they note, is not simply interpreted with a weaker quantifier over the same modal base.

(3) a. To go to Ashfield, you have to use Route 2.
b. To go to Ashfield, you ought to use Route 2.

Rather, (3b) is interpreted as universal quantification but over a modal base that has an additional layer of restriction, to ways of getting to Ashfield that are better in one or more respects. On this view, (3b) asserts that among all words in which you get to Ashfield, the most favored are ones in which you use Route 2.

It is claimed by von Fintel and Iatridou that the truth conditions for weak necessity reduce to those for strong necessity under the Limit Assumption for modal interpretation, by which there is a set of best worlds, maximally in accord with the ordering source. But there is reason to resist their view of the consequences of the Limit Assumption for the interpretation of weak necessity modals. As discussed in Chapter 1, their argument hinges on an assumption that the set of possibilities in the modal base is finite, so that the tails of sequences or worlds in the modal base that get progressively better in accord with the ordering source are eventually immersed in the set of "best worlds," Max ($\Omega(w)$), maximally in accord with the ordering source $\Omega(w)$. But if the ordering source is infinite and the modal base is dense with possible worlds, this need not be the case – possible worlds on the tail of a sequence can get progressively more in accord with the ordering source and yet never become immersed in the set of "best" worlds because they might never "reach" the best worlds. Nevertheless, the interpretation proposed by von Fintel and Iatridou is a plausible one for weak necessity modals such as (3b), and arguably captures the most everyday uses of these modal statements. Furthermore, when the Limit Assumption fails to hold, other root modals have plausible similar interpretations. For example, the ability modal in (1d) could be interpreted as asserting that there are worlds among those most favored for the demonstration of Max's mile-running prowess in which he runs a four-minute mile. But if the conditions favoring this are infinite in number, such a set of most favored worlds may not exist, and even if it did, if relevant possible worlds are multitudinous, tails of sequences consisting of worlds that are progressively more and more in accord with conditions for Max demonstrating his mile-running prowess are not necessarily immersed in such a set of most favored worlds – they could get progressively better without reaching that set.

This discussion, with the proviso about the additional restriction on worlds that von Fintel and Iatridou propose for the interpretation of weak necessity modal statements such as (3b), suggests that Kratzer's possible worlds semantics for root modality has the right degrees of freedom to capture the exact sorts of variability that exist among root modals, including their different strengths and different modal bases, which range from realistic to more idealistic. Furthermore, in Chapters 3 and 4, when we apply the apparatus of modal interpretation to the interpretation of attitude reports, it will be argued that this approach to the semantics of root modality gets us out of some of the difficulties with the interpretation of bouletic attitude ascriptions noted by Heim (1992). But before we can leave the topic of root modal interpretation and turn to propositional attitude ascriptions in Chapter 3, we first engage with challenges to possible worlds semantics for the interpretation of root modals given by Lassiter (2011) and Cariani (2011). This is taken up in Section 2.2. Lassiter (2011) will be the main reference point for these objections since Lassiter spells them out at length as part of an argument for a scalar approach to modal interpretation, and the present concern is not simply to evaluate objections to Kratzer's possible worlds semantics for modality, but to place it among contenders that have a claim to be appropriate for modal interpretation in some contexts of modal utterance.

## 2.2 Challenges to possible worlds semantics of root modals

In Chapter 5 of his dissertation, Lassiter (2011) outlines five problems for any analysis of teleological, deontic, or bouletic modality involving quantification over possible worlds. He argues that these particularly challenge any version of Kratzer's analysis, which controls the restriction on the quantifier with an ordering source or any other preference order on possible worlds, as in (2). Three of the challenges have been presented before in the literature, while the other two are concerned with issues of gradability and comparisons of degree that Lassiter thinks are better handled by his own scalar theory of root modality. Kratzer (2012) and von Fintel (2012) present responses to the first three objections. The present section airs the issues involved in all five of these challenges by outlining the objections and presenting responses to them. But rather than repeating responses given elsewhere in the literature, where they exist, it focuses on understanding them in terms of the expression of Kratzer's analysis of modal interpretation given in (2), as well as on constructing suitable contexts to determine whether the effects alleged in the challenges rely on pragmatic factors rather than on semantic ones. Some novel observations are made along the way, and a couple of the responses take a different direction from those in the literature; departures from responses in the literature will be noted as they arise.

### *2.2.1 Challenge to the upward monotonicity of Kratzer's analysis*

The first challenge concerns the property of upward monotonicity. Using the truth conditions in (2a), if $\mathbf{S}_1$ entails $\mathbf{S}_2$, then **must** $\mathbf{S}_1$ entails **must** $\mathbf{S}_2$. That is, **must** has the property of upward monotonicity. Likewise, using (2b,c), **might** and **should** are upward monotonic. (These are all due ultimately to the upward monotonicity of the universal and existential quantifiers.) Lassiter (2011), following Cariani (2011), argues that this prediction gives intuitively wrong results over a range of cases, where deontic modals fail to exhibit the monotonicity behavior of standard quantifiers.

One area in which this problem emerges is in inference schemas involving the interaction of disjunction with deontic or bouletic modal statements, such as in (4a,b,c), or bouletic attitude ascriptions, such as those in (4d), all adapted from Lassiter (2011: 130) and Cariani (2011). Since such schemas were discussed by Ross (1944), Lassiter uses the term "Ross's Paradox" to refer to such cases, the alleged paradox being that the monotonicity of the modal operators would render the inferences expressed in these examples valid, though they seem intuitively invalid.

(4) a. You must wash the dishes. So, you must wash the dishes or destroy them.
b. We are required to drive less than 70 mph. So, we are required to drive less than 70 mph or more than 100 mph.
c. Mary needs to work harder. So, Mary needs to work harder or quit her job.
d. The boss wants you to go to Atlanta. So, the boss wants you to go to Atlanta or Houston.

Lassiter (2011: 130–132) analyzes the problem with these as follows. Any quantificational analysis of them involves quantification over a set of deontically or bouletically accessible worlds, **Acc**. On this approach, the examples in (4) all involve an inference of the form given in (5), for propositions *p* and *q*, which is essentially Lassiter's (2011: 131) formula (5.18), which clearly expresses a valid inference.

(5) $[\forall w \in \mathbf{Acc}\text{: } [[p]]^{c,w} = 1] \models [\forall w \in \mathbf{Acc}\text{: } [[p \vee q]]^{c,w} = 1]$

The universal quantifier in (5) matches the strength of necessity modals, but since existential quantifiers are also upward monotonic, the objection holds for all of the quantificational analyses in (2). Relating (5) to the analyses in (2), **Acc** is just the result of filtering worlds in **W** according to an ordering source $\Omega(w)$ appropriate to the deontic modal – or in the case of (4d), bouletic modal – involved.

To talk this through for (2a), if *p* is the proposition that we wash the dishes, then the first sentence of (4a) asserts that every sequence of worlds that is weakly ordered by $\Omega(w)$ has a *p*-uniform tail. Let *q* be the proposition that we destroy the dishes. Since every world in which *p* is true is a world in which is $p \vee q$ is true, it follows that every sequence of worlds that is weakly ordered by $\Omega(w)$ has a $(p \vee q)$-uniform tail, since it has a *p*-uniform tail, which is, ipso facto, $(p \vee q)$-uniform. Thus, the interpretations of the two successive modal statements in (4a) using (2a) do indeed support an instance of the inference schema in (5).

Another such divergence in behavior elicited by Cariani and Lassiter arises in downward entailing contexts such as the complement *to doubt*. The example in (6) illustrates the fact that this context is downward entailing.

(6) I doubt he wore a tie or scarf. Therefore, I doubt he wore a tie.

In this context, the deontic modal *ought* diverges in behavior from the upward monotonic universal quantifier, as seen in the contrast between (7a) and (7b) (Lassiter 2011: 131) when each occurs in the scope of *doubt*.

(7) a. I doubt that Lynn ought to wear either a tie or a scarf. In fact, I'm reasonably certain that she ought to wear a scarf.
b. # I doubt that everyone is Italian or French. In fact, I'm reasonably certain that everyone is Italian.

This sort of divergence in the semantic behavior of deontic modality with universal quantification is also seen in the examples with retraction in (8)–(9), which Lassiter (2011: 131) adapts from Cariani (2013). The judgment in (9c) is Lassiter's.

(8) a. Sam: According to the rules, you have to follow suit or play the king of trumps.
b. Joan: No, the rules quite explicitly say you ought to follow suit, no matter what.
c. Sam: I guess what I said was wrong.

(9) a. Sam: Everyone followed suit or played the king of trumps.
b. Joan: Nobody had the king of trumps. Everyone followed suit.
c. Sam: # I guess what I said was wrong.

Intuitions in the modal examples (7a) and (8) diverge from intuitions regarding the universal quantifiers in (7b) and (9). Lassiter presents this as evidence that the modal statements are not semantically interpreted as universal quantification over possible worlds.

To critically address these arguments, we begin with the examples in (7) and (8)–(9), and then turn to the examples in (4). The paragraphs that follow make some of the same points as von Fintel's (2012) discussion of challenges to the monotonicity of Kratzer's analysis, except that von Fintel addresses Cariani's (2011) formulation of the challenge. Cariani's formulation contrasts judgments in (8) with judgments for a similar set of examples involving epistemic *must* instead of the universal quantifier in (9); as a result, the line of argument here diverges from that of von Fintel in several respects.

Lassiter's argument based on (7) is undermined by the following points. If you take out *either* in (7a), then (7a) is as unacceptable as (7b). Under *ought*, *either* creates free choice in (7a). That is what makes it acceptable, as von Fintel (2012) demonstrates at greater length. If I doubt that Lynn has a free choice between a tie and a scarf, I could nevertheless be certain that she should wear a scarf. It may also be worth adding to the debate here the observation that if we replace the disjunction of singulars, *a tie or a scarf* in (7a) with a disjunction of bare plurals, *ties or scarves*, as in (10), it becomes akin to (7b).

(10) # I doubt that Lynn ought to sell ties or scarves at her booth. In fact, I'm reasonably certain that she ought to sell scarves.

So the contrast between (7a) and (7b) hinges on other factors than any alleged semantic differences between deontic *ought* and universal *everyone*.

As for the contrast between (8c) and (9c), there are potential confounds in arriving at this judgment. First, the direct contradiction by Joan in (8b) may prime us to regard Sam's disavowal in (8c) as quite natural. In (9c), Sam's disavowal does not have the benefit of such priming. The change in strength of the modal operator from *have to* to *ought* in (8a,b) is not matched in (9a,b). Eliminating these and other differences, we have the examples in (11) and (12), with no judgments indicated.

(11) a. Sam: According to the rules, you must follow suit or play the king of trumps.
b. Joan: No, you must follow suit.
c. Sam: I guess what I said was wrong.

(12) a. Sam: Everyone followed suit or played the king of trumps.
b. Joan: No, everyone followed suit.
c. Sam: I guess what I said was wrong.

With the priming by *no* in (12b), and removing other differences, we seem to level the difference between Sam's (c) utterances in (8) and (9), making (11c) and (12c) equally acceptable (or equally unacceptable).

Turning to (4), Lassiter considers the possibility that we should regard (4a) as semantically valid, while we write off its oddness as the result of a Gricean implicature generated by the use of disjunction, or some other pragmatic infelicity. After all, that is what we do with the nonmodalized example in (13), which expresses a valid inference if we accept the standard truth conditional semantics for disjunction.

(13) We washed the dishes. So we washed the dishes or destroyed them.

Whatever oddness (13) has can be explained away on pragmatic grounds in normal contexts. We can even imagine a context in which it is acceptable. In a world with highly aggressive bacteria, there might be an imperative to clear all dirty dishes immediately, which can be accomplished by either washing them in antibacterial soap, or by breaking them up and throwing the pieces into a furnace to melt them down and recycle their material. In such a world, the speaker of (13) could be announcing that we washed the dishes, and thus met the imperative – that we not leave dirty dishes sitting around – in one of the two ways that exist for meeting that imperative.

Lassiter claims that the modalized example in (4a) is more irredeemably bizarre, giving us grounds to reject a quantificational semantics for deontic modals since it makes the inference in (4a) valid, even as we retain the standard truth conditional semantics for disjunction, which makes the inference in (13) equally valid. The heart of the argument is an alleged difference in intuitive acceptability between (13) and (4a), the greater susceptibility of (13) to Gricean explanation, and the correspondingly greater possibility of accommodating the second conjunct in (13) than in (4a).

The response to Lassiter's argument is to acknowledge that the semantics for modality in (2) predicts that the inferences in (4) are all valid, but to note that the examples in (4) and examples such as (13) seem to be more akin to one another than Lassiter admits since the oddness of both can be explained on the grounds of Gricean implicature, and the disjunction in both can be accommodated by a shift in background in suitable contexts. There is, therefore, no more reason to reject the quantificational semantics for deontic or bouletic modals in (2) on the grounds that it makes the inferences in (4) valid, than to reject the standard truth conditional semantics of disjunction on the grounds that it renders the inference in (13) valid. To support the argument that the oddness of the examples in (4) are due to Gricean implicature and not to the semantics of the modal, von Fintel (2012) employs various tests of Gricean implicature, and then presents Cariani's (2011) objections to this response and argues that the objections are not persuasive. Rather than recapitulating von Fintel's argument, the present discussion will invoke a wider range of contexts supporting von Fintel's response, and add an additional observation centered on the examples presented in (14)–(15).

To make sure we are making fair comparisons between (4a) and (13), consider the interaction of the disjunction with the respective backgrounds, modal and contextual, in the two types of examples. In the second sentence of (13), the use of the disjunction, in typical contexts, would implicate that the speaker doesn't know anything about which of the two disjuncts is actually realized – that the dishes were washed, or that they were destroyed, or the speaker is otherwise in a position where they cannot say which it is. In (13), this is odd since the first sentence asserts that the dishes were washed. The example can be redeemed in the "aggressive bacteria" context, since the disjunct, *washed the dishes or destroyed them*, in that case expresses the content of the imperative that is understood to be operative in the context; the use of the disjunction in this context does not carry the usual implicature about speaker inability to divulge the truth status of the individual disjuncts. In processing this redeemed version of the example, we shift from asserting simply that an action was performed, washing the dishes, in the first sentence, to asserting that the imperative was met. At the level of reference, we evidently shift from interpreting the simple VP with its ordinary reference, to interpreting the coordinated VP, which refers to the content of the imperative, and thus refers to its ordinary intension (in Fregean terms, it refers to its ordinary sense). This use of the disjunction to assert that the imperative was met is a form of semantic ascent – a shift in reference from normal denotation to normal intension, and does not generate the implicature, so the inference is, in this case, intuitively acceptable.

In the modal example in (4a), the second sentence has an attachment site ambiguity between VP coordination, with wide scope *must* over *or*, or S coordination with antecedent-contained deletion (*you must wash the dishes or* ~~*you must*~~ *destroy them*). The second is a disjunction of two modal statements, which normally generates an implicature of speaker inability to specify the truth status of individual disjuncts, just as in (13). In the first case, coordination of two VPs under *must*, there is normally an implicature that the deontic conditions do not specify whether you are to wash the dishes or destroy them.[3] This is odd since the first sentence asserts unequivocally that you must wash them. Here too, a shift in background between the two sentences can redeem the inference. But in this case, the shift is in the operative deontic criteria. Imagine that in a world with aggressive bacteria, it has been customary to wash dishes in antibacterial soap, radiate them in a microwave at a special high setting, or destroy them by throwing them in a furnace, a three-way range of options. We don't have a furnace at our restaurant, so our long-standing practice is to do one of the first two. A news story comes on reporting that the high microwave setting is a health hazard. So, with that option closed off, the boss tells us we must wash the dishes. Reflecting on the consequences, I utter (4a), generalizing from the deontic conditions at our restaurant (we must wash the dishes) to the new imperative in society at large (we must wash them or destroy them in a furnace).

Speakers typically see this possibility of accommodation more easily in (4c), where it can be true that Mary needs to work harder to satisfy the requirements of her demanding boss, and therefore, from the perspective of her friends who care about her, she needs to either work harder to meet those requirements, or quit her job to get away from the demands of her boss. Similar shifts serve to redeem the others. Is this cheating? It doesn't change the fact that the inference is valid, by the semantics of quantification in (2), when there is no shift in the deontic criteria. But as an accommodation that makes the inference involved intuitively acceptable, it is on a par with the accommodations we make in (13), shifting from a simple description of what we have done, to a characterization of what we have done relative to an operative imperative, by

[3] There is a further ambiguity in whether the second sentence in (4a), with VP coordination, poses the choice for each dish, or for the whole lot of them. But this ambiguity doesn't matter here.

which we reconcile its semantic validity with our pragmatic intuitions. The shift in background relative to which we report on an event in (13) (which arguably involves a shift in reference, as well) is no more innocent than a shift in deontic criteria relative to which we assert a modal statement in (4). In both cases, the shift gets us away from simple inference schemas to ones that are arguably much more sophisticated, involving either semantic ascent or a shift in deontic ordering source.[4] In both cases, the simple, semantically valid inference is intuitively bizarre.

The bizarreness of (13) is mitigated when we replace the disjunction with the statement that the disjunction is true. But this behavior is also matched by (4a), as (14) and (15) show.

(14) We washed the dishes. So, it is true that we washed the dishes or we destroyed them.

(15) You must wash the dishes. So, you must make it true that you wash the dishes or destroy them.

Prefacing *it is true* or (respectively) *make it true* may serve to key our intuitions to bare truth conditions, thereby keeping the implicature at bay. But it has this effect equally in (4a) and (13), making them, in this respect too, more alike than different.

In (4a–d), the shift in deontic criteria turns a logically simple and valid, but intuitively bizarre, inference, into a more sophisticated inference that is logically valid and intuitively acceptable. The shift is fairly natural in (4c), which is why many speakers find it intuitively acceptable in the first place, but the shift is more contrived in (4a,b,d), and may therefore be less helpful in fostering the intuitive acceptability of these examples. The full range of examples suggests that some form of pragmatic fit of an inference matters more to its intuitive acceptability than its semantic validity. This should warn us against drawing hasty conclusions regarding the semantics of an assertion based on whether logical inferences supported by a semantic interpretation seem intuitively valid.

Lassiter's (2011: 132) argument from the example of Professor Procrastinate, derived from Jackson and Pargetter (1986), calls for a similar response.[5] It is widely known that Professor Procrastinate finishes very little of what he

[4] In the "aggressive bacteria" context mentioned earlier, some speakers find (4a) to be on a par with (13), or even less odd, because they find that *wash the dishes* embedded under the modal in the first sentence of (4a) can be accommodated to be interpreted more generally, to go proxy for "get the dishes cleared away," which makes the second sentence a natural conclusion. This involves a shift of reference, making the accommodation here more parallel to that which was used previously to redeem (13).

[5] The response developed in the following page is essentially that given by von Fintel (2012), but articulated in terms of the expression of Kratzer's analysis given in (2).

undertakes. He has been invited to review an article for an academic journal, an article that he is more qualified to review than anyone else in the world. The best outcome would be that he accepts the assignment and writes the review. On this basis, a speaker asserts the deontic modal in (16). But (16) entails (17) under an upward monotonic quantificational semantics for deontic *ought*, such as (2a) or (2c).

(16) Professor Procrastinate ought to accept the assignment and write the review.

(17) Professor Procrastinate ought to accept the assignment.

But the worst outcome would be for the professor to accept the assignment and fail to finish the review, and this outcome is overwhelmingly likely if he does accept the assignment. If we are sufficiently mindful of this fact, our intuition is that (17) should be false in this context, even if (16) is true. Intuitively, (16) does not entail (17).

Again, let's think this through with attention to pragmatic factors that condition judgments of acceptability. If (16) is true, then every weakly ordered sequence of worlds, progressively improving in their accord with deontic conditions in $\Omega(w)$, has a $(p \wedge q)$-uniform tail. In really "good" worlds, according to $\Omega(w)$, $(p \wedge q)$ holds. And, therefore, $p$ holds in those worlds. (16) entails (17) because, in all the "best" worlds, according to $\Omega(w)$, the professor accepts the assignment and writes the reviews. But when we evaluate (17), if we know the facts of the professor's nature, we naturally shift the set of accessible worlds to exclude ones in which the professor writes the review, as too unrealistic to consider. We evaluate (17) for practical purposes, on a different set of possible worlds, thereby shifting in midstream between a set of worlds that include "best imaginable" ones in (16), to a more realistic set in (17). None of this changes the fact that (16) entails (17) when we keep the deontic criteria fixed. But we understand our intuition that the inference from (16) to (17) is unsound as the result of an accommodation we naturally make, shifting the accessible worlds from one modal statement to the next.

So Lassiter's explanation of the fact that (16) does not intuitively entail (17) is that the quantificational semantics in (2), on which (16) does entail (17), is wrong. But another explanation of the fact that (16) does not intuitively entail (17) is that, ordinarily, we habitually and unconsciously shift our deontic criteria between more ideal and more realistic sets in the interpretation of these two statements, given our understanding of the context, and we intuitively recognize the invalidity of the inference when we shift the deontic or bouletic criteria.

We are able to shift modal backgrounds with even more dexterity in another example from the literature, from Jackson (1985: 189), from which Lassiter (2011: 133) quotes the following setup.

Attila and Genghis are driving their chariots towards each other. If neither swerves, there will be a collision; if both swerve, there will be a worse collision (in a different place, of course); but if one swerves and the other does not, there will be no collision. Moreover if one swerves, the other will not because neither wants a collision. Unfortunately, it is also true to an even greater extent that neither wants to be "chicken"; as a result what actually happens is that neither swerves and there is a collision.

In this context, the following three are all intuitively true.

(18) a. Attila ought to swerve.
b. Genghis ought to swerve.
c. It is not the case that Attila and Genghis ought to both swerve.

(18a) is intuitively true against a backdrop of worlds in which Genghis does not swerve, and likewise, (18b) is true against a backdrop of worlds in which Attila does not swerve. Thus understood, (18a,b) do not support the inference to the proposition that Attila and Genghis ought to both swerve, so (18c) can be true as well.

### 2.2.2 *Challenge to the reliance on best worlds*

In an example from Goble (1996), Lassiter invites us to consider an ailing patient who can be treated by either of two mutually exclusive medications, medicine A and medicine B. If medicine A is administered, there is a small chance of a total cure, and a high likelihood of death. If B is administered, the patient will be spared and largely cured, but permanently debilitated. Our intuition is that the medicine B ought to be administered, since medicine A is too risky.

But Lassiter claims that Kratzer's semantics for root modals ensures the truth of the claim that medicine A ought to be administered since it resolves the truth value of a root modal statement on best available worlds. In the best available possible worlds (holding fixed the fact that the patient is afflicted with a condition requiring treatment), medicine A is administered and, against the odds, the patient does not die from the side effects, and recovers fully. Kratzer's semantic analysis, Lassiter claims, is essentially responsive only to what happens in these best of all worlds, and thereby predicts that it is true that medicine A ought to be administered, the wrong result.

According to Lassiter, this indicates that we need a semantic interpretation that takes account of relative probabilities. On such an account, the low probability of a tolerable outcome with medicine A renders false the proposition that medicine A should be administered, even though medicine A offers a slim chance of the perfect outcome, and such an account renders true the proposition that medicine B should be administered.

It will be argued here that Lassiter's claims about these examples confuse deontic modal statements, pertaining to obligations constraining action under current medical practice, with bouletic modal statements pertaining to wishes or desires about outcomes.

Let's carefully consider the possible worlds account of the interpretation of such modal statements, in (19a–c). For ease of discussion we will use the necessity modal statements in (19b,c), but parallel points hold for the weak necessity modal in (19a).

(19) a. Medicine B ought to be administered.
b. Medicine B must be administered, rather than A.
c. Medicine A must be administered, rather than B.

The possible worlds interpretation of a necessity modal is repeated in (20).

(20) [[**must** $\mathbf{S_1}$]] $^{c,\,w}$ = 1 iff every sequence of worlds in **Acc** that is weakly ordered by $\Omega(w)$ has an $\mathbf{S_1}$–uniform tail.

Lassiter's argument assumes that $\Omega(w)$ orders possible worlds according to the desirability of patient outcomes. In the best worlds for patient outcomes, the patient not only lives, but is free of the debilitating side effects of medicine B. These worlds are in the mix, Lassiter assumes, and in all of these worlds, the patient was given medicine A (since we know that medicine B always has permanently debilitating effects). So when $\mathbf{S_1}$ is the phrase marker for the prejacent, *The patient is given medicine A*, all tails of sequences totally ordered by $\Omega(w)$ have $\mathbf{S_1}$–uniform tails, and (19c) is therefore true. The prejacent in (19b), *The patient is given medicine B*, is not true in any of the best worlds, so (19b) is false.

This argument presumes that there is a fixed intuition regarding the truth conditions of (19a–c), which can be rendered independently of context. This is simply not the case. The truth conditions of these modal statements are relative to the goals and interests of the parties involved. Lassiter's observations are correct for the possible worlds interpretation of (19c), applied to a context in which it is understood as asserting the teleological modal in (21).

(21) In order for the patient to survive without debilitating side effects, medicine A must be administered.

In such a context, (19a,b) are false, and (19c) is true, and the possible worlds analysis, articulated along the lines given by Lassiter, correctly yields those results. But this isn't the use of (19a–c) in a more normal context. (19c) would not normally be interpreted as in (21) in the context given in the example. The modal statement in (21) does, indeed, specify the only way the patient could survive without debilitating side effects. But that's a long shot, and in the example given, neither (19b) nor (19c) would elevate that long-shot possibility to the goal at issue. In the more ordinary context of the example, (19a–c) are asserting what must be done, in full recognition of the slim chances of survival if medicine A is administered, and the certainty of the debilitating side effects of medicine B. In particular, in the more normal context answering to Lassiter's claims that (19a) is evidently true, (19b) would be making a claim along the lines of (22).

(22) In order for the patient to have the best overall prognosis, medicine B must be administered.

In computing this interpretation using possible worlds semantics, worlds get successively better under $\Omega(w)$ in accord with the extent to which the treatment of the patient in each world accords with best practices, from the point of view of doing what is most likely to secure the best prognosis for the patient based on what is known when the medicine is administered.

Of course, the notion of best prognosis can depend on probability calculations of the sort that Lassiter invokes for the interpretation of these examples. But that doesn't mean that these sentences have implicit probability calculations as their meanings. Probability calculations are just a precise way of characterizing the stakes involved in the choice of treatment.

In the more normal contexts invoked by Lassiter's argument, the modal operators in (19) can be regarded as teleological or deontic – to obtain the desirable result of patient recovery, it is incumbent upon a doctor, as an upholder of the Hippocratic oath, to give the best available treatment, based on what the doctor knows at the time of treatment. The ordering source $\Omega(w)$ ranks worlds as better and better according to whether the doctor's actions conform to best practices, in order to maximize the likelihood of a positive outcome. It does not rank worlds in terms of desirability of the outcomes themselves. Probability is implicated in the notion of likely positive outcomes, but it is not clear that a quantitative theory of probability is involved in the interpretation of (19), however useful it might be in making the estimation of the chances of success mathematically precise.

Just as epistemic modal statements are interpreted based on what is known to the speaker and a community to which the speaker belongs (see Chapter 1), so too, teleological and deontic modal statements are interpreted based on what the relevant party hopes or expects to achieve, or is required to do, in the given context. The modal statements in (19a–c) are not expressions of the speaker's "best case" wishes. In Chapter 4, we will see that pure bouletic attitude ascriptions do, in fact, have an idealized interpretation where they reference the best imaginable worlds. But (19a–c) are not bouletic statements; they are statements of obligation to pursue best practices to achieve a desired outcome in the context of the constraints of the situation. In teleological or deontic modal statements such as (19a–c), better and best worlds are determined by their accord with best practices as a means to achieve goals, or to meet obligations to achieve goals, as understood in the context, based on what is known to relevant parties in the context.

In the interpretation of bouletic modal statements, our intuitions tolerate unrealistic expectations – we accept purely bouletic modal statements that shoot for pie in the sky. A protester arrested by the police and under a three-day lockup, can, at the end of the first day, say, "I must have a cigarette," even if the police have made it clear that the realization of this desire is an impossibility. Even teleological modals, which tend to have a realistic modal base, can be interpreted relative to an idealized modal base. A speaker can assert (23) to describe the steps in the preparation of a perfect guacamole, even in postapocalyptic conditions where all limes and all lime juice have disappeared from the face of the earth.

(23) To get the perfect guacamole, you must blend in the juice of one whole, fresh lime.

Evaluating this in a world in which fresh limes are a thing of the past, unrealistic counterfactual worlds in which we still have fresh limes can figure as better or best worlds in the interpretation. We tolerate slipping the bounds of realism in statements of unfettered preference, most readily in bouletic modal statements, but sometimes even in statements of teleology and obligation.

Lassiter reviews other examples from the literature to make the same point. One involves miners trapped in a flooding mine. The setup by Lassiter (2011: 136), quoted here, is from Kolodny and MacFarlane (2010) with slight adaptations.

Ten miners are trapped either in shaft A or in shaft B, but we do not know which. Flood waters threaten to flood the shafts. We have enough sandbags to block one shaft, but not both. If we block one shaft, all the water will go into the other shaft, killing all miners inside it. If we block neither shaft, both shafts will fill halfway with water, and just one miner, the lowest in the shaft, will be killed.

They continue that the teleological or deontic modal in (24) is obviously true in this case.

(24) We ought to block neither shaft.

We don't rank worlds according to outcomes, where the best worlds are ones in which all miners survive. Rather, we rank worlds according to best practices, based on a probability calculation, to maximize the expected outcome. The probability calculation is not intrinsic to the meaning of the modal statement, but it is a means to establish the best outcome, in terms of expected value, and thus the best practice.

Kolodny and MacFarlane (2010) considered this problem and argued for a role of information states, specifically, argued that the modal judgment should be relativized to the information state of the speaker in (24). The present account also appeals to information states. Lassiter dismisses this as implausible on the grounds that it would allow our preferences to shift as our information state shifts. But, he points out, our preferences wouldn't shift if we were to gain information. We prefer to maximize the number of miners who survive. Furthermore, Lassiter critiques an account of root modality that makes reference to what we know as too inexplicit. He notes (2010: 139) that Kolodny and MacFarlane (2010), who propose such an account, "say nothing about how information gain actually affects the deontic selection function."

But on the present account, it isn't our preference that would shift if we had more information – it is our judgment of what outcomes are possible in the case at hand. The concern with shifting preferences gives away the main problem with Lassiter's critique of possible worlds semantics as applied to these examples. Throughout these discussions, including in his discussion of Kolodny and MacFarlane (2010), Lassiter continues to take it for granted that best worlds are worlds with the most desirable outcomes, essentially conflating teleological and deontic modals with bouletic modals. (Lassiter [2011: 140] makes this conflation explicit.) But teleological and deontic modals are critically different from bouletic modals. In epistemic modals, we order worlds in accord with their normality for what the speaker (and community) knows: the best worlds are those that are most realistically in accord with what is known. In deontic modals, we order worlds in accord with what behaviors are required by law or obligation: the best worlds are those in which the subject's behavior is most in accord with the law. To evaluate (25), the best worlds are not those in which Max is in the courtroom at the specified hour. That is, rather, the content of the prejacent of the modal.

(25) Max must be in court at 8 am tomorrow.

In interpreting (25), the best worlds are those in which Max acts most fully according to the requirements of law or obligation. Likewise, in the miner example, the best worlds are not those in which the number of miners that survive is the maximum possible across all contexts, whether the action toward that end is discernible to those involved or not. The best worlds are those in which a course of action is pursued that is best suited to achieve the highest "expected value" in terms of number of survivors, based on what is known in the context. The expected value of blocking one shaft is five miners saved: zero miners saved multiplied by 50% (if the wrong shaft is blocked), summed with ten miners saved multipled by 50% (if the correct shaft is blocked). The expected value of leaving both shafts open is nine miners saved.

Only bouletic modal statements order worlds according to the desirability of outcomes. And bouletic modal statements exhibit exactly the behavior Lassiter capitalizes on – they allow a modal statement to be true by virtue of unobtainable ideal worlds.

As for the inexplicitness, this is simply Lassiter's lament that possible worlds semantics does not do the work of probability calculation, as his account does. This reverts to the question of how detailed the semantics of modality should be. Lassiter's view seems to be that the semantics of the modal operators should encode whatever we might use to explicate the probabilities involved. This has the intriguing consequence that, if we were to devise a better theory of probability tomorrow, then the meanings of our modal statements might be said by Lassiter to change, in light of the better theory. The view implicit in possible worlds semantics is that the meanings of the modal operators should suffice to reveal the source of their truth conditions and to distinguish operators of different types and strengths from one another, but that ancillary notions are "plugged in" to the application of modal meaning, depending on whether modal statements are being used in casual, everyday contexts, or quantitatively precise technical contexts, or imaginable future contexts in which our notions of technical theories of probability and statistics have substantially changed.[6]

### 2.2.3 *Challenge based on examples of modal gradability*

Lassiter (2011: 141–142) notes that there can be comparative statements of deontic, teleological, and bouletic modality. The example in (26), example (5.41a) of Lassiter (2011: 141), will suffice for illustration and discussion.

(26) Constance ought to help George more than she ought to help anyone similarly situated.

Lassiter (2011: 148) claims that the interpretation of examples such as (26), in any semantic theory, requires an ordering on objects to be compared, and the establishment of a threshold for comparison. In particular, for comparative deontic and bouletic modality, it requires an ordering of propositions according to which is most in accord with relevant obligations or desires.

[6] To acknowledge the literature, von Fintel (2012) ends up with a solution like the one proposed in this section for the miner's paradox discussed in the second half of this section. But he doesn't apply it to the moral dilemmas discussed in the first half of this section, and he does not identify this challenge to Kratzer's analysis as resulting from a confusion of the differences between bouletic and deontic modals, as it is analyzed here. In short, I see this challenge as resulting more fundamentally from an underlying failure to appreciate the distinction between bouletic and deontic ordering sources. Charlow (2013) and Cariani, Kaufmann and Kaufmann (2013) present solutions to the miner's paradox that give up the classical semantics of modals. This isn't necessary, as von Fintel argues. Goble (1996) analyzes deontic modals in terms of expected value, as does Levinson (2003) for bouletic attitude ascriptions. That is on the right track; but it is worth pointing out that expected value can figure into ordering sources in Kratzer's semantics for modals (which will be extended to bouletic attitude ascriptions in Chapter 6).

But possible worlds semantics orders worlds, not propositions, according to deontic or bouletic criteria. The ordering on worlds doesn't tell us whether the proposition that Constance helps George is more highly rated by deontic criteria than the proposition that she helps $x$, for any person $x$ other than George.

In possible worlds semantics, we are working with truth conditions for deontic *ought* as given in (27).

(27) $[[\textbf{ought S}_1]]^{c,w} = 1$ iff
 a. every sequence of worlds in **Acc** that is weakly ordered by $\Omega(w)$ has an $\mathbf{S}_1$–persistent tail, and
 b. there is a sequence of worlds in **Acc** that is weakly ordered by $\Omega(w)$ that has an $\mathbf{S}_1$–uniform tail.

$\Omega(w)$ is a preorder on possible worlds, ordered according to how many obligations in the set $O_1$ are satisfied. In better worlds, more obligations in $O_1$ are satisfied.

But this, in fact, provides a fairly straightforward interpretation of comparative *ought* in examples such as (26). In this example, the obligations for Constance helping George are similar to, or parallel to, the obligations she might have to help anyone else in a similar situation. This provides an interpretation of (26) as asserting that the set of obligations by virtue of which Constance ought to help George is larger, and therefore more stringent and binding, than the set of obligations for her to help anyone similarly situated to George. In general, we can formulate truth conditions for comparative deontic modality as in (28), where $O_2$ is the set of obligations by virtue of which **ought** ($\mathbf{S}_2$) is true, for a sentential phrase marker $\mathbf{S}_2$, and $O_1$ is the set of obligations by virtue of which **ought** ($\mathbf{S}_1$) is true, for a sentential phrase marker $\mathbf{S}_1$.

(28) $[[\textbf{ought S}_2 \textbf{ more than S}_1]]^{c,w} = 1$ iff
 $[[\textbf{ought S}_1]]^{c,w} = 1$ and $[[\textbf{ought S}_2]]^{c,w} = 1$ and $O_2 \supset O_1$.

There are more obligations that require the truth of $\mathbf{S}_2$ than there are that require the truth of $\mathbf{S}_1$. That makes the obligation for $\mathbf{S}_2$ stronger, or more stringent, than the obligation for $\mathbf{S}_1$. We don't have to rank the propositions expressed by $\mathbf{S}_1$ and $\mathbf{S}_2$ in order to interpret (26).

Lassiter notes that deontic *may* and *must* are exceptions to the availability of comparative deontic modality. His example for deontic *may* is given in (29). The interpretation of deontic *may* for our purposes is given in (30).

(29) ?* You may leave more than John may.

(30) $[[\textbf{may S}_1]]^{c,w} = 1$ iff there is a sequence of worlds in **Acc** that is weakly ordered by $\Omega(w)$, with an $\mathbf{S}_1$–persistent tail.

We can assume a context in which the obligations that pertain to you leaving are parallel to those that pertain to John leaving, making this a contextually good candidate for comparative deontic modality. Why then is it unacceptable? It may just be confusing. Does *more than* signal that the obligation set for you leaving is larger and more stringent than for John leaving, which makes your permission more restricted than John's? Or does *more than* signal that your license is greater, because your obligation set is smaller and less stringent? The idea that this uncertainty may be responsible for the infelicity in (29) is all the more plausible when we see that comparative modality is acceptable when the obligation sets are not inversely related to an associated notion of license, as they are in (29).

To this end, consider Lassiter's other claimed exception, with *must*, in (31), presented with Lassiter's judgment, for example (5.48) from Lassiter (2011: 144).

(31) ?* Sam must stay home more than Bill must go out.

The judgment in (31) is unsurprising since it is hard to compare the obligation set of Sam, pertaining to his staying home, with the obligation set of Bill, with regard to his going out. But if we make the actions and obligation sets parallel, we get the example in the last sentence of (32).

(32) I think Bill has to stay here. Now, I know that Sam really wants to leave right away. But Sam must stay here (even) more than Bill must.

Here we can have quite similar sets of obligations on Sam and Bill, allowing that Sam has a slightly larger and more stringent set. Furthermore, there is no license to leave for either Sam or Bill, so confusion about the direction of the comparison is not a factor – a larger and more stringent set of obligations is a tighter, more binding necessity. Thus, comparative deontic modality is, in fact, possible with *must*, and the fact that we can naturally obtain it when the obligation sets are comparable provides indirect support for the explanation of the judgment in (29). Across a range of examples, deontic comparatives are acceptable when the obligation sets are comparable, and when the relation between the obligation sets is not confounded by a confusingly inverse relation between the degrees of license involved.

The remainder of the challenge from Lassiter (2011: 144–147) based on gradability and scales involves interactions between deontic modals and Neg-Raising, and is mainly designed to target a single-quantifier analysis of deontic modal operators. This does not bear on the Kratzer-style analyses adopted here, presented in (2), (27), and (30). Even so, Chapter 7 of the present work gives reasons to substantially reconceptualize Neg-Raising in the terms of possible worlds semantics, and to do so in a way that captures the facts about the interaction of Neg-Raising with the relative strengths of modal operators, explored in detail by Horn (1978, 1989), and cited by Lassiter on this point. The idea that possible worlds semantics is not equipped to handle these facts is thus refuted by the account in Chapter 7.

### 2.2.4 *Challenge based on deontic and bouletic comparatives*

In an effort to construct a degree semantics for root modality on top of Kratzer's possible worlds semantics for modality, Lassiter (2011: 147) claims that the examples in (33a,b) [his (5.59a,b)] end up not having a truth value.

(33) a. It is better to trespass than it is to murder.
b. It is better to murder than it is to trespass.

Lassiter assumes that we have incomparable scales for murder and trespassing. For the murder scale, the best worlds are ones in which there is no murder (but there may be trespassing), and for the trespassing scale, the best worlds are ones in which there is no trespassing (but there may be murder).

But murder and trespassing pertain to deontic modality in conjunction with all pertinent legal or moral issues, not taken singly, as they pertain to the content of the prejacent in (33a,b). Compare with epistemic modality. Epistemic modal statements are not interpreted in possible worlds semantics with individualized scales dependent on the content of the prejacent. They have a single ordering of worlds, appropriate for the speaker and the context, in which worlds are ordered based on how fully and realistically they accord with what the speaker knows in the context. Likewise, deontic and bouletic modal statements are not interpreted in possible worlds semantics with individualized scales dependent on the content of the prejacent. The deontic modals in (33a,b), pertaining to a given set of social circumstances, have a single ordering of worlds based on the degree to which the actions of pertinent actors are in accord with legal, social, or moral obligations. The better or best worlds are ones in which people act highly in accord with those obligations. As we trace the tails of totally ordered sequences of such worlds, it is plausible that uniformly trespass-free tails properly contain uniformly murder-free tails – that is, the worlds become murder-free before they become trespass-free, with some persistence of trespassing in worlds that are too good to contain any instances of murder. This is a basis for evaluating (33a) as true, and (33b) as false.

Likewise, Lassiter (2011) claims that a bouletic comparative such as that in (34a) [Lassiter's (5.60)] is without a truth value since there are incompatible scales for the speaker's lunch preferences and the speaker's desire to watch a movie. The best worlds for the speaker's lunch preferences may not feature the speaker watching a movie, and the best worlds for the speaker's nonlunch activities may not feature a sandwich for lunch.

(34) a. I want to go to a movie more than I want to have a sandwich for lunch.
b. I want Romney to win more than I want Palin to win.

But in fact, we look at sequences of worlds ordered according to the degree to which the speaker's desires as a whole are met. The main problem in stating the correct truth conditions for (34a) is that (34a) could be true even if it is not true that the speaker wants to go to a movie or to have a sandwich for lunch.[7] If the speaker is confronted by a set of unpalatable options, (34a) can be used simply to state that going to a movie is less unpalatable than eating a sandwich for lunch. It is just required that, as worlds get more in accord with the speaker's preferences, worlds in which she has a sandwich for lunch give out before worlds in which she goes to a movie. Or the other way around, worlds in which the speaker goes to a movie persist longer than worlds in which she has a sandwich for lunch. If the two options are compatible with one another, this means that tails of sequences that start out with worlds in which the speaker goes to a movie and has a sandwich for lunch are persistent in worlds in which the speaker goes to a movie even after they "run out" of worlds in which the speaker eats a sandwich for lunch. This becomes more complicated when the two options are inconsistent, as in example (34b) when Romney and Palin are running against one another for the same office, but the idea works out essentially the same. In this case, there are no worlds (respecting the basic structure of the election process) that are in accord both with Romney winning and Palin winning. But we can still look at sequences of worlds that get better in accord with the speaker's desires, and note that tails persist in worlds with a Romney victory after running out of worlds with a Palin victory. In sum, (34a,b) are both true if and only if, as we trace the tails of sequences of worlds weakly ordered by the speaker's preferences, worlds in which the speaker's greater preference (or lesser abhorrence) is met persist longer than worlds in which the lesser preference (or greater abhorrence) is met.

This approach provides some elucidation of the meaning of *p is much better than q*, for propositions *p* and *q*. (E.g., Lassiter's example, "It is much better to give your money to charity than to gamble it away on sports.") Looking at the tail of a sequence of worlds ordered according to what one ought to do, worlds uniform in *p* would be attained much earlier than worlds uniform in *q*. However, this provides a basis for interpreting only very imprecise comparability. It would not suffice for the interpretation of *p is twice as good as q*, or *p is three times as good as q*. But these interpretations are arguably not produced by the theory of modal interpretation alone. This is the sort of problem for which room should be made for semantic pluralism, allowing more precise

[7] I am grateful to Jerry Weltman (p.c.) for this observation and for providing the example in (34b).

mathematical interpretations to be superimposed on more basic semantic interpretations. (See Section 2.2.6 for elaboration of this comment.)

### 2.2.5 *Challenge based on deontic conflicts*

Suppose that I care for two frail people in a house in the countryside. One is ominously late in returning from a walk, and nobody else is available to search for him. The other is ill and would be imperiled without constant help, and nobody else is available to help. There is no way for me to meet the needs of both of these people, and an intuitive weighting of the associated risks and likely outcomes rates them as equal. I have a deontic conflict: it seems that I ought to do A, search for the missing person, but equally, I ought to do B, stay and care for the ill person, and there is no way for me to do both. The situation is summarized in (35).

(35) a. One ought to do A.
b. One ought to do B.
c. A and B are incompatible; one cannot do both A and B.

Lassiter (2011: 150) claims that deontic conflicts pose a problem for quantificational semantics of deontic modality. He frames the problem as follows: "Deontic conflicts are real, and yet, if standard quantificational theories are correct, it is an *a priori* truth that they cannot exist." He specifically cites the analysis of Kratzer (1991: 647–649), in which all obligations are factored into the ordering source, but there is no guarantee that the best worlds will be ones in which all deontic obligations are met. The problem, according to Lassiter, is that under such a quantificational semantics for modality, the *ought*-claims in (35a,b) cannot both be true, so the conflict is not modeled by the semantics.

To simplify the discussion, we frame the problem in terms of the necessity modal, interpreted as in (36), rather than weak necessity.

(36) $[[\mathbf{must}\ \mathbf{S}_1]]^{c,w} = 1$ iff every sequence of worlds in **Acc** that is weakly ordered by $\Omega(w)$ has an $\mathbf{S}_1$–uniform tail.

Let $\mathbf{S}_1$ be the phrase marker for the sentence asserting that one does A, and let $\mathbf{S}_2$ be the phrase marker for the sentence asserting that one does B. Using two symmetric proofs by contradiction, we can demonstrate that, using (36), **must** $\mathbf{S}_1$ and **must** $\mathbf{S}_2$ are both false in the situation at hand. To do so, suppose that **must** $\mathbf{S}_2$ is true in this situation. Then there are sequences weakly ordered by $\Omega(w)$ that have $\mathbf{S}_2$–uniform tails. They get better and better in one's meeting of the obligation to do B (to care for the ill person at home) until eventually they consist of worlds in which that obligation is uniformly met. But since A and B are incompatible, these worlds are worlds in which one fails to meet the obligation to do A, so these sequences do not have

$S_1$–uniform tails. So **must** $\mathbf{S_1}$ is false. By a symmetric chain of reasoning, **must** $\mathbf{S_2}$ is also false.

And this, Lassiter takes it, is a failure of the semantics to model the deontic conflict. He claims that we could abide **must** $\mathbf{S_1}$ and **must** $\mathbf{S_2}$ both being true (or **ought** $\mathbf{S_1}$ and **ought** $\mathbf{S_2}$ in his discussion), even though $\mathbf{S_1}$ and $\mathbf{S_2}$ express incompatible actions. But the result that **must** $\mathbf{S_1}$ and **must** $\mathbf{S_2}$ are both false is one that Lassiter interprets as effectively asserting that the conflict does not exist, and thus, Lassiter asserts, a failure of the theory.

Lassiter's discussion makes clear that a semantic theory, for him, successfully models the deontic conflict only if it interprets the deontic modal statements so that **must** $\mathbf{S_1}$ and **must** $\mathbf{S_2}$ (or **ought** $\mathbf{S_1}$ and **ought** $\mathbf{S_2}$) are both true. Lassiter's own analysis is that **ought** $\mathbf{S_1}$ is true if and only if the expected value of doing A is greater than the average expected value computed across all possible courses of action, including A and B as possible actions together with all alternative actions. Since most of the alternatives will not involve doing either A or B, and none of them can involve doing both A and B, the average expected value of all possible courses of action will be lower than the expected value of doing A (or likewise, B), so **ought** $\mathbf{S_1}$ is true, and so, likewise, is **ought** $\mathbf{S_2}$. At the same time, the expected value of **ought** [$\mathbf{S_1}$ **and** $\mathbf{S_2}$] is zero since $\mathbf{S_1}$ **and** $\mathbf{S_2}$ is not true in any possible world, and thus **ought** [$\mathbf{S_1}$ **and** $\mathbf{S_2}$] is false. von Fintel (2012) argues that Lassiter's analysis gives deontic *ought* truth conditions that are too weak to be fully plausible and he sketches possible revisions of Kratzer's analysis such that it would also give the result that **ought** $\mathbf{S_1}$ and **ought** $\mathbf{S_2}$ (or **must** $\mathbf{S_1}$ and **must** $\mathbf{S_2}$) are both true.

For a semantic theory of deontic modality to model the deontic conflict, Lassiter assumes that the theory must interpret **ought** $\mathbf{S_1}$ and **ought** $\mathbf{S_2}$ (or **must** $\mathbf{S_1}$ and **must** $\mathbf{S_2}$) in the context of the example as both true. The deontic conflict then has a semantic locus: **ought** $\mathbf{S_1}$ and **ought** $\mathbf{S_2}$ are both true, while **ought** [$\mathbf{S_1}$ **and** $\mathbf{S_2}$] is false. But Lassiter does not explain why this counts as the definitive semantic "model" of the deontic conflict, or why the deontic conflict should have an exclusively semantic model at all. Obviously, this isn't a conflict that anything will resolve; what we are looking for is simply a good explanation of the conflict, one that might respond to features of the conflict in ways that seem right for an apt account of it, at whatever level that might reside. We should be suspicious of any declaration prior to an inquiry of what tools or ideas must be utilized by an explanation of something.[8]

With the semantics of deontic modality expressed as in (36), or an equivalent, we can explain the conflict as a mismatch between our naive expectations and what the semantics says. I must do A, and I must do B, so we feel that **must**

[8] von Fintel (2012) talks more realistically of deontic conflicts being adjudicated by the semantics, not modeled by the semantics.

$S_1$ and **must** $S_2$ should both be true, but, as a result of the logical structure of the situation, the semantics says that neither one is true. That is a reflex of the conflict, one that Kratzer's analysis captures very well. Another way to express or explain the conflict is that if **must** $S_1$ is true, then in any world in which I do what is right, I do A, which, of course, prevents me from doing B, which means I am not doing what is right after all, so there is a problem if **must** $S_1$ is true, and likewise, by symmetric reasoning, if **must** $S_2$ is true.

But neither way of expressing or explaining the deontic conflict necessarily reflects on the semantics of the deontic modal statements. We would like our semantics of deontic modality to reveal the presence of a deontic conflict by leading to trouble, one way or another. And Kratzer's theory, as formulated in (36) for deontic necessity, does that. It isn't clear that the best explanation of a deontic conflict demands any particular result from a semantic theory of deontic modality, other than to reveal the presence of the conflict. Kratzer's theory may or may not be a guide to the proper conceptual analysis of the conflict. To say that Kratzer's theory says that the conflict does not exist is to impose a particular notion of the level at which the explanation must be given, and to impose it by fiat, pretheoretically.

### 2.2.6 *Summary and discussion*

In arguing against quantificational analyses of modal interpretation based on the upward monotonicity of the quantifiers, Lassiter (2011) does not allow for the fact that, in examples such as that of Professor Procrastinate, discussed in Section 2.2.1, the deontic modal base and ordering source naturally shifts between two successive modal statements, so that semantically valid entailments are not pragmatically warranted. In other cases, Lassiter makes empirically flawed comparisons to support the claim that modal operators behave differently from quantifiers.

With the critique of reliance on best worlds discussed in Section 2.2.2, Lassiter misconceives the teleological or deontic ordering source as yielding sequences of worlds that converge on most desirable worlds, rather than worlds in which an agent's actions conform to applicable laws, conventions, or other obligations, or to best practices to achieve a desired outcome, thus confusing deontic or teleological modality with bouletic modality. This misconception may underlie Lassiter's failure to see the shift in modal base and ordering source in some of the examples that are central to the monotonicity challenge.

The same misconception about the modal base and ordering source of root modality infects Lassiter's discussion of deontic and bouletic comparatives, discussed in Section 2.2.4.

Finally, as discussed in Section 2.2.5, Lassiter may be demanding more than we can expect in insisting that the semantics of deontic modals should be a definitive guide to the analysis of deontic conflicts.

It is not clear that Lassiter has shown that there is any principled problem with a possible worlds semantics for the interpretation of deontic, teleological, or bouletic modal statements. This is not to say that possible worlds semantics alone would serve to meet all needs for conceptual analysis, or even semantic analysis, of these root modals. The sixth chapter of Lassiter (2011) develops a scalar account of deontic and bouletic modals based on weighted preferences, integrating information about preferences and probability. This permits Lassiter to develop a notion of weighted preference for propositions that provides a quantitative probabilistic purchase on the issues addressed in the previous sections, for which Lassiter claims that possible worlds semantics is inadequate.

In favor of resisting the idea that one theory must be chosen over the other, consider the semantics of sentences with the preposition *on*, such as, *The marble ball is on the table*. We can let *on* be a curried function that takes an individual to a function from individuals to truth values, and enter it into compositional semantics with a lexical representation as a double-lambda abstract. For many semantic purposes, that will do just fine. But if we look at what happens when we evaluate the truth of that sentence across a range of cases where the angle of the table to the horizontal increases gradually from 0 degrees to 90 degrees, and why at some point, as the angles become less acute and the marble ceases to rest on the table top, the marble can no longer be truthfully said to be on the table, the curried function account will not be of much use. An account of opposing forces, and support against gravity, will be useful, and this might be given to us "for free" from an image schema in cognitive semantics. Does it follow that curried functions are inadequate to the task of preposition interpretation, and cognitive semantics should be adopted instead? We could resist that conclusion by insisting that the two theories are both useful, each in its respective domain. The formal semantic representation of preposition meaning is useful in precisely delineating the composition of its meaning with other meanings in a variety of syntactic contexts, whereas the cognitive semantic account is useful in representing information about the physical embodiment of the *on*-relation. A stance of semantic pluralism, being open to the use of different semantic theories for representing different types of semantic information, seems to be merited.

The examples discussed here, including those presented from the literature in support of objections to possible worlds semantics of root modal statements, may also recommend semantic pluralism. Possible worlds semantics captures the idea that modal statements are true by virtue of how things play out in situations that differ from the actual situation along particular dimensions. Furthermore, as will be argued in the remaining chapters of this work, possible worlds semantics for modality does useful work in distinguishing two types of propositional attitude report, and the interaction of these two types with negation that determines whether Neg-Raising is possible. Lassiter's accounts of epistemic and root modality are useful in explicating the meanings of modal statements in certain cases when quantificational precision is warranted, both in scientific and technological contexts and in daily life. As a society becomes more saturated with scientific and technological notions, it is plausible that the role of Lassiter's semantics for modality will grow. When technically informed people talk about exponential growth, they may be simply appropriating technical terminology for convenience, or to sound impressive, or they may lend some actual precision in co-opting this term even for an essentially metaphorical use. In a similar development, everyday speech about modality may be more quantitatively informed, on some occasions of use, and less so on others. There is reason to be wary of the idea that there is a single, correct truth conditional semantics for modality for all modal statements.

In the previous two chapters, we have examined Kratzer's theory of possible worlds semantics for the interpretation of epistemic and root modal statements. In Chapter 1, some emendations to Kratzer's theory were proposed, and we saw reason to take the epistemic modal base to be knowledge of the speaker or of a community to which the speaker belongs. That is, following von Fintel and Gillies (2008), there is a cloud of contexts for the evaluation of epistemic modal statements. We countered arguments by Lassiter (2011) that possible worlds semantics is inadequate for the semantics of epistemic modality. In the present chapter, we examined possible worlds semantics for teleological and deontic modality, and found that it typically employs modal bases of more or less realistic worlds ordered by the extent to which a relevant actor (speaker, agent of the action described, or other agent) acts in accord with relevant laws, obligations, or conventions, or follows best practices toward a desired teleological end. This contrasts with bouletic modals, which more readily tolerate a counterfactual, idealistic base, and ability and circumstantial modals, which employ a more realistic base. The attention to differences among bouletic, deontic, and teleological bases provides grounds, in addition to those in the literature, on which to counter Lassiter's (2011) objections to possible worlds semantics for deontic and bouletic modality.

# 3 Bare attitude ascriptions

## 3.1 Introduction

In a widely adopted analysis developed by Hintikka (1969), and elaborated by Heim (1992), a belief ascription of the form **Alex believes that $\mathbf{S_1}$**, where $\mathbf{S_1}$ is the phrase marker of a complement clause, is interpreted as true at a world of evaluation $w$ just in case $[[\mathbf{S_1}]]^{v} = 1$ for all possible worlds $v$ in an appropriate set of alternatives compatible with what Alex knows or believes in $w$. Letting $A_3(w)$ be the alternative set for Alex, the individual $a_3$ in the domain, this approach proposes the truth conditions in (1) for this ascription in context $c$.

(1) $[[\textbf{Alex believes that } \mathbf{S_1}]]^{c,w} = 1 \text{ iff } \forall v \in A_3(w)\text{: } [[\mathbf{S_1}]]^{c,v} = 1$

What is desirable about such an analysis, if something along these lines can be maintained, is that the subordinate clause is interpreted in terms of the truth conditions of its phrase marker $\mathbf{S_1}$, directly parallel to the interpretation of $\mathbf{S_1}$ when it occurs as a main clause.

There are two problems with this analysis, however. The first problem is that the use of the universal quantifier in (1) precludes a parallel analysis for attitude ascriptions with predicates that are similar to *believe,* but stronger, such as *be certain* and *be sure*. The standard logical quantifiers ($\forall$, $\exists$), used by themselves as in (1), turn out to be inadequate to the task of modeling the strengths of natural language attitude ascriptions. The second problem concerns the identification of the alternative set. The present chapter examines candidates for the alternative set in (1), and discusses problems with taking it to be either an epistemic modal context of possible worlds consistent with what Alex knows, derived from an epistemic modal base, or a doxastic context of possible worlds consistent with what Alex believes, derived from a doxastic base of propositions. It will be argued that neither is adequate in itself, that a doxastic base is potentially inconsistent, and that therefore an ordering source is required for the doxastic ascription. This result opens the door to the modal quantificational schemas of Chapter 1, and thus provides truth conditions for a range of attitude ascriptions of different types and strengths. The effort expended working out the formulation of Kratzer's analysis of epistemic modal interpretation in Chapter 1 will pay off since standard formulations will not serve the purpose of modeling attitude ascriptions along the lines called for in the present chapter. In Sections 3.3 and 3.4, this analysis will be applied to the interpretation of examples in which attitude predicates embed modal statements or other attitude ascriptions.

Chapter 6 will explore the question of whether the ascribed propositional content in an attitude ascription is related to an eventuality or mental state that might be semantically represented as a Davidsonian eventuality variable. The interpretation of bouletic attitude ascriptions with predicates such as *want* will be taken up in Chapter 4.

## 3.2 Interpreting attitude ascriptions

To fix terminology and notation, where Alex is the individual $a_3$ in the domain of interpretation, let $E_3(w)$ be the set of propositions that Alex knows in a world of evaluation $w$, each proposition equal to a set of possible worlds. If the set of propositions that Alex knows is finite or denumerable, we could list these propositions using natural number indices, $p_1^K$, $p_2^K$ and so on, and then $E_3(w) = \{p_i^K: i \in I\}$, where I is the set of natural numbers, or a finite initial segment of that set. But this representation of $E_3(w)$ allows equally that I could be a nondenumerable index set (such as the set of real numbers) if it were argued or established that Alex knows a nondenumerably infinite number of propositions. However big this set is, and whatever notation we use to represent it, this is a version of the epistemic modal base, and epistemic modal statements are interpreted by quantifying over possibilities in $\cap E_3(w) = \cap\{p_i^K: i \in I\}$, which we call the epistemic alternative set. For epistemic modality, as discussed in Chapter 1, this might have to be generalized to $\cap\{\cap K_n(w): x_n \in G\}$ for a group G of participants in a discourse. In contrast, let $D_3(w) = \{p_j^B: j \in J\}$ be the set of propositions that Alex believes in $w$, each proposition equal to a set of possibilities. (The index set J might be the natural numbers, or a finite initial segment of them, or an infinite set such as the real numbers.) We will call this the doxastic base of Alex in $w$. The set of all possibilities compatible with each of these propositions is $\cap D_3(w)$, which we will call the doxastic context of Alex in $w$.

### *3.2.1 Doxastic base and ordering source*

Reconsider, for a moment, the rationale for Kratzer's approach to epistemic and root modal interpretation. Kratzer's theory keeps the modal base and the ordering source separate, even though they are both just sets of propositions, because they function differently in the theory. The modal base makes a categorical distinction, restricting quantification to possibilities compatible with what the speaker knows (epistemic modality), or restricting to possibilities compatible with one or more things that the referent of the subject of the sentence wants (bouletic modality), or more or less realistic worlds (ability, circumstantial, teleological, and deontic modals) . In contrast, the ordering source effectively sorts possibilities in the modal context, selecting ever better possibilities for the evaluation of the modal statement, approaching (but perhaps never reaching) an ideal.

Turning to propositional attitude ascriptions, such as those in (2), the first thing we note is a modal-style variation in flavor. The ascriptions in (2a,b) pertain to knowledge or beliefs of the attitude experiencer (the referent of the subject), while those in (2c,d) pertain to goals, preferences, or desires of this experiencer.

(2) a. Alex believes that the man in the green hat stole the artifact.
b. Sally suspects that she is being followed by the CIA.
c. I hope that we find the ring before Tuesday.
d. Maria wants Bill to write the report.

Following the lead of Kratzer's approach to modal interpretation, this suggests that ascriptions with *believe* and *suspect* quantify over the set of possibilities consistent with what the experiencer subject knows or believes, with an ordering source characterizing the prototypicality of possibilities for the generalization stated by the ascription, while ascriptions with *want* and *hope* differ from this either in the set of possibilities over which they quantify, or in how those possibilities are ordered. (Exactly what this difference is will be taken up in Chapter 4.) For now, we will discuss ascriptions such as (2a,b), which pertain to knowledge or beliefs.

*3.2.1.1 Doxastic base* What Alex knows underdetermines what he believes; he believes things he does not know. Otherwise, he would be either infallible, or abnormally cautious, committing to believe only what he definitely knows. Thus, the epistemic and doxastic bases of Alex are in the proper subset relation given in (3a).[1] As a result, the doxastic base puts more restrictive conditions on possibilities that are consistent with all of its member propositions than the epistemic base: there are worlds that are consistent with everything that Alex knows, but not with everything that he believes. Thus, the epistemic alternative set and the doxastic context of Alex are in a reverse proper subset relation, as indicated in (3b).

(3) a. $\{p_i^{\mathrm{K}}: i \in \mathrm{I}\} \subset \{p_j^{\mathrm{B}}: j \in \mathrm{J}\}$
b. $\cap\{p_j^{\mathrm{B}}: j \in \mathrm{J}\} \subset \cap\{p_i^{\mathrm{K}}: i \in \mathrm{I}\}$

[1] This requires tacit knowledge to be accorded whatever degree of self-awareness is necessary for belief.

To see that the subset relation in (3b) is, in general, proper, suppose that *p* is a true proposition that Alex does not believe because he thinks, erroneously, that it is inconsistent with what he knows. Then there are possibilities consistent with *p* and with what Alex knows that are not consistent with his beliefs, as diagrammed in (4).

(4)

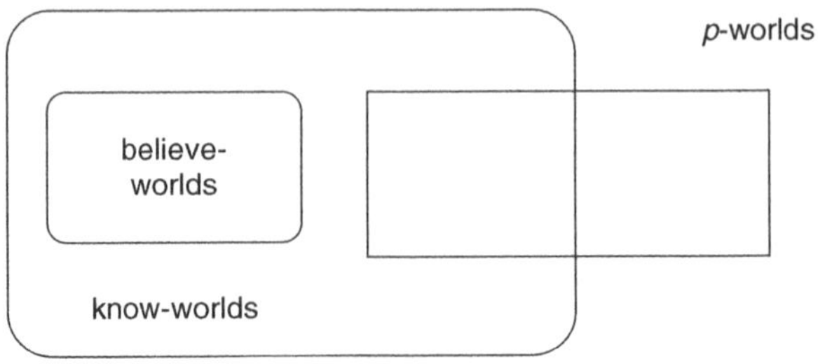


If Alex is convinced of the truth of a false proposition *q* expressed by a subordinate clause S, we would not want to interpret the true ascription *Alex is sure that* S by quantifying over possibilities consistent with what Alex knows (see the diagram in (5)), since these could include possibilities at which *q* is false, which would then falsify the report, if the quantifier is universal or near-universal. And the quantifier is likely to be universal, or near-universal in such a case to express the force of the predicate *be sure*. Thus, quantifying over the larger set of worlds consistent with what Alex knows is problematic for the interpretation of false beliefs and other belief-like attitudes in relation to false content.

(5)

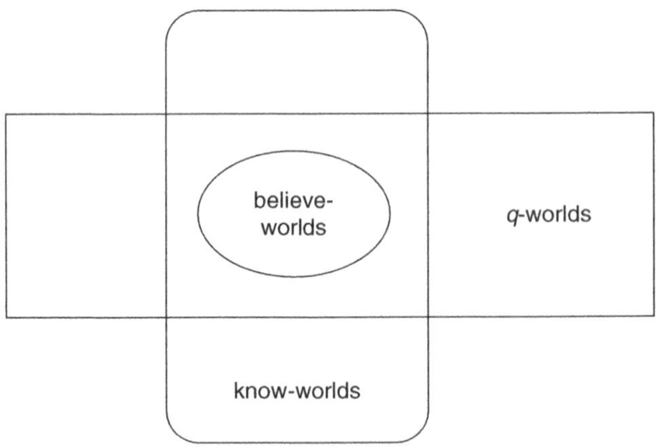

For the reasons just discussed, the appropriate quantification domain for these reports seems to be the set of worlds consistent with what the experiencer believes, not the set of worlds consistent with what the experiencer knows. Belief reports have a doxastic base. This has, in fact, been assumed by most researchers at least since Heim (1992).

*3.2.1.2 Ordering source* Quantifying over the set of possibilities consistent with what Alex believes, we have a problem similar to that with epistemic modality – that this set, without further constraint, contains possibilities that are not realistic or relevant for the attitude being reported.

Consider (2a), and suppose that Alex is a sober-minded museum guard with quite normal knowledge and perceptions of the world around him. Suppose that Alex, as well as the speaker and addressee, has perceptual contact with the man in the green hat, and that the speaker has some normal grounds to assert, truly or falsely, that Alex has the belief reported. Let S be the subordinate clause, the complement to the verb *believe*. In this example, it would not matter if S is false in some possible world in which the man in the green hat is a robot installed by Martians to fool humans, and capable of vaporizing the artifact, but not of stealing it, even if this is logically consistent with what Alex believes. Alex may be open to the possibility of extraterrestrial life, and may have never thought about, and never disavowed, the idea that aliens would populate the world with robots with such capacities. But if he is not willing to actually entertain any such possibilities in the case at hand, they would normally be excluded from consideration in the interpretation of (2a).

However, possibilities in which the history of the world before the present day is different in a number of respects from what unfolded in the actual world should be countenanced to accommodate gaps or mistakes in Alex's knowledge of history without having an impact on the truth of the belief ascription in (2a). The possibilities so entertained can vary within bounds, so long as they are consistent with those gaps and mistakes, but some of the resulting possibilities will comport better with Alex's knowledge and beliefs than others.

So the set of possibilities compatible with what the experiencer Alex believes in (2a) admits diversity among these possibilities, as well as divergence from how the world actually is, in various ways, to varying degrees, and at various levels of detail. They reflect the full range of possibilities consistent with what Alex takes the world to be, and the varying degrees of divergence of relevant possibilities from one another or from an ideal corresponding to what Alex takes the world to be. These observations underscore the need for an ordering source. On this view, (2a) is true just in case S becomes true at more possibilities, the closer we get, among possibilities consistent with what Alex believes, to what he actually takes the world to be.

These considerations provide grounds to employ an ordering source on the doxastic base in the interpretation of propositional attitude ascriptions. The relative strengths of attitude ascriptions will be considered in the following pages, but if we provisionally take the truth conditions for **Alex is sure that $S_1$** to have the strength of a necessity modal, then we conjecture that these truth conditions can be explicated as in (6), quantifying over worlds in all the propositions in Alex's doxastic base $D_3(w)$, ordered relative to one another by an ordering source, $\Omega_3(w)$, according to their suitability for evaluating whether Alex is doxastically committed to the truth of $\mathbf{S_1}$ in them.

(6) [[**Alex is sure that $S_1$**]] $^{c,w} = 1$ iff

a. $\forall u \in \cap D_3(w)$: $(\exists v \in \cap D_3(w)$: $v \leq_{\Omega 3(w)} u)$ $(\forall z \in \cap D_3(w)$: $z \leq_{\Omega 3(w)} v)$ $[[\mathbf{S_1}]]^{z} = 1$, iff

b. every sequence of worlds in $\cap D_3(w)$ that is weakly ordered by $\Omega_3(w)$ has an $\mathbf{S_1}$–uniform tail

c. $=_{df}$ MUST ($\mathbf{S_1}$ | $D_3(w)$ || $\Omega_3(w)$).

This uses the modal quantificational schema of necessity, "living on" possibilities in $\cap D_3(w)$, ordered by $\Omega_3(w)$. Appropriating the idea at the end of Chapter 1 that there is an autonomous quantificational structure for necessity that can be identified apart from the epistemic base, and letting "MUST" designate that quantificational schema, the truth conditions in (6a,b) can be abbreviated in the metalanguage of the semantics as in (6c).

### *3.2.2 Rethinking the role of the doxastic base*

The problem with the truth conditions in (6) is that Alex's doxastic base, $D_3(w)$, the set of all propositions he believes, could be inconsistent. If it is, then the doxastic context $\cap D_3(w)$ is empty, and the phrase marker **Alex is sure that $S_1$**, evaluated as in (6), is vacuously true.

The problem here is not with using the doxastic base $D_3(w)$. The problem lies in confining the interpretation to possible worlds in the doxastic context $\cap D_3(w)$, which could be empty. The doxastic base $D_3(w)$ could still guide the interpretation of **Alex is sure that $S_1$**, provided we don't require the statement of truth conditions to "live on" the doxastic context, $\cap D_3(w)$.

In (6), the ordering source $\Omega_3(w)$ is used to rank possible worlds according to their suitability and prototypicality for the ascribed belief. At the outer levels (on the "greater than" end of the order $\leq_{\Omega 3(w)}$), are wilder possibilities that should not be considered accessible for the belief ascription, which are excluded as we proceed to better ranked possible worlds on the "less than" end of the order. In the "mid-range" and beyond, the truth conditions in (6) make a vital distinction between possibilities that are consistent with the content of $\mathbf{S}_1$, and ones that aren't, while usefully ignoring differences among possibilities that don't pertain to Alex's doxastic state with respect to the truth of $\mathbf{S}_1$, such as different ways of filling in gaps in Alex's knowledge of irrelevant aspects of history. As possibilities become progressively better ranked under $\leq_{\Omega 3(w)}$, they become more and more relevant to Alex's belief $\mathbf{S}_1$. We would be in business if just we could ensure that the condition that possibilities accord all the better with Alex's beliefs is one of the conditions that are progressively better met in possibilities as they improve under the ordering of worlds.

We can't ensure that so long as the interpretation "lives on" the doxastic context, $\cap D_3(w)$. But we can if we rank worlds, additionally, by how large a subset of $\cap D_3(w)$ they are compatible with, so that one world is higher ranked than another if it is compatible with a larger subset of $\cap D_3(w)$.

That is, we could interpret **Alex is sure that $\mathbf{S}_1$** by looking at the truth of $\mathbf{S}_1$ as possible worlds improve in ranking based on how they accord with Alex's beliefs, as well as with their relevance to evaluating the truth of **Alex is sure that $\mathbf{S}_1$**. To do this, we add the doxastic base as an additional set of conditions to be met within the ordering source, to form what we will call the doxastic ordering source, $\Omega{:}D_3(w)$.

(7) $\Omega{:}D_3(w) = D_3(w) \cap \Omega_3(w)$

Let **W** be the set of all possible worlds. The resulting truth conditions for **Alex is sure that $\mathbf{S}_1$** are given in (8), where the double line in MUST ($\mathbf{S}_1 \parallel \Omega{:}D_3(w)$) is meant to emphasize that $\Omega{:}D_3(w)$ is the ordering source, not the doxastic context (which we separate off by a single line).

(8) $[[$**Alex is sure that $\mathbf{S}_1$**$]]^{c,w} = 1$ iff

a. $\forall u \in \mathbf{W}$: $(\exists v \in \mathbf{W}$: $v \leq_{\Omega:D3(w)} u)$ $(\forall z \in \mathbf{W}$: $z \leq_{\Omega:D3(w)} v)$ $[[\mathbf{S}_1]]^{z} = 1$, iff

b. every sequence of worlds in **W** that is weakly ordered by $\Omega{:}D_3(w)$ has an $\mathbf{S}_1$–uniform tail, $\text{iff}_{\text{def}}$

c. MUST ($\mathbf{S}_1 \parallel \Omega{:}D_3(w)$).

This says that, as possibilities get better for the evaluation of **Alex is sure that $\mathbf{S}_1$**, $\mathbf{S}_1$ is true more uniformly on those possibilities, and it is that density or uniformity of possible worlds in which $\mathbf{S}_1$ is true, systematically realized as we go further down on the tails of sequences defined with respect to $\Omega{:}D_3(w)$, that bears out or constitutes the truth of the ascription.

### *3.2.3 Variation in strength*

What is it that makes the report *Alex believes that* S an ascription of belief, in particular, rather than of a different doxastic attitude? One dimension of difference is the "flavor" of the ascription in (2a) relative to other examples in (2). This will be addressed in later sections. Another dimension of difference is the particular strength of the ascription relative to others of the same "flavor." This distinguishes the doxastic ascriptions in (9a–c) from one another.

(9) a. Alex is sure/certain/convinced that Sam will resign as president.
b. Alex believes/thinks/expects that Sam will resign as president.
c. Alex suspects/conjectures/imagines that Sam will resign as president.

The gradations in (9) are lexically salient, and are clearly reminiscent of the gradations among necessity, weak necessity, and possibility modals. The present view is that they are not merely reminiscent of these three grades of modal strength, but parallel to them (with one significant point of divergence to be discussed in the pages that follow), because the quantificational mechanisms of modality are autonomous, and show up in attitude ascriptions as well as in modality.

The truth conditions of attitude ascriptions with *sure* were already given in (8) using the quantificational schema of modal necessity. Since *believe*, *think*, and *expect* have a lower order of strength than *sure*, we can conjecture that they have the strength of a weak necessity modal, as spelled out in (10) .

(10) $[[\textbf{Alex believes that S}_1]]^{c,w} = 1$ iff
a. $(\forall u \in \mathbf{W}: [[\textbf{not-S}_1]]^{c,u} = 1)\ (\exists v \in \mathbf{W}: v \leq_{\Omega:D3(w)} u)\ [[\mathbf{S}_1]]^{c,v} = 1 \wedge (\exists u \in \mathbf{W}: [[\mathbf{S}_1]]^{c,u} = 1)\ (\forall v \in \mathbf{W}: v \leq_{\Omega:D3(w)} u)\ [[\mathbf{S}_1]]^{c,v} = 1$, iff
b. every sequence of worlds in **W** that is weakly ordered with respect to $\Omega$:$D_3(w)$ has an $\mathbf{S}_1$–persistent tail, and there is such a sequence with an $\mathbf{S}_1$–uniform tail
c. $=_{df}$ SHOULD $(\mathbf{S}_1 \parallel \Omega{:}D_3(w))$.

The interpretations in (8) and (10) distinguish *believe* from *be sure* in that (10), but not (8), allows that there is at least one sequence of weakly ordered possible worlds in $\Omega$:$D_3(w)$ with a (**not-S**$_1$)–persistent tail, that is, a tail persistent in worlds in which $\mathbf{S}_1$ is false. This plausibly captures the speaker's hedging of bets in using *believe* rather than *be sure*. The ascription with *believe* expresses the experiencer's sense of the fallibility of the truth of $\mathbf{S}_1$, that no matter how good worlds get with respect to the experiencer's criteria in $\Omega$:$D_3(w)$, the experiencer recognizes that it is still possible that $\mathbf{S}_1$ is false. The ascription with *be sure* attributes to the experiencer a mindset that closes the door on $\mathbf{S}_1$ being false on worlds that are sufficiently good according to $\Omega$:$D_3(w)$.

Empirical support for this will be given in the following pages, when we examine the interactions of these predicates with negation in the context of a scale of modal strength, which will reveal relationships of duality. To do that, we have to fill the associated scale with doxastic attitude predicates.

It might be thought that we can fill out the paradigm by identifying a weak ascription using *suspect*, *imagine*, or *conjecture* with the strength of a possibility modal, with the truth conditions given in (11), for (9c).[2]

(11) [[**Alex suspects that S$_1$**]] $^{c,w} = 1$ iff
  a. there is a sequence of worlds in **W** that is weakly ordered by $\Omega$:D$_3$($w$), which has an **S$_1$**–persistent tail, iff$_{\text{def}}$
  b. MIGHT (**S$_1$** || $\Omega$:D$_3$($w$)).

But Kai von Fintel (personal communication) provides a compelling reason against this proposal for the interpretation of the *suspect*-class of ascriptions. The possibility modal has the property that might ($p$) is compatible with might (~$p$). But predicates in the *suspect*-class do not have this property; the coordinated ascriptions in (12) are not consistent with one another, evaluated at the same time index.

(12) (Alex suspects that S) and (Alex suspects that not-S).

In fact, as von Fintel reports, it seems hard (if not impossible) to find any simply lexicalized attitude predicate in any language that, substituted for *suspect* in (12), would make it true.

There is, however, an instantiation of this pattern, given in (13), just not with a simply lexicalized predicate.

(13) (Alex is not sure that S) and (Alex is not sure that not-S).

As we saw in Chapter 1, negation is well-behaved with Kratzer's quantificational structures. As a result, we can define the dual of **sure** with a clausal complement as in (14), and compute its truth conditions by writing out the truth conditions of the negation of a *sure*-ascription in (15), and then replacing **S$_1$** by **not-(S$_1$)**. The result, given in (16), exhibits the quantificational structure of the possibility modal, MIGHT, on Kratzer's analysis.

(14) [[**Alex is sure$^{\perp}$ that S$_1$**]] $^{c,w} = 1$ iff [[**Alex is not sure that not-S$_1$**]] $^{c,w} = 1$

(15) [[**Alex is not sure that S$_1$**]] $^{c,w} = 1$ iff
  $\exists u \in \mathbf{W}$: $(\forall v \in \mathbf{W}$: $v \leq_{\Omega:D3(w)} u)$ $(\exists z \in \mathbf{W}$: $z \leq_{\Omega:D3(w)} v)$
  $[[\mathbf{S_1}]]^{z} = 0$

[2] This was, in fact, the proposal put forth in previous versions of the present work, including Hegarty (2006b).

(16) [[**Alex is sure$^{\perp}$ that $\mathbf{S_1}$**]] $^{c, w}$ = 1 iff

a. $\exists u \in \mathbf{W}$: $(\forall v \in \mathbf{W}$: $v \leq_{\Omega:D3(w)} u)$ $(\exists z \in \mathbf{W}$: $z \leq_{\Omega:D3(w)} v)$ [[$\mathbf{S_1}$]] $^{z}$ = 1, iff

b. there is a sequence of worlds in **W** that is weakly ordered by $\Omega$:$D_3(w)$, with an $\mathbf{S_1}$–persistent tail

c. $=_{df}$ MIGHT ($\mathbf{S_1}$ || $\Omega$:$D_3(w)$).

The present claim is that **sure**$^{\perp}$ is not simply lexicalized in English, and possibly not in any language, and in particular, is not lexicalized as any of the *suspect*-class predicates. But it has the quantificational force of a possibility modal, and this is the effective strength of a negated ascription with **sure**.

This successfully captures the behavior of *sure* under negation in actual examples. The naturally occurring example in (17a) was uttered by tennis announcer John McEnroe during his commentary before the tennis championship match between Venus Williams and Marion Bartoli played at Wimbledon on July 7, 2007; the examples in (17b,c) are constructed to provide other intuitive touchstones.

(17) a. I'm not sure Venus will crumble like Henin did yesterday.
b. I'm not sure we can get in there.
c. Alex isn't convinced that Sam will resign the presidency today.

Bartoli had played consistently well throughout the Wimbledon fortnight. But McEnroe was warning against the presumption that Bartoli would win. In the context, his utterance of (17a) could be paraphrased as, "A Bartoli victory is not certain; Venus might not crumble like Henin did yesterday." Under negation, the strong doxastic predicate *be sure* in (17a) has a strength comparable to that of a possibility modal. The same is true of the constructed examples in (17b,c). This substantiates the truth conditions in (16).[3]

Now consider negation with *imagine* and *suspect*, as in (18).

(18) a. I don't imagine/suspect you can lend me a hundred dollars.
b. I don't imagine/suspect we can get in there.
c. Alex doesn't imagine/suspect that Sam will resign the presidency.

It is instructive to consider why (18a) is more polite as a request than the variant, "Can you lend me a hundred dollars?" The interpretation of (18a) can be

[3] An anonymous reviewer asks why (ia) is infelicitious and not parallel to (ib), if negated *sure* is interpreted as a weaker predicate on the same scale as *sure*. The answer to this will be given in Chapter 7.

(i) a. # I'm not sure Venus will win; in fact, I'm sure she will.
b. It's possible Venus will win; in fact, it's certain she will.

paraphrased as, “I rather expect that you cannot lend me a hundred dollars,” a formulation that, in asserting that the addressee probably cannot make the loan, gives the addressee an easy way out if the addressee decides not to comply with the request.[4] The predicate *imagine* is intuitively somewhat weaker than the midpoint of the scale for belief ascriptions, so it is not surprising that, as the paraphrase of (18a) just given shows, when we negate it as in (18a), we get an ascription with strength somewhere above the midpoint of that scale. Likewise, (18b) would normally be used to assert that the speaker expects or believes that we cannot get in (say, to a fancy restaurant or trendy nightclub), and (18c) to assert that Alex expects or believes that Sam will not resign.

This suggests that *believe* is the dual of *imagine* and *suspect*, which means that *imagine* or *suspect* is the dual of *believe*. The truth conditions in (10) provide grounds for the observation that *believe* has strength above the midpoint of the scale, but below *sure* and *certain*. The discussion of modal interpretation in Chapter 1 established that Kratzer's truth conditions for $[[\textbf{should S}_1]]^{c,w}$ support the definition of $[[\textbf{should}^{\perp}\ \textbf{S}_1]]^{c,w}$ such that $[[\textbf{should}^{\perp}\ \textbf{S}_1]]^{c,w}$ = iff $[[\textbf{not (should not-S}_1\textbf{)}]]^{c,w}$. Transposing those results from modality to attitude ascriptions, the dual to **believe** has the truth conditions in (19) such that the equivalence in (20) holds.

(19) $[[\textbf{Alex believes}^{\perp}\ \textbf{that S}_1]]^{c,w} = 1$ iff

a. $(\exists u \in \textbf{W}: [[\textbf{S}_1]]^{u} = 1)(\forall v \in \textbf{W}: v \leq_{\Omega:D3(w)} u)\ [[\textbf{S}_1]]^{v} = 1 \lor (\forall u \in \textbf{W}: [[\textbf{not-S}_1]]^{u} = 1)(\exists v \in \textbf{W}: v \leq_{\Omega:D3(w)} u)\ [[\textbf{S}_1]]^{v} = 1$, iff

b. there is a sequence of possible worlds weakly ordered with respect to $\Omega{:}D_3(w)$ that has an $\textbf{S}_1$–uniform tail, or every such sequence has an $\textbf{S}_1$–persistent tail, $\text{iff}_{\text{def}}$

c. $\text{SHOULD}^{\perp}\ (\textbf{S}_1 \parallel \Omega{:}D_3(w))$.

(20) $[[\textbf{Alex does not believe that S}_1]]^{c,w} = 1$ iff $[[\textbf{Alex believes}^{\perp}\ \textbf{that not-S}_1]]^{c,w} = 1$

Note that truth conditions for **Alex believes**$^{\perp}$ **that** $\textbf{S}_1$ are specified in (19) in terms of the same two conditions as for **Alex believes that** $\textbf{S}_1$ in (10), but are weaker by virtue of having the connector “or” rather than “and” between the two conditions. By the argument given here, **suspect that** $\textbf{S}_1$ is interpreted as **believe**$^{\perp}$ **that** $\textbf{S}_1$, thus, with the truth conditions given in (21).

(21) $[[\textbf{Alex suspects that S}_1]]^{c,w} = 1$ iff $\text{SHOULD}^{\perp}\ (\textbf{S}_1 \parallel \Omega{:}D_3(w))$

In summary, predicates such as *believe* have strength somewhere above the midpoint of the doxastic scale, but below the top of the scale, which is occupied

[4] Some speakers do not obtain this interpretation as readily with *suspect* as with *imagine*.

by *be certain* and *be sure*. The predicates *suspect* and *imagine* have the force of the dual of *believe*, which is correspondingly weaker than the midpoint. The resulting scale of predicates is shown in (22); the predicates increase in strength from left to right, and each predicate on the top line is matched with an identification of it on the bottom line as the dual of another predicate, except that the dual of **be-sure** is not lexicalized as a predicate; it emerges only as the interpretation of **sure** under negation, and then only when negation is understood to be interpreted with the lower clause (i.e., as an instance of Neg-Raising).[5] This will be discussed at length in Chapter 7.

(22) ∅ suspect believe be-sure
|-----------|----------|----------|---------|
be-sure$^{\perp}$ believe$^{\perp}$ suspect$^{\perp}$

Attitude ascriptions of want, preference, or desire involve a bouletic component to the ordering source, which is easy to add. But they also typically involve another aspect of interpretation that will not be taken up until Chapter 6. So discussion of these attitudes will be deferred until that chapter.

## 3.3 Iterated attitude ascriptions

Since attitude reports are interpreted as quantificational schemas, and quantificational statements readily iterate, allowing one to be embedded within another, it is easy to interpret iterated attitude ascriptions such as that in (23).

(23) Jill is sure that Alex suspects that Josh stole the artifact.

The ascription in (23) is about possibilities consistent with Jill's doxastic base, $D_4(w)$, as favored by her ordering source $\Omega_4(w)$ on all possibilities in $\mathbf{W}$. As discussed in Section 3.2, these are worlds favored by the derived ordering source $\Omega{:}D_4(w) = \Omega_4(w) \cap D_4(w)$. Let $\mathbf{S}_2$ be the phrase marker **Josh stole the artifact**, and $\mathbf{S}_3$ be the phrase marker **Alex suspects that** $\mathbf{S}_2$. Then (23) asserts that $\mathbf{S}_3$ is true in any possible world $v$ that ranks sufficiently high, on tails of sequences of possible worlds weakly ordered by $\Omega{:}D_4(w)$. Then (23) has the truth conditions in (24), to be further spelled out.

[5] Reinforcing this point, an anonymous reviewer notes that *not sure* does not have the scalar implicature cancelling properties it would have if it were regarded as a lexicalization of the low end of the scale in (22). Thus, *possible* implicates *not certain (to be the case)*, an implicature that can be cancelled as in, *It's possible Venus will win; in fact, it's certain she will.* In contrast, *not sure* does not behave like a weak lexical predicate in this respect, as demonstrated by the following example provided by the anonymous reviewer: *#I'm not sure Venus will win; in fact, I'm sure she will.* The appropriate implicature cancelling scalar relationship goes instead in the other direction, given that the negation of *sure* has strength near the lower end of the scale, but only with negation interpreted below it. Thus, *I'm not sure Venus will win; in fact, I'm quite sure she won't win*, where the follow-up, interpreted with lower negation, strengthens the predicate, in line with the discussion of (17).

(24) [[**Jill is sure that** $\mathbf{S_3}$]] $^{c,w} = 1$ iff MUST ($\mathbf{S_3} \parallel \Omega{:}D_4(w)$)

Recall that the right-hand side of (24) means that every sequence of possibilities that is weakly ordered by $\Omega{:}D_4(w)$ has an $\mathbf{S_3}$–uniform tail. It is an abbreviation for the universally quantified condition in (25).

(25) $\forall u \in \mathbf{W}: (\exists v \in \mathbf{W}: v \leq_{\Omega:D4(w)} u)\ (\forall z \in \mathbf{W}: z \leq_{\Omega:D4(w)} v)\ [[\mathbf{S_3}]]^{z} = 1$

Using (19) and (21), $\mathbf{S_3}$ = **Alex suspects that** $\mathbf{S_2}$, evaluated at $z$, has the truth conditions in (26), the right-hand side of which we symbolize as SHOULD$^{\perp}$ ($\mathbf{S_2} \parallel \Omega{:}D_3(z)$), where $\Omega{:}D_3(z)$ is the doxastic ordering source of Alex.

(26) [[**Alex suspects that** $\mathbf{S_2}$]] $^{z} = 1$ iff
a. $(\exists\, u' \in \mathbf{W}: [[\mathbf{S_2}]]^{u'} = 1)\ (\forall v' \in \mathbf{W}: v' \leq_{\Omega:D3(z)} u')\ [[\mathbf{S_2}]]^{v'} = 1$ $\vee$ $(\forall u' \in \mathbf{W}: [[\textbf{not-S}_2]]^{u'} = 1)\ (\exists v' \in \mathbf{W}: v' \leq_{\Omega:D3(z)} u')$ $[[\mathbf{S_2}]]^{v'} = 1$, iff
b. there is a weakly ordered sequence with respect to $\Omega{:}D_3(z)$ that has an $\mathbf{S_2}$–uniform tail, or every such sequence has an $\mathbf{S_2}$–persistent tail, iff
c. SHOULD$^{\perp}$ ($\mathbf{S_2} \parallel \Omega{:}D_3(z)$).

Plugging the right-hand side of (26) into (25), we obtain (27a), which expands as (27b).

(27) [[**Jill is sure that** $\mathbf{S_3}$]] $^{c,w} = 1$ iff MUST ($\mathbf{S_3} \parallel \Omega{:}D_4(w)$).
a. $\forall u \in \mathbf{W}: (\exists v \in \mathbf{W}: v \leq_{\Omega:D4(w)} u)\ (\forall z \in \mathbf{W}: z \leq_{\Omega:D4(w)} v)$ $[[\mathbf{S_3}]]^{z} = 1$, iff
b. $\forall u \in \mathbf{W}: (\exists v \in \mathbf{W}: v \leq_{\Omega:D4(w)} u)\ (\forall z \in \mathbf{W}: z \leq_{\Omega:D4(w)} v)$ $(\exists\, u' \in \mathbf{W}: [[\mathbf{S_2}]]^{u'} = 1)\ (\forall v' \in \mathbf{W}: v' \leq_{\Omega:D3(z)} u')\ [[\mathbf{S_2}]]^{v'} = 1$ $\vee$ $(\forall u' \in \mathbf{W}: [[\textbf{not-S}_2]]^{u'} = 1)\ (\exists v' \in \mathbf{W}: v' \leq_{\Omega:D3(z)} u')$ $[[\mathbf{S_2}]]^{v'} = 1$, iff
c. MUST (SHOULD$^{\perp}$ ($\mathbf{S_2} \parallel \Omega{:}D_3(z)$) $\parallel \Omega{:}D_4(w)$).

This is tedious to write out, but perfectly straightforward. Other combinations, with different choices of attitude predicates, can be formed and interpreted similarly. On this view, there is nothing special about interpreting iterated attitude ascriptions; we simply iterate their interpretations.

## 3.4 Prospects for semantically reduced interpretations

### *3.4.1 Some views from the literature*

On the interpretations formulated in Section 3.2, attitude ascriptions report on the doxastic state of the experiencer subject, by means of the ordering source. It has also been claimed that they may be interpreted as evidential or modal qualification of the assertion of the content of the complement clause. This interpretation has been discussed by Urmson (1952), Bolinger (1968), Cattell (1973), Ross (1973), R. Lakoff (1969), Lindholm (1969), G. Lakoff (1974), Hooper (1975), Prince (1976), Horn (1978), and more recently by Stephenson (2007), Cappelli (2007), Simons (2007), and Hacquard (2010). Some prospective examples of this are given in (28), where the capitalization in (28a) indicates nuclear stress, focusing the positive polarity of the content of the complement clause (the bill does have enough votes to pass) according to Sam.[6]

(28) a. Senate staffer: Does the bill have enough votes to pass?
Colleague: Sam thinks it DOES.
b. I suppose/think/am sure that the bill has enough votes to pass.

On such a use, the "main point," in Simons' (2007) terminology, of the colleague's response in (28a) or of the whole utterance in (28b) is to assert the content of the complement clause with appropriate qualification, whether very confidently (*am sure*), somewhat less confidently (*think*), or quite tentatively (*suppose*), according to the doxatic state of Sam or the speaker.

Simons (2007) and Cappelli (2007: 117–139) argue that the main clause verb in these ascriptions expresses an evidential condition on the content of the complement clause. Simons, citing Rooryck (2001), appeals to the analysis of evidentiality as featuring two components, one pertaining to the source of information, the other to its reliability or veridicality. In the use described previously, the main clause verb expresses an evidential condition of the second type. A similar point is made throughout Cappelli's (2007) book. Hooper calls this the assertive use of the ascription (although Simons [2007] takes issue with that characterization), and Urmson characterizes the main clause as parenthetical in these cases, since the main clause subject and verb are merely included to provide the evidential comment on the content of the complement clause. However, even if we were to adopt Urmson's terminology, the utterances must be recognized as instances of hypotaxis, unlike (29), where the attitude verb and experiencer subject would normally be set off by pauses and pronounced at a lowered pitch relative to the surrounding material.

(29) The bill has, I suppose, enough votes to pass.

We can admit the force of these claims, and in particular, admit that the attitude predicates in (28a,b), in the uses imagined, do provide a form of commentary

[6] Stress on the subject or verb in the colleague's utterance in (28a) is much more favorable to the interpretation of the utterance as primarily an assertion about Sam's doxastic state, rather than as primarily about the bill's prospects of passage. In (28b), nuclear stress within the complement clause, on *pass* or *enough*, most strongly favors the interpretation of the whole utterance as primarily about the bill's prospects.

on the reliability of the content of the complement clause. Nevertheless, it remains open to note that they may do this through the standard quantificational semantics of the attitude predicate. The qualification in (28a) can be construed as the claim that the bill passes in worlds that comport with propositions in Sam's doxastic base (realized as tails of sequences weakly ordered by Sam's doxastic base that are uniform or persistent in the truth of the complement clause). In the first person, in (28b), the qualification can be regarded as the degree to which the complement clause comports with the speaker's own doxastic base, which does express a speaker judgment of reliability or veridicality, but does so through the quantificational mechanisms of doxastic ascription. The ascriptions in (28a,b) contrast with attitude ascriptions discussed in most of the literature on the semantics of propositional attitudes, for reasons noted by the authors cited, but the difference seems to be attributable to informational prominence, rather than to the employment of truth conditions other than those of doxastic attitude ascription.

One could claim that the higher attitude predicate is merely a strengthener or weakener of the assertion made by the complement to this predicate. In (30), the responses of the parent might seem to be likely candidates for this, as providing qualifications of the strength of the deontic permission modal statements that they embed. The idea is that they can be used to give permission, with different degrees of permissive strength, and say nothing about the doxastic state of the speaker.

(30) Child: Can I go on the boat with Sam?
Parent: I suppose you can./I think so. Be careful!/You sure can!

Note that one can't actually say, *I'm sure you can!*, with a purely permissive interpretation. But for the sake of the argument, it could be granted that *You sure can!* is a variant of the others with strength at the top of the associated scale.

But the attitude predicates here aren't just strength modifiers. When the embedded deontic condition has a source other than the parent, such as the law or the will of the boat owner, the parent's utterance has a primary doxastic function that it achieves by expressing a doxastic judgment regarding the deontic state of affairs. When the parent's own will is the source of the embedded deontic modal, the doxastic judgment is understood to reflect the parent's will, but that is available as an immediate inference from the doxastic interpretation, as the parent "reading" his or her own deontic judgment. Through that inference, *I suppose you can* means that the parent rather reluctantly grants permission, *I think so* means that the parent quite readily grants permission, and *You sure can!* means that the parent enthusiastically grants permission. So the observation is correct that the higher predicate provides strength modification, but it isn't interpreted as a pure, domain-neutral, strength modifier.

In a different vein, Stephenson (2007) argues that doxastic ascriptions that embed modal statements semantically reduce to pure doxastic ascriptions. Look at *Sam thinks it might be raining*, for which Stephenson adopts the truth conditions in (31).

(31) In all of Sam's doxastic alternative worlds $z$ at $w$, it is the case that in some epistemic alternative to $z$, it is raining.

But Stephenson (2007: 46) assumes that "to believe something is to believe that one knows it, which means that the epistemic alternatives of a person's doxastic alternatives are the same as the person's doxastic alternatives." Under this assumption, (31) becomes (32) (assuming that the universal quantifier over doxastic alternatives is not vacuous).

(32) There is some doxastic alternative world for Sam at which it is raining.

That is, *Sam thinks it might be raining* is interpreted essentially as, "Sam suspects it is raining."

Specifically, Stephenson (2007: 122–123) argues for (33), which states that when you start at $w$ and unite the epistemic alternatives computed at all the doxastic alternatives to $w$, you end up with just the set of doxastic alternatives at $w$.

(33) $\cup_z [\cap K_4(z): z \in \cap D_4(w)] = \cap D_4(w)$

In (33), $\cup_z$ corresponds to the strength of the doxastic predicate, the universal quantifier Stephenson assumes for *think*.

What happens if the doxastic predicate has a lesser strength? Taking account of differences in attitude strength, we can change the quantifier over doxastic alternatives. Sticking with the semantically simpler doxastic predicates, *be sure* and *suspect*, and the possibility and necessity modals, we have the four possibilities in (34).

(34) a. Sam is sure it might be raining.
b. Sam suspects it might be raining.
c. Sam is sure it must be raining.
d. Sam suspects it must be raining.

Assume that the quantifier for *be sure* is the universal quantifier. Then (34a) is the sentence that Stephenson actually analyzed, resulting in the reduced truth conditions in (32), essentially, "Sam suspects that it is raining." For *suspect*, the quantifier over doxastic alternatives will be "some" (which we will interpret here with its scalar quantity implicature, "but not too many"). This is one of the mixed strength cases. Next, consider the other mixed strength case, (34d). It will have the full truth conditions given in (35).

(35) In some (but not too many) of Sam's doxastic alternative worlds, it is the case that in all epistemic alternatives to those worlds, it is raining.

We start at *w* and take the epistemic alternatives at each doxastic alternative *z* to *w*. The set of Sam's epistemic alternatives at *z* is $\cap K_4(z)$. If we do this for all of Sam's doxastic alternatives $z \in \cap D_4(w)$, and unite all the resulting worlds, we get all and only the worlds in $\cap D_4(w)$, according to Stephenson. But if we do this for just some of the doxastic alternatives *z*, then we get a collection of epistemic alternatives for a partial set S of doxastic alternatives. On Stephenson's assumption, the epistemic alternatives are still doxastic alternatives, as stated in (36).

(36) For any subset S of $\cap D_4(w)$, $\cup_S$ $[\cap K_4(z){:}\ z \in \cap D_4(w)] \subseteq \cap D_4(w)$

But we have no real purchase on how much of $\cap D_4(w)$ is composed of $\cup_S$ $[\cap K_4(z){:}\ z \in \cap D_4(w)]$. As a result, we don't know what doxastic predicate might result from such a reduction. Thus, it is not clear how Stephenson's approach would generalize to strengths of doxastic predicates other than what is modeled by the universal quantifier over doxastic alternatives, $\cap D_4(w)$, which Stephenson takes to be *think* and *believe*, but which should properly be identified as *be certain* and *be sure*.

The "same strength" cases of embedding in (34b,c), repeated here, are intuitively more amenable to reduction.

(34) b. Sam suspects it might be raining.
c. Sam is sure it must be raining.

(34b) might be interpreted as asserting that Sam has a weak or tentative suspicion that it is raining, and (34c) might be interpreted as asserting that Sam is very confidently certain that it is raining. In these cases, it does seem that the embedded modal might effectively provide a strength modification of the higher doxastic attitude predicate.

It is easy to see that if we follow Stephenson and interpret *be sure* and *suspect* as universal and existential quantification over doxastic alternatives, respectively, and likewise *must* and *might* as universal and existential quantification over epistemic alternatives, and if we adopt Stephenson's assumption "that the epistemic alternatives of a person's doxastic alternatives are the same as the person's doxastic alternatives," then (34b) reduces to "Sam suspects that it is raining," and (34c) to "Sam is sure that it is raining." But then, with these straightforward provisions for semantic reduction, we don't have a semantic account of the intuitive strength modification.

This suggests that these interpretations don't actually reduce semantically, but have "shortcut" renditions arrived at through contextual inference. That they are actually semantically unreduced would explain why we seem to be able to retain the full semantic embedding of the modal, even in (34a), the one that actually has the truth conditions that Stephenson semantically reduced.

(34) a. Sam is sure it might be raining.

The example (34a) can mean that all of Sam's doxastic alternatives allow the possibility of rain, keeping the unreduced semantic structure. Sam might even enumerate the possible scenarios and show that the possibility of rain exists in each of them. This is not the same thing as suspecting that it is raining, understood as enumerating all doxastically possible scenarios and showing that enough of them are rainy to support a suspicion.

### 3.4.2 *Exploring the reduction of doxastic-epistemic embeddings*

This section explores doxastic ascriptions that embed epistemic modal statements, the type of example Stephenson discussed, and argues that semantically reduced interpretation as a simple doxastic ascription is not obtained in general, but that a reduced interpretation is obtained through nonsemantic accommodation in a special case. At this point, the discussion reverts to the interpretation of epistemic modals presented in Chapter 1, following Kratzer, and the interpretation of doxastic attitude ascriptions based on them, given earlier in this chapter.

Yalcin (2007) noted a potential problem with the analysis of reductions of iterated ascriptions. This problem arises in (34a,b,c) if they are taken to mean that rain is actually compatible with Sam's beliefs. Referential opacity would normally be at work here, so that the examples in (34) should merely mean that Sam believes that rain is compatible with her beliefs (with the appropriate modal strength). As noted by Hacquard (2010: 105), who also discusses this problem, the interpretation of the examples in (34) should not rely on an assumption "that whatever [Sam] *believes* to be compatible with [her] beliefs *is* actually compatible with [her] beliefs."

To investigate this on the present approach, assume that (34b) is true, *Sam suspects that it might be raining*. How is it interpreted? Given a doxastic alternative $z$, we look at possible worlds consistent with $K_4(z)$ and check whether there is a weakly ordered sequence with an $\mathbf{S}_3$–persistent tail in the order defined by the ordering source $\Omega{:}K_4(w)$ for Sam's epistemic modal judgment. The complement clause has the truth conditions in (37), with the epistemic alternative set given in (38).

(37) $[[\textbf{might } \mathbf{S}_3]]^{z} = 1$ iff
$\exists u \in \cap K_4(z){:}\,(\forall v \in \cap K_4(z){:}\, v \leq_{\Omega:E4(z)} u)(\exists z' \in \cap K_4(z){:}\, z' \leq_{\Omega:E4(z)} v)$
$[[\mathbf{S}_3]]^{z'} = 1$

(38) $\cap K_4(z)$ = possibilities consistent with what Sam knows at $z$.

As discussed, if we interpret "know" opaquely in (38), then Sam "knows" at $z$ whatever she believes she knows, which suffices to make (34b) true. The modal in (34b) is interpreted relative to what Sam suspects or believes, not what she actually knows. What Sam knows at $z$, within the scope of the attitude predicate, is quite different from what we grant that Sam knows at $w$, which is constrained to be veridical. In contrast, if we interpret "know" transparently in (38), then $K_4(z) = K_4(w)$. But what is this? If Sam glances out the window and concludes that it might be raining based on signs of water on the street that, unbeknownst to Sam, has flowed from a neighbor's sprinkler, then Sam knows (transparently) some number of things in $w$, but nothing Sam knows is about it raining, since the rain is a mistaken apprehension on her part. So in $w$, it is consistent with what Sam actually knows, that it is raining out, and equally consistent with that it is not. Thus, we lose all purchase on the task of reporting Sam's suspicions or beliefs if we interpret "know" transparently in (38).

So "know" must be interpreted opaquely in (38), within the scope of an attitude predicate. And within the scope of the attitude predicate, $K_4(z)$ is what Sam thinks she knows at $z$. As a result, the interpretation of **Sam suspects that might ($S_3$)**, in (39a), does not reduce to (39b), which would be the quantificational schema for **Sam suspects $S_3$**.

(39) a. MIGHT (MIGHT ($\mathbf{S_3}$ | $K_4(z)$) || $\Omega$:$D_4(w)$)
b. MIGHT ($\mathbf{S_3}$ || $\Omega$:$D_4(w)$)

That is, we do not obtain the counterpart to a semantic reduction to a quantifier over the doxastic alternative set, as proposed by Stephenson.

Now consider the first person ascription in (40), which has the interpretation represented in (41), using modal quantifier schemas, which expands as in (42).

(40) I suspect it might be raining.

(41) MIGHT (MIGHT ($\mathbf{S_3}$ | $K_1(z)$)) || $\Omega$:$D_1(w)$)

(42) [[**I suspect might ($\mathbf{S_3}$)**]] $^{w}$ = 1 iff
$\exists u \in \mathbf{W}$: $(\forall v \in \mathbf{W}: v \leq_{\Omega:D1(w)} u)$ $(\exists z \in \mathbf{W}: z \leq_{\Omega:D1(w)} v)$
$\exists u' \in \cap K_1(z)$: $(\forall v' \in \cap K_1(z): v' \leq_{\Omega:E1(z)} u')$ $(\exists z' \in \cap K_1(z)$: $z' \leq_{\Omega:E1(z)} v')$ $[[\mathbf{S_3}]]^{z'} = 1$.

Consider a possible world $z$ by virtue of which (40) is true, and consider the epistemic alternative set in (43).

(43) $\cap K_1(z)$ = possible worlds consistent with what I know at $z$.

The specification of possibilities in (43) would normally be transparent in what I know at $z$. It would be very unusual for a present tense, first person attitude ascription to report an attitude based on the assumption that the doxastic base is anything other than veridical.

(44) $K_1(z) = K_1(w)$

Replacing $K_1(z)$ with $K_1(w)$ in (42), and $E_1(z)$ with $E_1(w)$, we obtain (45).

(45) $[[\textbf{I suspect might } (\mathbf{S_3})]]^{w} = 1$ iff
$\exists u \in \mathbf{W}$: $(\forall v \in \mathbf{W}$: $v \leq_{\Omega:D1(w)} u)$ $(\exists z \in \mathbf{W}$: $z \leq_{\Omega:D1(w)} v)$
$\exists u' \in \cap K_1(w)$: $(\forall v' \in \cap K_1(w)$: $v' \leq_{\Omega:E1(w)} u')$ $(\exists z' \in \cap K_1(w)$:
$z' \leq_{\Omega:E1(w)} v')$ $[[\mathbf{S_3}]]^{z'} = 1$.

Furthermore, any proposition I know is one I believe (under our usual self-awareness assumption), but not vice versa, so the belief base, which constrains possibilities sufficiently far down on tails of sequences ordered by $\leq_{\Omega:D1(w)}$, is more restrictive than the knowledge base; see (3). It follows that any possible world consistent with enough of what I believe is consistent with what I know, but not vice versa.

(46) $\cap D_1(w) \subset \cap K_1(w)$

Thus, (45) is not quite equivalent to might (might ($\mathbf{S_3}$)), which under the interpretations adopted here would reduce to might ($\mathbf{S_3}$). However, to the extent to which the speaker is confident that his or her own beliefs are veridical, the speaker would think that (40) should be interpreted with set equality in (46), in which case (45) would reduce to the interpretation of might ($\mathbf{S_3}$). This explains why a speaker would utter (40) as a complex way of expressing might ($\mathbf{S_3}$), with the embedding attitude predicate *suspect* used to situate the utterance relative to evidence that the speaker has, or to situate it on a scale of strengths pertaining to the claim that it might be raining, if that is a claim at issue in the discourse. It also explains why the statement in (40) does not semantically reduce to the assertion of might ($\mathbf{S_3}$).

For example, suppose that Jill and George are sitting reading together indoors, and George gets up and says, "Let's go take a walk in the park." Although they are not within the direct sight of a window, Jill, mindful of the weather forecast and the fact that daylight from the adjoining room with a window has dimmed even though sunset is hours away, shakes her head and responds with, *I'm sure it must be raining*, to convey that they can't walk because it must be raining. Jill might intend her utterance to reduce to a simple modal statement that it must be raining, but the embedding attitude nevertheless signals that she and George are at a further remove from direct evidence for rain than they normally would be for a direct modal statement. And the additional evidential layer expressed by the attitude verb allows for a further layer of error, even in the first person. In line with this, George could respond, "You're wrong; I don't think it must be raining at all, although it could be."

Likewise, the interpretation of (40) does not reduce to the interpretation "It might be raining." However, to the extent that a speaker takes his or her beliefs to be veridical, a speaker could use (40) with the intention that the interpretation reduce to might ($\mathbf{S_3}$). But addressees are free to resist this reduction, and will be inclined to resist it to the extent that they think that the speaker's beliefs are not veridical. The interpretation of (40) does not semantically reduce to the truth conditions for might ($\mathbf{S_3}$), but they can be accommodated to this interpretation.

But there is reason to think that we might sometimes interpret an attitude ascription with embedded espistemic modal transparently. In (47), Sally responds to this interpretation, keeping the embedded modal responsible to conditions in the actual world.

(47) Joe: John believes that it might be raining.
Sally: He's wrong!

Sally can say this even though the modal is true at worlds on the doxastic sequences. Alternatively, Sally can make allowances for the fact that John is enmeshed in fantasy, preserving full opacity, as in (48).

(48) Joe: John believes that it might be raining.
Sally: And so it might be in the fantasy-land he lives in.

Whether the embedded modal is interpreted with full opacity or not depends on the purposes of the ascription in the context of use.

### *3.4.3 Other complement clauses interpreted using modal quantificational schemas*

The interpretation of propositional attitude ascriptions using modal quantificational schemas can be applied to interpret instances of clausal arguments of other predicates, including predicates of perception and probability, such as *seem*, *look like*, *hear*, *possible*, *likely*, and *certain*, as in (49).

(49) a. It is possible/likely/certain [that the legislature will pass the bill].
b. It seems [that the legislature will pass the bill].
c. It looks like [the legislature will pass the bill].
d. I hear [(that) Max is being considered for chair of that committee].
e. I see [(that) Max has been appointed as chair of the committee].

The ascriptions in (49a) are straightforward modal statements, interpretable as discussed in Section 3.2. The modal adjective *certain* would be interpreted as a necessity modal, and *possible* as a possibility modal; *likely* would plausibly have the strength of *should*.

The statements in (49d,e) are a different matter. Recalling Rooryck (2001), they are evidential qualifiers according to the source of information, not just according to its prospects for veridicality. The schemas of modal quantification could be employed to capture the strengths of these ascriptions, as one aspect of their interpretation. But they have an additional component of their interpretation pertaining to the source of the information. This part of the interpretation must be deferred to the theory of that part of evidentiality pertaining to the source of information, which is beyond the scope of the present work.[7]

[7] See Faller (2002) for a fuller picture of evidentiality in natural language, and one view of how this might go.

# 4 Presupposition filtering and bouletic ascriptions

## 4.1 Introduction

This chapter addresses issues that arise with coordinated pairs of doxastic ascriptions, with bouletic ascriptions, and finally, coordinations of doxastic and bouletic ascriptions, all of which were investigated by Heim (1992). Heim solved problems in this area by formulating her solutions in dynamic semantics. Heim also addressed the interpretation of counterfactual bouletic ascriptions in possible worlds semantics, which she solved by recourse to a "maximally similar" relation on possible worlds. It is argued here that these problems can be solved by iteratively taking tails of weakly ordered sequences of possible worlds, ordered according to joint doxastic and bouletic ordering sources. On this approach, there remains a need for some dynamic interpretive processes, but it is not necessary to replace truth conditional semantics by a context update formulation of semantics. Furthermore, the appeal to maximally similar possible worlds is argued here to be an essential part of one fairly restricted interpretive option for bouletic ascriptions, but not an intrinsic part of the interpretation of bouletic ascriptions in general.

## 4.2 Presupposition filtering

Consider successive doxastic ascriptions, following Heim (1992).

(1) Alex believes Mary is on the balcony, and he believes Susan is there too.

The use of *too* in the second of the two conjoined ascriptions introduces the presupposition that somebody other than Susan is on the balcony for the interpretation of the second conjunct. The content of this presupposition is furnished by the first of the two conjoined ascriptions; it is not a presupposition of the entire sentence in (1). In this sense, the presupposition of the second conjunct is introduced by the first conjunct, and "filtered out" from being a presupposition of the whole.

Heim argues that a dynamic semantic analysis of the ascription captures this presupposition filtering behavior. In a possible worlds semantics for interpreting attitude ascriptions, in order to interpret the second conjunct, we must first update the doxastic context of Alex with the content of the first conjunct. To follow Heim's discussion, assume for now that a belief ascription is interpreted in terms of universal quantification over the doxastic context of the experiencer subject. If $\mathbf{S_2}$ is the phrase marker **Mary is on the balcony**, then the first conjunct in (1), *Alex believes that* $\mathbf{S_2}$, is dynamically interpreted as reflecting the update of the doxastic context of Alex, $\cap D_3(w)$, by the Context Change Potential (CCP) of $\mathbf{S_2}$, which means that when we apply the CCP of $\mathbf{S_2}$ to $\cap D_3(w)$, the doxastic context remains unchanged.[1] Using a standard notation, let $[[\mathbf{S_2}]]$ be the CCP of $\mathbf{S_2}$, and for a set of possibilities X, let X + $[\mathbf{S_2}]$ be the result of applying the CCP $[[\mathbf{S_2}]]$ to X. Then, if we were to start with an interpretation of the belief ascription in the first conjunct as universal quantification over a doxastic context, as in (2), the dynamic counterpart of (2) can be written as in (3), where Con($c$) is the set of possibilities constituting the speaker-addressee context in $c$.

(2) $[[\textbf{Alex believes that } \mathbf{S_2}]]^{c,w} = 1$ iff $\forall v \in \cap D_3(w)$: $[[\mathbf{S_2}]]^{v} = 1$

(3) Con($c$) + [**Alex believes that** $\mathbf{S_2}$] is defined iff $\cap D_3(w) + [\mathbf{S_2}]$ is defined for each
$w \in$ Con($c$). Where defined,
Con($c$) + [**Alex believes that** $\mathbf{S_2}$] = $\{w \in \text{Con}(c)$: $\cap D_3(w) + [\mathbf{S_2}] = \text{unchanged}\}$.

That is, we update Con($c$) by the Context Change Potential of **Alex believes that** $\mathbf{S_2}$, just in case $\cap D_3(w)$ is in the domain of the Context Change Potential of $\mathbf{S_2}$. When it is, we get the set of possibilities $w$ in which the doxastic context of Alex, $\cap D_3(w)$, is unchanged by the Context Change Potential of $\mathbf{S_2}$, which means that $\mathbf{S_2}$ is consistent with every possibility in $\cap D_3(w)$.

To interpret (1), where $\mathbf{S_3}$ is the phrase marker for *Susan is there too*, using the normal dynamic semantics of conjunction, we update successively by $[[\mathbf{S_2}]]$ and by $[[\mathbf{S_3}]]$, so that (1) is interpreted as the set of possible worlds in Con($c$) in which (4) holds.

(4) $(\cap D_3(w) + [\mathbf{S_2}]) + [\mathbf{S_3}] = \text{unchanged}$

The presence of Mary on the balcony before the update with $[[\mathbf{S_3}]]$ is ensured by the prior update of $\cap D_3(w)$ with $[[\mathbf{S_2}]]$.

[1] In pursuing this discussion, we temporarily put aside our concern that the doxastic base may be inconsistent, and adopt the simplifying assumption that it is consistent. In the pages that follow, we will reconsider the problem without this simplifying assumption.

If the doxastic base, $D_3(w)$, happens to be inconsistent, the doxastic context will be empty, and this approach will not get off the ground. However, we can try to implement the idea behind (4) in the update of the conversational background, in a way that can tolerate the doxastic context being inconsistent. Several ways of doing this will be explored here.

In general, using Kratzer's semantics for modality to interpret attitude ascriptions, as in Chapter 3, we interpret a belief ascription in terms of the truth of the complement clause on the tails of sequences of possible worlds weakly ordered by the ordering source, as in (5).

(5) [[**Alex believes that** $\mathbf{S_2}$]] $^{c,w}$ = 1 iff
every sequence of worlds in **W** that is weakly ordered with respect to $\Omega$:$D_3(w)$ has an $\mathbf{S_2}$–persistent tail, and there is such a sequence with an $\mathbf{S_2}$–uniform tail.

But for subsequent belief ascriptions, or subsequent ascriptions of any doxastic attitude, Kratzer's conversational background must be updated. For doxastic ascriptions, on the approach developed in Chapter 3, what matters is the update of the ordering source, $\Omega$:$D_3(w)$.

One solution to consider is the following. For (1), when we complete the interpretation of the first belief ascription, **Alex believes that** $\mathbf{S_2}$, we could update the ordering source for any subsequent ascription of a belief to Alex, by adding the proposition expressed by $\mathbf{S_2}$ to the ordering source. For the presupposition of *too* to be satisfied, we have to make sure that the proposition expressed by the first sentence, that Mary is on the balcony, is satisfied in the worlds we look at for the evaluation of the second conjunct. The interpretation process is iterative. The first ascription looks at sequences ordered according to Alex's doxastic ordering source, and asserts that their tails are uniform (or at least persistent) with worlds in which Mary is on the balcony. The interpretation of the second conjunct looks at sequences of worlds weakly ordered by the updated ordering source, including the proposition that Mary is on the balcony, and checks whether Susan is on the balcony in worlds that are progressively better in accord with the proposition that Mary is on the balcony and all other propositions in the updated ordering source.

This approach appeals to something that is implicit in Kratzer's analysis of modal interpretation, which is the need for context update, particularly for the update of the modal conversational background, extended here to the doxastic ordering source. The approach is dynamic, but doesn't require replacing truth conditional semantics with context update semantics. The iterative interpretation is one interpretation of the coordinate structure, alongside a semantic interpretation as a "bare logical" conjunction (e.g., a list of things Alex believes, in no particular order). Conversational implicature (Grice [1975]) and considerations of coherence (Longacre [1983], Hobbs [1990], Kehler [2002], *inter alia*) will suffice to interpret coordinate structures in richer terms of temporal succession, causal relations, and other contextual relations. But the coordinated clauses in (1) exhibit a logical dependence that requires an account in the semantics. The presence of *too* in the second conjunct relies on the process of context update in the interpretation of modal statements and propositional attitude ascriptions. If this approach is right, it is not necessary to appeal to a completely different semantic value for clauses in the semantics, but the process of context update, implicit in truth conditional semantics with a contextual index, is essential to account for the interpretation of (1).

Since this interpretation adds the proposition expressed by $\mathbf{S_2}$ to the doxastic ordering source for ascriptions of belief to Alex, but not to the speaker-addressee context, it does not become a general presupposition to the discourse. This feature carries over from Heim's analysis, that presuppositions of belief complements are filtered out since they are added to a set of propositions that pertain strictly to doxastic ascription, the doxastic context for Heim, and the doxastic ordering source on the account proposed here.

As formulated earlier, this approach unfortunately does not yield a tenable analysis of the presupposition filtering that we observe in (1). After interpreting the first conjunct of (1), we add to the doxastic ordering source of Alex, $\Omega{:}D_3(w)$, the proposition $p_2$ expressed by the complement clause $\mathbf{S_2}$. The updated ordering source may be inconsistent, and so, as we take sequences of worlds weakly ordered by it, we may have some sequences that become progressively better in accord with propositions within $\Omega{:}D_3(w)$, but that have tails that are uniform in consisting of possible worlds in which $p_2$ is false. In particular, if $p_2$ is inconsistent with one or more other propositions in $\Omega{:}D_3(w)$, then some sequences that become progressively better in accord with $\Omega{:}D_3(w)$ will be ones with $p_2$–uniform tails, but others will "miss" $p_2$ altogether, and become progressively more in accord with $\Omega{:}D_3(w)$ by virtue of becoming more uniform in worlds that accord with propositions that are incompatible with $p_2$. Thus, the second conjunct, that Alex believes $\mathbf{S_3}$, will not necessarily be evaluated on sequences with $p_2$–uniform tails.

We could remove this difficulty by defining the update of the doxastic ordering source more strictly. Instead of just adding $p_2$, the proposition that Mary is on the balcony, to $\Omega{:}D_3(w)$, suppose instead that, in addition to adding $p_2$ to the ordering source, we delete from the ordering source any other propositions that are incompatible with $p_2$. Any subsequent ascription thereby effectively "lives on" possible worlds in which $\mathbf{S_2}$ is true. This ensures that we cannot "miss" $p_2$ in evaluating the second conjunct, that Alex believes Susan is on the balcony too. This has the fortunate consequence that it ensures that the presupposition of *too* is satisfied in the evaluation of the second conjunct of (1).

But this also has an unfortunate consequence – it predicts that the coordinated ascriptions in (6) should be felicitous.

(6) # Alex believes that Mary is on the balcony, and he is sure that Susan is there too.

If the second conjunct in (6) lives on worlds in which $p_2$, the proposition expressed by $\mathbf{S_2}$, is true, then as sequences of worlds get progressively better in accord with the updated ordering source, there is no inconsistency that will prevent them from improving their accord with $p_2$, along with all other propositions remaining in the ordering source. So we could evaluate the truth of the claim that all such sequences have $\mathbf{S_3}$–uniform tails, and be sure thereby that such tails are eventually $\mathbf{S_2}$–uniform as well, thereby satisfying the presupposition introduced by *too*. But, in fact, we want to draw a rather fine distinction, whereby *believe-believe* coordinations such as (1) support the presupposition of *too*, as do *sure-sure* coordinations such as that in (7), which can be usefully interpreted along the lines just described.

(7) Alex is sure that Mary is on the balcony, and he is sure that Susan is there too.

But the ascription with *believe* in the first conjunct should not, as a matter of semantic intuition, support the presupposition of *too* when the subsequent ascription uses *sure*, as in (6). The fact that the update of the ordering source sketched earlier makes (7) interpretable as true is therefore a problem.

It's not clear that any intuitively motivated definition of the update of the doxastic ordering source will suffice to account for this richer paradigm of facts. However, we can consider the possibility that the presupposition imposed by *too* is realized at a different level. With *believe-believe* coordinations, as in (1), the second conjunct must be interpreted as asserting the $\mathbf{S_3}$–persistence of tails of sequences that get uniformly better in accord with the doxastic ordering source $\Omega{:}D_3(w)$. If we look at the same sequences that were used to evaluate the truth of the first conjunct, they will have $\mathbf{S_2}$–persistent tails, but there is no guarantee that the worlds that make a tail $\mathbf{S_2}$–persistent will be the same ones that make it $\mathbf{S_3}$–persistent. This opens the way to the conjecture that the presupposition imposed by *too* is that the $\mathbf{S_2}$–persistent tails, by virtue of which the first conjunct is true, are persistent with worlds in which $\mathbf{S_2}$ and $\mathbf{S_3}$ are both true.

To see this, consider (1) without *too*, as in (8).

(8) Alex believes that Mary is on the balcony and he believes that Susan is on the balcony.

The coordination in (8) could be true by virtue of conditions that don't link the $\mathbf{S_2}$–worlds that make the first ascription true to the $\mathbf{S_3}$–worlds that make the second ascription true. It could be true that Alex believes that Mary is on the balcony (by virtue of something he was told an hour ago), and also true that Alex believes that Susan is on the balcony (by virtue of Alex just having heard her voice there), but if Alex never puts the two pieces of evidence together, because the situation is complicated or he is inattentive, there may be no worlds in Alex's doxastic base in which Mary and Susan are both on the balcony. In such a case we can add *too* to (8), but *too* then operates on the propositional level – it is true that Alex believes that Mary is on the balcony, and it is true that Alex believes that Susan is on the balcony. The presupposition of that use of *too* is "Alex believes something other than that Susan is on the balcony." But with the original interpretation of (1), where the presupposition of *too* is that Alex believes that someone other than Susan is on the balcony, *too* is inappropriate for the situation of complexity or inattention just described for (8).

So when *too* in (1) is interpreted as introducing the presupposition that Alex believes that someone other than Susan is on the balcony, this cannot be a case of cognitive dissonance, where the first conjunct of (1) is true by virtue of $\mathbf{S}_2$–persistent tails, and the second conjunct is true by virtue of $\mathbf{S}_3$–persistent tails that do not necessarily "live on" the $\mathbf{S}_2$–worlds that make the first conjunct true. We must instead have $\mathbf{S}_3$–persistent tails in which the $\mathbf{S}_3$–worlds are also $\mathbf{S}_2$–worlds.

To get this, we interpret *too* as imposing the condition that the $\mathbf{S}_3$–worlds on $\mathbf{S}_3$–persistent tails that make the second conjunct true are also $\mathbf{S}_2$–worlds. They are worlds in which Mary and Susan are both on the balcony. The coordinated ascriptions in (1) then assert that every sequence of worlds in $\mathbf{W}$ that is weakly ordered with respect to $\Omega{:}D_3(w)$ has a tail persistent in $\mathbf{S}_2$ and $\mathbf{S}_3$, and there is such a sequence with a tail that is uniform in $\mathbf{S}_2$ and $\mathbf{S}_3$.

Returning to (7), we have a comparable interpretation using the quantificational conditions for *be sure*, namely, that every sequence of worlds in $\mathbf{W}$ which is weakly ordered by $\Omega{:}D_3(w)$ has a tail that is uniform in $\mathbf{S}_2$ and $\mathbf{S}_3$.

Finally, what about (6)? The first conjunct asserts that every sequence of worlds in $\mathbf{W}$ which is weakly ordered by $\Omega{:}D_3(w)$ has an $\mathbf{S}_2$–persistent tail, and there is such a sequence with an $\mathbf{S}_2$–uniform tail. The interpretation of the second conjunct would assert that every such sequence has an $\mathbf{S}_3$–uniform tail. The presupposition of *too* would be that all the worlds that make the tails $\mathbf{S}_3$–uniform are worlds in which $\mathbf{S}_2$ is true. But the first conjunct does not make enough $\mathbf{S}_2$–worlds available on these sequences for this purpose. We cannot satisfy the presupposition of *too* when the second ascription has greater quantificational strength than the first.

## 4.3 Bouletic ascriptions

On the basis of presupposition filtering, Heim (1992) argues that ascriptions of preferences and desire through clausal complements to predicates such as *wish*, *want*, *be glad* are interpreted with quantification over possible worlds consistent with what the experiencer believes, ordered according to the experiencer's preferences or desires, rather than quantification over worlds consistent with what the experiencer prefers, wants, or desires. This argument will be recapitulated here. We will find that some of the difficulties encountered by Heim are consequences of having the interpretations "live on" the doxastic context of the experiencer. These difficulties can be more easily removed when we partly "step off" the doxastic context of the experiencer and combine the doxastic base with the bouletic ordering source, parallel to what we did in the previous section for the interpretation of doxastic attitude ascriptions. The interpretations thus obtained are ones that hold even if the doxastic base is inconsistent, or if the bouletic preferences are inconsistent with the doxastic base. Section 4.3.1 sketches Heim's account. Section 4.3.2 will develop an account using Kratzer's quantificational schemas for modality, which apply even if the doxastic base is inconsistent.

### *4.3.1 Heim's (1992) account of bouletic ascriptions and presuppositions*

Heim starts with the idea that an initially plausible interpretation of an ascription with *want*, such as that in (9), would quantify over a set of possibilities consistent with what the experiencer Alex wants.

(9) Alex wants Mary to come to the party.

This can be expressed as in (10), where *Alex* refers to the indivdual $x_3$ in the domain, $\mathbf{S_4}$ is the phrase marker for the complement clause, *Mary to come to the party*, and $B_3(w)$ is the set of propositions, each a set of possibilities, for the operative bouletic conditions of Alex at a world of evaluation $w$, expressing those preferences or desires of Alex that are implicitly referenced by the bouletic ascription. The ascription then quantifies over possible worlds in $\cap B_3(w)$, as in (10).

(10) $[[\textbf{Alex wants } \mathbf{S_4}]]^{c,w} = 1 \text{ iff } \forall v \in \cap B_3(w)\text{: } [[\mathbf{S_4}]]^{v} = 1$

But in (11), similarly to the interpretation of (1) in Section 4.2, the interpretation of the second conjunct relies on the presupposition that Alex wants someone other than Mary to come to the party, and this presupposition comes from the interpretation of the first conjunct. This presupposition is not a presupposition of the entire sentence, so it counts as filtered out.

(11) Alex wants Mary to come to the party, and he wants Susan to come too.

The situation is parallel to that in (1), and Heim considers a parallel dynamic semantic account, interpreting the first conjunct in (11) by updating the speaker-addressee context Con($c$), as in (12).

(12) Con($c$) + [**Alex wants S$_4$**] is defined iff $\cap B_3(w)$ + [**S$_4$**] is defined for each
$w \in$ Con($c$). Where defined,
Con($c$) + [**Alex wants S$_4$**] = {$w \in$ Con($c$): $\cap B_3(w)$ + [**S$_4$**] = unchanged}.

This says that if the presuppositions of **S$_4$** (if any) are satisfied so that $\cap B_3(w)$ + [**S$_4$**] is defined, then the CCP of **Alex wants S$_4$** updates the speaker-addressee context Con($c$) such that it reflects update of $\cap B_3(w)$ by the CCP of **S$_4$**, more literally, by keeping possibility $w \in$ Con($c$) just in case **S$_4$** is consistent with all of Alex's wants and desires at $w$. Using the dynamic semantic rule for the interpretation of conjunction, (11) will then be interpreted in terms of the successive updates in (13), where **S$_5$** = **Susan to come to the party**.

(13) ($\cap B_3(w)$ + [**S$_4$**]) + [**S$_5$**] = unchanged

The update $\cap B_3(w)$ + [**S$_4$**] ensures that all possibilities left in $\cap B_3(w)$ are ones in which Mary comes to the party. This satisfies the presupposition of *too* in the application of the CCP of **S$_5$** to $\cap B_3(w)$ + [**S$_4$**].

But, as Heim notes, the use of the bouletic modal base $B_3(w)$ in (12) is problematic in a couple of respects. First, suppose that **S$_4$** presupposes a proposition $p$. Then, in order for $\cap B_3(w)$ + [**S$_4$**] = unchanged to hold, it must be that $\cap B_3(w)$ + [$p$] = unchanged. It follows that if Alex wants **S$_4$**, then Alex wants $p$. This is useful for the interpretation of examples like (11) and likewise for Heim's example in (14), where the resolution of pronominal reference in the second conjunct depends on the satisfaction of the presupposition of the definite nominal in the bouletic alternative set $\cap B_7(w)$, provided for by the first conjunct (where *Patrick* refers to individual $x_7$ in the domain). This works out just like it did previously for (1).

(14) Patrick wants to buy a cello. He wants to play it in the school orchestra.

But this very property of the interpretation in (12) is a problem for an example like (15), where the *want*-complement presupposes that I have been sick.

(15) I want to get well.

Again, the interpretation in (12) has the property that if I want **S$_7$**, and **S$_7$** presupposes $p$, then it must be the case that I want $p$. This has the undesirable consequence that (15) entails the truth of *I want to be sick now*. Furthermore, continuing to follow Heim's reasoning, (12) is problematic for the interpretation of *believe-want* sequences such as (16), a minimally altered version of an example of Heim. Interpreting (16) in isolation, the presupposition of *too*, occurring in the *want* complement, is that Alex believes that someone other than Susan is coming to the party, and that presupposition is satisfied by the first conjunct, a belief ascription, with the result that the whole sentence is felicitous, interpreted as filtering out the presupposition that Alex believes someone other than Susan is coming to the party.

(16) Alex believes Mary is coming to the party, and he wants Susan to come too.

Interpreting the first conjunct with a doxastic base as Heim does, and the second with a bouletic base according to (12), the first conjunct effects the condition $\cap D_3(w)$ + [$\mathbf{S_4}$], that Alex's doxastic context reflects update by the CCP of $\mathbf{S_4}$, and thus reflects Alex's belief that Mary is coming to the party. The interpretation of the second conjunct then involves ($\cap B_3(w)$ + [$\mathbf{S_5}$]). But the bouletic context $\cap B_3(w)$ does not reflect update by the CCP of $\mathbf{S_4}$, and therefore does not satisfy the presupposition of *too*. Thus, the interpretation of a *want*-clause such as (9), given in (12), together with the dynamic interpretation of belief ascriptions, fails to interpret (16). For this reason, Heim argues, ascriptions of desire and preference should be interpreted using a doxastic base, just as belief ascriptions are. The ascription in (9), *Alex wants Mary to come to the party*, would then be interpreted as in (17), similar to (12) except that the CCP of $\mathbf{S_4}$ must comport with Alex's doxastic context $\cap D_3(w)$, in order to assert that Alex's bouletic context $\cap B_3(w)$ reflects update by the CCP of $\mathbf{S_4}$.

(17) Con($c$) + [**Alex wants $\mathbf{S_4}$**] is defined iff $\cap D_3(w)$ + [$\mathbf{S_4}$] is defined for each
$w \in$ Con($c$). Where defined,
Con($c$) + [**Alex wants $\mathbf{S_4}$**] = {$w \in$ Con($c$): $\cap B_3(w)$ + [$\mathbf{S_4}$] = unchanged}.

Continuing to follow Heim's reasoning, the interpretation in (17) solves problems with the bouletic ascriptions encountered so far. But there remains a problem with what we might call counterfactual bouletic ascriptions. Suppose that (9), *Alex wants Mary to come to the party*, is used to assert that Alex wants Mary to attend the party, even though he knows that she won't attend. Then $\cap D_3(w)$ + [$\mathbf{S_4}$] is undefined since $\mathbf{S_4}$ is incompatible with what Alex knows, the propositions in $D_3(w)$. Then the update condition in (17) will never hold, and (9) will evaluate as false. But (9) could be truly asserted in the circumstances described. We need to amend our conception of what is going on in the interpretation of (9). Heim invites us to do so, as follows. The utterance in (9) is used to assert that, if $u$ is a possible world in which Mary attends the party, then $u$ is more desirable to Alex, ceteris paribus, than possible worlds in which Mary doesn't attend the party, but which are otherwise as similar to $u$ as they possibly could be.

Toward this end, Heim defines a selection function on propositions dependent on possibility $u$, which, given proposition $p$, produces the set of possibilities that maximally resemble $u$, among those in which $p$ is true. This selection function is written as $\text{Sim}_u$, and its definition is recorded in (18). To paraphrase Heim, $\text{Sim}_u$ maps a proposition $p$ to the $p$-possibilities that maximally resemble $u$.

(18) $\text{Sim}_u\ (p) = \{v \in p$: $v$ resembles $u$ no less than other possibilities in $p\}$

Then, extending the order $\leq_{\Omega:B3(w)}$ to sets of possible worlds, Heim (1992: 197) proposes that **Alex wants $\mathbf{S_4}$** is interpreted dynamically as in (19), asserting that the CCP of **Alex wants $\mathbf{S_4}$** updates the context $c$ by retaining those possibilities $w$ at which Alex's doxastic context $\cap D_3(w)$, updated by the CCP of $\mathbf{S_4}$, is satisfied in possibilities $v$ that are more desirable than ones at which $\cap D_3(w)$ is updated by **not-$\mathbf{S_4}$**.

(19) $\text{Con}(c)$ + [**Alex wants $\mathbf{S_4}$**] =
$\{w \in \text{Con}(c)\text{: } \forall u \in \cap D_3(w)$:
$\text{Sim}_u\ (\cap D_3(w) + [\mathbf{S_4}]) \leq_{\Omega:B3(w)} \text{Sim}_u\ (\cap D_3(w) + [\textbf{not-}\mathbf{S_4}])\}$

Heim notes that (19) opens a new door to the problem with the presupposition of (15), that *I want to get well* presupposes that I am sick. In an appropriate context for (15), I am, in fact, sick, so if $\mathbf{S_6}$ = **PRO to be (currently) sick**, with PRO controlled by the speaker, who is the individual $x_1$ in the domain, then $(\cap D_1(w) + [\textbf{not-}\mathbf{S_6}])$ is empty. So then $\text{Sim}_u\ (\cap D_1(w) + [\textbf{not-}\mathbf{S_6}])$ is empty, the universal quantification over $u$ in a first person counterpart to (19) for the interpretation of **I want $\mathbf{S_6}$** is vacuously true, and that interpretation therefore automatically updates Con($c$) by the CCP of **I want $\mathbf{S_6}$**, counting *I want to be (currently) sick* as true. A similar problem arises from (19) itself whenever the experiencer believes either $\mathbf{S_4}$ or **not-$\mathbf{S_4}$**, since then one or the other of the Sim sets ordered in (19) is empty, making the ascription **I want $\mathbf{S_4}$** vacuously true, whether or not $\mathbf{S_4}$ is something that the speaker actually wants.

To address these problems, Heim amends (18) so that $\text{Sim}_u\ (p)$ is undefined when $p$ is empty. Thus, when Alex knows or firmly believes that $\mathbf{S_4}$ is true, $(\cap D_3(w) + [\textbf{not-}\mathbf{S_4}])$ is empty, so the update in (19) is undefined. Similarly, when Alex knows or firmly believes that **not-$\mathbf{S_4}$** is true, $(\cap D_3(w) + [\mathbf{S_4}])$ is empty, so the update in (19) is undefined. And if $\mathbf{S_4}$ presupposes $\mathbf{S_5}$, the update by the CCP of **Alex wants $\mathbf{S_4}$** using (19) effects update of $\cap D_3(w)$ by the CCP of $\mathbf{S_5}$, so that $(\cap D_3(w) + [\textbf{not-}\mathbf{S_5}])$ is then empty, making (19) undefined for **Alex wants $\mathbf{S_5}$**, and thus preventing **Alex wants $\mathbf{S_4}$** from entailing **Alex wants $\mathbf{S_5}$**.

Nevertheless, even with these emendations, there are two remaining problems with Heim's account, which will prevent us from adopting it. The first is an internal problem. Heim defines Sim ($p$) to be undefined when $p$ is empty, in order to handle what we might call "boundary value" problems when the bouletic attitude pertains to the content of a sentence with phrase marker **S**, when **S** expresses something that the experiencer antecedently believes to be true, by presupposition or hypothesis, or antecedently believes to be false. But there are attitude ascriptions, even with the same predicate *want*, such as (15), which have perfectly normal, contingent interpretations at the boundary values. Therefore, there are good reasons not to build a "fix" for the boundary values problem into the semantic interpretation of bouletic ascriptions. Second, from our own perspective, Heim's account is too limited since it implicitly rests on the assumption that the doxastic base is consistent. If the doxastic base is inconsistent, then $\cap D_3(w)$ is empty, so the quantifier over possible world $u$ in (19) is vacuous, so the dynamic condition in (19) automatically updates Con($c$) by **Alex wants** $\mathbf{S_4}$, whether or not $\mathbf{S_4}$ is more desirable to Alex than **not-**$\mathbf{S_4}$.

### *4.3.2 Bouletic ascriptions and joint doxastic-bouletic ordering sources*

Let's return to Heim's example in (15), repeated here.

(15) I want to get well.

In the most typical context, (15) would be interpreted with respect to a realistic doxastic base that includes the proposition that the speaker is sick, along with attendant details. The ordering source would typically be a small set of bouletic conditions – that the speaker gets well, sooner rather than later, fully and completely rather than partially, and so on. If the doxastic base is a consistent set, and the eventuality of the speaker's recovery is consistent with the propositions in the doxastic base, then (15) can be interpreted on the doxastic context in Kratzer's theory as asserting that sequences of worlds in the doxastic context that progressively improve with respect to the bouletic ordering source have tails that are uniform, from some point onward, in being composed of worlds in which the speaker gets well. That is usually presented as the "typical" use of (15), presupposing that the speaker is sick and that the prospect of getting well is not inconsistent with what the speaker knows.

If the doxastic base is an inconsistent set, the doxastic context is empty, so this analysis won't work. We have the propositions in the doxastic base, nevertheless, and a set of propositions in the ordering source (that the speaker gets well, sooner rather than later, fully and completely rather than partially). But, in general, we cannot interpret by taking the intersection of all the propositions (sets of worlds) in the doxastic base.

Instead, we will formulate an effective ordering source for the interpretation of (15) obtained by intersecting the normal bouletic ordering source, $\Omega{:}B_1(w)$, described at the beginning of this section, with all, part, or – in an extreme case – none of the doxastic base $D_1(w)$, with the result used as an ordering source on the set **W** of all possible worlds, in a way that is parallel to the interpretation of belief ascriptions in Chapter 3.

The rationale for this is that bouletic ascriptions can be interpreted at various levels of realism, and the interpretation of the ascription in a particular context has to choose the level of realism that is appropriate for that use, in that context. In a context where (15) is intended to be realistic, the speaker could be asserting that she wishes to be well only under conditions compatible with her doxastic base. In this case, conditions in the bouletic ordering source are presumed to be compatible with what the speaker knows, and their intersection, $D_1(w) \cap \Omega{:}B_1(w)$, is the effective ordering source for the interpretation of the ascription, obtained by restriction from the omnibus source, the set union.

The speaker of (15), in a counterfactual use of the ascription, might intend to assert that she wishes to be well under counterfactual conditions that are as realistic as possible. Or, the speaker of (15) might simply be expressing a preference for members of the broad class of more or less fanciful alternative realities, some of them perhaps very fanciful indeed, in which she gets well. When recovery is known to be impossible, the utterance of (15) could be intended as a statement that the speaker wishes she could recover under maximally realistic counterfactual conditions. Or it could be a despairing outburst, that she would grasp at recovery under just about any circumstances, including doxastically inaccessible possible worlds that may be more than minimally different from worlds that are compatible with what is known. Nothing requires expressions of clearly counterfactual desire to "live on" worlds that are compatible with what is known. In this extreme case, the effective ordering source is obtained by intersecting $\Omega{:}B_1(w)$ with none of the doxastic base, resulting in a pure bouletic order on all possible worlds.

If the propositions in the ordering source are incompatible with propositions in the doxastic base, then (15) is in some measure the expression of a counterfactual desire – the speaker knows she will not get better, or at least not as soon or as fully and completely as she would wish. It won't do to look at tails of sequences of worlds in the doxastic context, or worlds that accord progressively better with propositions in the doxastic base, since worlds that accord with doxastic conditions will not, in general, be worlds in which the speaker's desire to get better is realized.

In Chapter 1, we saw that von Fintel and Gillies (2008a, 2008b) proposed that epistemic modal statements are evaluated relative to a context chosen from a cloud of contexts (in their terms) provided by different choices of the community whose knowledge is used to compose the epistemic modal base. Under the proposal made earlier, the effective bouletic ordering source for a bouletic modal ascription such as that in (15) is variable, and can be chosen from a "cloud of bouletic ordering sources," appropriate in different contexts. The effective bouletic ordering source can be highly factual (to the best of the speaker's knowledge), if the possibility of recovery is not ruled out by what is known. When recovery is known to be impossible, the effective bouletic ordering source can be counterfactual in a highly restricted way (anticipated by Heim), confining the ordering source to operate on possible worlds that are maximally similar to what is known (but not identical to what is known, since that would preclude the possibility of recovery). Or, when recovery is known to be impossible, the effective ordering source can be quite loosely counterfactual, admitting possible worlds that are more fanciful than what is maximally similar to what is known. This gives us reason to invoke a cloud of effective bouletic ordering sources for the interpretation of the bouletic ascription. In present terms, the contexts correspond to the bouletic ordering source taken on its own (for an ascription of a bouletic attitude unconstrained by what is known about the world), at one extreme, to the bouletic ordering source intersected with the complete doxastic base, for a maximally realistic bouletic ascription, at the opposite extreme. And effective ordering sources corresponding to contexts with intermediate levels of realism in bouletic modal interpretation can be obtained by choosing an ordering source consisting of just those possibilities in the doxastic base that are compatible with the ordinary bouletic ordering source $\Omega{:}B_1(w)$.

The role of this cloud of contexts will be discussed further in Chapter 8, in connection with proposals in the literature to address the problem of modal determination – how a modal comes to be interpreted as epistemic or root, and the role that the syntactic position of a modal auxiliary plays in that determination.

A third person bouletic ascription, such as Heim's (1992) example in (20), is open to the same range of interpretations, from fully realistic, to highly fanciful.

(20) Patrick wants to buy a cello.

Suppose that Patrick really wanted to be a piano player, but doesn't have room for a piano in his small apartment, and therefore chose to play cello instead. Then worlds in which Patrick can, counter to fact, buy and play a piano would be ordered higher than most worlds in which he buys and plays a cello (higher than any such worlds, ceteris paribus), so sequences of worlds ordered by Patrick's desires and preferences would have tails uniform with worlds in which Patrick buys a piano, with no worlds in which he buys a cello instead. If we evaluate (20) using those sequences, (20) would evaluate as false. Thus, to interpret (20), we can't just take all possibilities, with a weak order on them determined by Patrick's preferences.

But now suppose that Patrick is wrong about not being able to fit a piano into his apartment. Perhaps he has overlooked something, and speaker and addressee(s) know this. In this case, worlds consistent with what speaker and addressee know include lots of worlds in which Patrick has a piano in his apartment, and when we order them with respect to Patrick's preferences, we would get sequences whose tails are uniform in worlds in which Patrick has a piano but not ones in which he has a cello. So (20) would evaluate as false under this interpretation procedure. Clearly, (20) must be interpreted relative to Patrick's doxastic base, not that of the speaker and addressee(s). As with previous examples considered, the doxastic base may be internally inconsistent, or inconsistent with those of Patrick's bouletic conditions that are operative in interpreting (20). So we combine the operative bouletic conditions with Patrick's doxastic base, but throwing out the base propositions that are inconsistent with the bouletic conditions, to form a joint doxastic/bouletic ordering source to apply to the set of all possible worlds, **W**. We can do this in a way that keeps worlds that are maximally similar to what Patrick knows, for a more realistic interpretation, or we can allow counterfactual worlds that are quite fanciful, if Patrick's wish for a cello is a fantasy that is completely beyond the realm of possibility, given the conditions of Patrick's existence, and therefore more divorced from reality. The interpretation of (20) is entirely parallel to the interpretation of (15) in this respect.

A purely bouletic counterpart to (1) would be Heim's example in (21).

(21) Patrick wants to buy a cello. He wants to play it in the school orchestra.

The pronoun *it* in the second conjunct refers to the discourse entity for the cello, introduced in the interpretation of the bouletic ascription of the first conjunct. This signals that the interpretation of the second conjunct should live on the worlds that compose the tails of sequences of worlds weakly ordered by the bouletic condition of the first conjunct, together with a collection of compatible doxastic conditions (if any), reflecting an interpretation that might be more realistic or more fanciful, depending whether the attitude being reported is more or less realistic. This ensures that *it* has a referent in the second conjunct. It also ensures that the second conjunct cannot be more realistically interpreted than the first conjunct, since a more realistic interpretation could eliminate all the possible worlds in which Patrick buys a cello. This matches the facts of (21). The first conjunct cannot be interpreted less realistically than the second, for the reason just given, but the second conjunct could be less realistic than the first, if the prospect of Patrick playing in the school orchestra is even more remote than his prospects for purchasing a cello.

So we interpret the first conjunct of (21) as was discussed in connection with (20). Tails of sequences weakly ordered by Patrick's operative bouletic criteria, and those of his doxastic criteria that survive interaction with his bouletic ones, are uniform in possible worlds in which Patrick buys a cello. (Or, if *want* is interpreted as weaker than a necessity modal, tails matching the appropriate quantificational condition are persistent or uniform, as appropriate, in worlds in which Patrick buys a cello.) The second conjunct introduces an additional set of bouletic conditions. These are added to the ordering source, with purging of any incompatible propositions in Patrick's doxastic base. If there are such incompatibilities, that is, if the second bouletic ascription in (21) is counterfactual, then the purge of doxastic conditions can be done loosely, for an interpretation corresponding to a fanciful wish on Patrick's part, or it can be done with care to cultivate worlds that are maximally similar to ones that are fully compatible with Patrick's doxastic base. The more restricted ordering source is imposed on worlds within the tails produced under the first ascription, and the second conjunct asserts that resulting sequences, weakly ordered by the enlarged bouletic preference set, are uniform in worlds in which Patrick plays in the school orchestra (or whatever the appropriate condition is for the strength of the ascription).

Finally, turn to Heim's example in (22), a variant of the examples in (1) and (20).

(22) John believes that Mary is coming to the party, and he wants Susan to come too.

The doxastic ascription in (22) asserts that sequences totally ordered by John's doxastic ordering source have tails that are uniform with worlds in which Mary is coming to the party. In line with examples discussed earlier, the follow-up bouletic ascription in (22) will piggy-back on this by taking only worlds on those tails in which Mary is coming to the party, and will look at sequences of those worlds, ordered by John's bouletic ordering source, for the interpretation of the bouletic ascription. In line with the discussion of previous examples, to interpret the second conjunct in (22), we don't just intersect the doxastic base and bouletic ordering source to form a new ordering source, and take sequences that are ordered by both of them, since John's doxastic base could be inconsistent with the prospect of Susan coming to the party. And we don't just take all worlds in which Mary is coming to the party, since they may not satisfy lots of other doxastic conditions of John's.

Instead, we first take tails of sequences weakly ordered by John's doxastic base to get worlds that are good to start with, with respect to John's doxastic base and ordering source. The interpretation of the second conjunct depends on whether Susan's attendance at the party is realistically in accord with John's doxastic base, or a counterfactual desire. If it is realistic, we can take sequences of worlds weakly ordered by John's doxastic conditions alone, in which Mary comes to the party, and then order the worlds on those tails by John's operative bouletic conditions for the interpretation of the second conjunct. The second conjunct will then assert the existence of tails of sequences weakly ordered by those bouletic conditions that are uniform (or of an appropriate condition) in worlds in which Susan also attends. If the desire for Susan's attendance is counterfactual, the doxastic base must be purged of propositions that are incompatible with the bouletic conditions. But critically, the proposition that Mary attends the party must be dynamically added to the doxastic base, and this proposition must not be purged. The purge must preserve worlds in which Mary comes to the party, even if they are not realistic. The resulting set of worlds, ordered by John's post-purge doxastic base together with his bouletic ordering propositions, will be more realistic or less realistic, depending on the circumstances, but they will all be worlds in which Mary comes to the party. The second conjunct then asserts that tails of sequences weakly ordered by the post-purge doxastic conditions, together with John's operative bouletic conditions, are uniform in worlds in which Susan comes to the party as well as Mary.

### *4.3.3 Bouletic ascription interpretation*

Stepping back, the previous sections suggest an approach to the interpretation of bouletic ascriptions, such as those in (23).

(23) a. Alex wants Bill to be in the room.
b. Alex wishes Bill were in the room.
c. Alex is glad that Bill is in the room.

The interpretations suggested involved quantifying over all possible worlds using Kratzer's modal quantificational schemas, with an ordering source obtained by intersecting the bouletic ordering source with some, all, or none of the doxastic base, as appropriate. This contextually dependent, derived bouletic ordering source will be notated as $\Omega^*{:}B_3(w)$, for individual $a_3$ (Alex) at world of evaluation $w$. The quantificational schema will be MUST or SHOULD, depending on the intended force of the assertion, yielding the truth conditions of the form given in (24) for (23a,b,c), where $\mathbf{S_6}$ is the phrase structure for the infinitival complement clause, *Bill to be in the room*, with differences stemming from different degrees of intersection of the bouletic ordering source with the doxastic base, together with some other differences to be discussed in Chapter 6.

(24) MUST/SHOULD ($\mathbf{S_6} \parallel \Omega^*{:}B_3(w)$)

For (23b), the intersection with the doxastic base is minimal, permitting sequences ordered by $\Omega^*{:}B_3(w)$ to become uniform or persistent in worlds in which Bill is in the room, even though the speaker and addressee(s) know or assume Bill is not in the room. For the factive ascription in (23c), there is full intersection with the doxastic base since the speaker and addressee(s) know or assume that Bill is in the room. For *want* in (23a), as discussed in Section 4.3, the extent of the intersection is variable, depending on what is known or assumed regarding Bill's presence in the room. The ascription in (23a) can be used to assert that Alex prefers Bill to be present in the room, when it is known or assumed that Bill is in the room, known or assumed that he isn't, or when nothing is known or assumed about Bill's presence in the room.

In an account of the distribution and interpretation of the subjunctive in Spanish, Villalta (2008) independently arrived at an analysis of bouletic ascriptions that bears several points of similarity to this one. Villalta quantifies over contextually relevant alternatives to the proposition that Bill is in the room (at a time resolved by the higher tense and context). She captures the differences among the three verbs in (23) by laying down conditions relating the doxastic base to this proposition or its alternatives; these conditions serve as definability conditions for the truth conditions. Thus, the ascription with *want* is defined only if alternatives for *want* are consistent with the doxastic base; it isn't felicitous to report that Alex wants Bill to be in the room if any relevant alternative to Bill being in the room is inconsistent with the doxastic base.[2] The ascription with *wish* is defined only if the proposition that Bill is in the room is inconsistent with the doxastic base; that is, wishes are counterfactual. And the ascription with *be-glad* is defined only if Bill is in his room in all worlds in the doxastic base, since *be-glad* is factive.

Villalta's truth conditions of the three ascriptions in (23) are, as in the present account, the same. For Villalta, to say that Alex wants or wishes Bill to be in the room, or is glad that Bill is in the room, is to say that the proposition that Bill is in the room is preferred by the ordering source over any relevant contextual alternatives, where we use the term "preferred by the ordering source" as shorthand for Villalta's (2008: 479) condition on the right-hand side of (25), expressed here in the notation of the present work.

(25) For any propositions (sets of worlds) $p$ and $q$, $p$ is preferred over $q$ by the ordering source $\Omega{:}B_3(w)$ iff $(\forall u \in p, \exists v \in q{:}\ v \leq_{\Omega:B3(w)} u)$ and not vice versa (swapping $p$ and $q$).

For an ascription with *want* with the force of MUST, the present account, (24), states that the tails of all sequences weakly ordered by $\Omega^*{:}B_3(w)$ are $p$-uniform. That will match Villalta's truth conditions in (25) provided that the contextually relevant alternatives to $p$ are all incompatible with $p$, in that when $p$ is true, they are all false, and when any of them is true, $p$ is false. (The simplest case of this, of course, is when there is only one alternative, $\sim p$, as assumed by Heim, but Villalta's account is designed to handle more complex cases than this.) The second part of this condition for the equivalency of the two accounts may be too strong, since it is not clear that the contextually relevant alternatives for Villalta's account would have this logical

[2] This may be too strong. The ascription in (23a) seems infelicitous indeed if alternatives are impossible, but not if they merely happen to be known to be false.

property of inconsistency with the proposition expressed by the bouletic complement. But they will be assumed to have this property for purposes of comparison of the two accounts. If (23a) is true on the present account, then all weakly ordered sequences are *p*-uniform; it is easy to confirm that *p* is therefore preferred over any contextually relevant alternative *q* by the ordering source since the failure of that condition would establish the existence of a weakly ordered sequence whose tail was not *p*-uniform. Conversely, if *p* is preferred over all alternatives *q* on a weakly ordered sequence of worlds, then that sequence must be *p*-uniform. There is, then, no difference in truth conditions between Villalta's account and the present one when the bouletic predicate has the force of MUST.

When the ascription in (23a) is interpreted with the force of SHOULD, the truth conditions allowed here can be the weaker ones given in (26).

(26) [[**Alex wants** $\mathbf{S}_6$]] $^{c,w}$ = 1 iff
every sequence of worlds in **W** that is weakly ordered with respect to $\Omega^*{:}B_3(w)$ has a $\mathbf{S}_6$–persistent tail, and there is such a sequence with a $\mathbf{S}_6$–uniform tail.

A sequence could have an $\mathbf{S}_6$–persistent tail on which worlds in which $\mathbf{S}_6$ is true alternate or interweave with worlds in which one of the contextually relevant alternatives is true. Thus, Villalta's account captures the stronger of the interpretations allowed by (24), but not the weaker one, spelled out in (26).

The same is true of an account of bouletic ascription offered by Rubinstein (2012: 116). Let *p* be the proposition (set of possible worlds) expressed by the complement $\mathbf{S}_6$, Rubinstein's account states that **Alex wants** $\mathbf{S}_6$ is true just in case the intersection of *p* with the doxastic context, $\cap D_3(w)$, is preferred by the ordering source over the intersection of ~*p* with the doxastic context. This doesn't allow that *p* and ~*p* worlds might alternate or interweave with one another on the tails of sequences weakly ordered by $\Omega{:}B_3(w)$.

One of the motivations for Villalta's account is to block the invalid inference in (27), from Villalta (2008: 478), which Villalta argues that Heim's (1992) account supported.

(27) a. I want to teach Tuesdays and Thursdays next semester.
b. I believe that I will teach Tuesdays and Thursdays next semester if and only if I work hard now.
c. (Invalid inference) Therefore, I want to work hard now.

Since truth conditions of a *want*-ascription are defined only if the alternatives to the proposition expressed by the complement are consistent with the doxastic base, the belief expressed in (27b) confines the two *want*-ascriptions in (27a) and (27c) to the same set of worlds. But the inference does not go through on Villalta's account since the set of contextually relevant alternatives to the speaker teaching on Tuesdays and Thursdays in (27a) is not necessarily the same (and typically would not be the same) as the set of alternatives to the speaker working hard now, invoked in the interpretation of (27c).

On the account developed in Section 4.3.2, the inference in (27) is inescapable when the interpretation adheres strictly to the intersection of the bouletic ordering source with the doxastic base since, evaluated in worlds in the doxastic base, the two complements in (27a) and (27b) are coextensive. But there is an available option to evaluate partly or entirely off of the doxastic base (to restrict the intersection of the bouletic base to part or none of the doxastic base), and in this case, the worlds in which the speaker teaches an easy schedule next semester are not restricted to be ones in which she works hard.

## 4.4 Conclusion

Interpreting attitude ascriptions by tracing possible worlds on sequences that are weakly ordered by an ordering source that includes some, all, or none of a doxastic base, we are able to account for the phenomena addressed by Heim (1992) while avoiding two theoretically troubling or problematic moves that Heim makes, namely, resort to general dynamic semantic interpretation, and appeal to a "maximally similar" relation among possible worlds. By iteratively refining tails of weakly ordered sequences of worlds, we can "piggy-back" the interpretation of a second attitude ascription conjoined with another, accomplishing a sort of *sui generis* dynamic interpretation without invoking dynamic semantics. And bouletic ascriptions can be interpreted in a simple and precise fashion when the preferred or desired condition is not incompatible with the subject's doxastic base. Unrealistic or counterfactual bouletic ascriptions can be interpreted with the same mechanisms, applied to adhere as closely to realistic conditions as is appropriate for the context of use. If the context of interpretation favors a maximally realistic (but still counterfactual) interpretation, then something like a maximal similarity relation can be invoked, but this may be more rightly seen to be part of the conceptual toolkit for talking about maximally realistic counterfactual states of affairs, rather than as part of the semantics of general bouletic ascriptions or other propositional attitude ascriptions.

## Conclusion to Part I

As Part I unfolded, we saw how Kratzer's quantificational schemas for modality can be used to solve a variety of problems in the semantics of modals and propositional attitude reports. Lassiter's (2011) objections to quantification over possible worlds in the semantics of modality and attitude ascriptions were examined; some were found not to gain purchase with the facts at issue, while others were seen to be appropriate to certain uses of modality and attitude ascriptions, uses that demand the richness or precision of mathematical formulations of probability theory. This observation prompted the recognition of a need for semantic pluralism, whereby different semantic approaches coexist, answering to different empirical and theoretical needs. Interactions between modal statements and an embedding attitude predicate were examined, and it was argued that these do not reduce semantically to simple attitude ascriptions, although there seem to be grounds for accommodation that would interpret such an embedding as a modified attitude ascription. Finally, it emerged that truth conditions for attitude ascriptions involving quantification over a doxastic context of worlds compatible with a doxastic base could be reformulated when the doxastic base is inconsistent (as it realistically might be). This was achieved by quantifying over all possible worlds, and intersecting the doxastic base with the ordering source, which we can think of as effectively using the doxastic base as a filter on possible worlds, parallel to the ordering source. Under this formulation, the account readily extends to counterfactual or unrealistic bouletic attitudes by limiting the intersection with the doxastic base.

Yet there is something missing. Some propositional attitude reports don't simply report on what possible worlds are compatible with a belief or other propositional attitude of an experiencer. Some also correlate this report with a mental state, or an act of assertion, or some other eventuality (event or state) in the world. To maintain that this is so, and to say how it is so, requires us to confront arguments in recent literature that purport to show that ascriptions of stative attitudes do not describe any eventuality in the world. These ideas, and the further applications they yield, will be taken up in Part II of this work.

# *Part II*

# Events and propositional attitudes

The interpretations of modal statements and propositional attitudes entertained in Part I were purely intensional, in that they were conditions on those possible worlds that constitute epistemic, deontic, bouletic, and doxastic (etc.) alternatives to the actual world. Since modality and propositional attitude ascriptions are traditionally intensional phenomena, this isn't surprising. But to identify them as intensional leaves open whether they are intensional by virtue of having an intensional component of meaning, or by virtue of being wholly intensional, so it is notable that, in these analyses, world-grounding is achieved primarily through the context parameter, *c*, which was implicitly taken to be responsible for the conversational backgrounds.

But much of the rest of language is overtly about some aspect of the world around us, and statements with intensional meanings must integrate with the rest. In Part I, that integration would be achieved through the relation of Kratzer's modal conversational backgrounds (base and ordering source) to the context *c*, and the knowledge of the speaker and associated community (for an epistemic modal) or the circumstances in the context of utterance (for a realistic root modal). Likewise, in the application of Kratzer's theory to attitude ascriptions, the integration of intensional content with the world would be achieved in the relation of the doxastic (or other) ordering source to the person to whom a propositional attitude is ascribed, as well as to the circumstances in which it is ascribed.

But it is reasonable to seek a more explicit semantic world-grounding for intensional statements. If Sally points out that Max beat George in a tennis match, and Alex replies that Max might have cheated, Sally is characterizing an event in the actual world, and Alex is saying something about how this event unfolded in possible worlds compatible with what they know (Sally and Alex, and perhaps others more widely). That raises questions about the individuation of the event of Max beating George in the actual world, and its identity with its counterparts in other possible worlds. Likewise, if Alex believes that Max's victory over George was attained by cheating, then questions of the individuation and identity of events across worlds arise regarding this event, just as they do regarding the cross-world identity of an individual specified with a definite description.

Questions of event identity arise at a different level. If, in the actual world, Alex announces to Sally that Max cheated in his tennis match with George, and in another world, at the same time, Alex whispers to Sally that Max cheated in his tennis match, we have no trouble saying that these events of Alex communicating with Sally are counterparts, one in which Alex speaks normally, and the other in which he whispers. But what do we say when, in the actual world, Alex announces to Sally that Max cheated in his tennis match with George, and in the other, he announces that Max did not cheat in his tennis match with George? Are the two events counterparts in this case?

Efforts have been made to give epistemic modals a more explicit semantic world-grounding, in the eventuality (state or event) of the speech context. Hacquard (2006, 2010) is a notable case of this, where interactions between modals and tense and aspect are addressed by interpreting modals as a condition on a Davidsonian event variable; see also Anand and Hacquard (2008).

Along similar lines, both of these sources interpret an ascription of belief or other propositional attitude as a condition on an associated eventuality. Meanwhile, Katz (2003, 2008) has argued that stative attitude ascriptions do not have an associated Davidsonian event or state variable. These contending views deserve consideration in relation to one another. But, in fact, research has proceeded along parallel, essentially independent lines, on the basis of each of these contrary premises, yielding useful results without resolving the underlying tension between them.

Part II will argue that each of these positions regarding propositional attitude ascriptions is partially right, in that there are two types of attitude ascription, one (primarily stative) type that states the ascribed content as a purely intensional condition on possible worlds, and a second type of attitude ascription in which the content of the ascription is used to individuate an eventuality (a cognitive state, or an event of assertion). Furthermore, the basis for this resolution will provide an account of Neg-Raising (which will be called the Lower Interpretation of Negation [LIN] to avoid implicating an archaic transformational syntactic analysis), and provide an account of predicates that permit it and those that don't, which, it will be argued, has particular advantages over ones in the literature.

Returning to modal statements, the literature has addressed the question of modal determination at considerable length – correlating different syntactic positions of epistemic and root modals within clausal phrase structure with their divergent interpretations in a way that is as coherent and explanatory as possible, and the prospect of positing a unitary semantics for a modal operator that will realize an epistemic modal interpretation if it occurs in one syntactic context, and a root modal interpretation if it occurs in another. Recent literature has productively addressed this issue by relating modal statements to an eventuality in the context of an utterance; see especially Hacquard (2006, 2010) and Anand and Hacquard (2008). The developments regarding two types of belief ascription in Part II will provide a perspective for critical evaluation of these treatments of modal determination.

Part I was an investigation of modality and propositional attitude ascriptions in which the analyses were not fundamentally new. That investigation addressed tensions in the literature, primarily between possible worlds semantics of modality using Kratzer's quantificational schemas and critiques of that approach. The resulting formulation of the theory, bearing some modifications to overcome challenges, was reapplied to uses to which it had been put before, but with advantages derived from the alterations made along the way. Part I maintained a place for a formulation of a possible worlds semantics for modality and developed that formalism to overcome challenges: permitting it to be applied to the analysis of doxastic ascriptions when the doxastic base is inconsistent; accounting for the duals we find in modality and attitude reports; and accommodating presuppositional properties of successive doxastic-doxastic and doxastic-bouletic attitudes in the framework of a truth conditional semantics.

Part II expands the reach of these analyses to new domains. As with Part I, the discussion begins with a review of ideas in the literature. However, as with Part I, the ideas reviewed must be given an original cast if they are to serve the purposes of the present work. Rather than blandly reviewing the existing literature (which comes in mutually incompatible formulations and notations), the "literature review" will proceed by rethinking and reconstituting the underlying ideas, giving credit where it is due to the extent possible.

Chapter 5 reviews the theory of Davidsonian events and states, and addresses challenges to those views in the literature. But in order to address those challenges, and for its own theoretical adequacy, the Davidsonian view must be developed with a particular perspective on the individuation and cross-world identity of events. This development is similar to proposals in the literature, but is developed here for our own use. Chapter 5 also presents and evaluates the arguments of Katz (2003, 2008) against Davidsonian events for stative predicates, including stative ascriptions of propositional attitudes.

In Chapter 6, a view of how propositional content can individuate an event is developed, and shown to be appropriate for some attitude ascriptions, including some stative ones, but not others.

Chapter 7 uses these results to propose mechanisms for the Lower Interpetation of Negation, which will account for which matrix predicates permit it. Chapter 8 takes up proposals in the literature to address the problem of modal determination, and its intersection with issues raised by the distinction between two types of attitude ascription developed in Chapters 6 and 7. Chapter 9 argues that the distinction between two types of attitude ascription shows up in child language development and in the deficits associated with forms of fluent jargonaphasia.

# 5 Events and states

The present work is founded on a view of events as individuals in the world that participate in multidimensional part-whole relationships. These relationships are sensitive to the spatial and temporal extents of events, and to the nature and structure of their participants, and the delineation of these relationships is dependent on the interests and purposes that motivate or inform the delineation. Events can have counterparts across possible worlds, but which events participate in the counterpart relation – and thus count as "the same event" – depends on our interests and purposes, just as with individuals. Later, in Section 5.3, we consider to what extent, and how, the present discussion of events should extend to states.

The view of events adopted here has its roots in Davidson (1967) and other works reproduced in Davidson (1980), and has been developed further, especially with regard to its application to semantic theory, by Higginbotham (1983, 1985, 1989, 2000), Parsons (1990), Herburger (2000), Landman (2000), and Pietroski (2005), as well as works assembled in the collections edited by Rothstein (1998), Higginbotham, Pianesi, and Varzi (2000), Maienborn and Wöllstein (2005), and Dölling, Heyde-Zybatow, and Schäfer (2008). The ontology of events assumed here is essentially that outlined in the chapter "Events" of Lewis (1986b), with the role of part-whole relationships on events as elaborated by Moltmann (1997). The present discussion will take into account critiques of the Davidsonian view, especially by Asher and Bonevac (1985), and the limits drawn for this approach by Katz (2000, 2003, 2008).

Rather than review this entire literature, which is ably done in the introductions to the edited collections mentioned earlier, the present chapter will motivate and formulate a particular variety of Davidsonian event theory by talking it through from its conceptual beginnings. This view differs from that of Davidson and Davidson's closest followers in ways that will be explained along the way. It reflects the idea that events can be individuated at multiple levels of granularity, just as physical objects are. Among views in the literature, it is closest to those of Lewis (1986a), Moltmann (1986b, 1997), and Carlson (1998), but detailed comparisons with their views will not be undertaken, and the present view should not be conflated with theirs. But to the extent that this view is useful or correct, the credit goes to them. For present purposes, we need to distill this view out and let it face challenges to Davidsonian semantics generally, in order to establish what it can do for the theory of modality and propositional attitudes.

Thus, what follows in Sections 5.1 and 5.2 is a distillation, modification, and discussion of ideas in the literature. From the perspective of the present work, the value added is to separate out and to underscore a formulation of event-based semantics that integrates well with propositional content. For event-based semantic theory, the value added is a demonstration that major challenges to the approach can be met. Section 5.3 outlines Katz's arguments against Davidsonian events for stative attitudes and other stative predicates. This effort is justified in part because Katz's conclusions, if accepted at face value, would completely undermine recent attempts in the literature to integrate events with propositional content, and in part because, as will be argued in the next few chapters, Katz is partly right.

## 5.1 Part-whole structure of events

### *5.1.1 Multidimensionality and multiple levels of granularity*

Physical objects can be individuated, and their part-whole structure delineated, along various dimensions and at various levels of granularity, which may be hierarchically related to one another. If I pour out the contents of a drawer, which include an ordinary screwdriver with metal shaft and plastic handle, and then count the tools in the drawer, the entire screwdriver will count as one object. But a count of plastic things in the drawer would count the handle only, while a count of metal things in the drawer would count only the shaft. Adding the latter two numbers together to obtain the cardinality of the union of the set of purely metal things with the set of purely plastic things in the drawer, the screwdriver would count as two things rather than one. What I count, and therefore how many things there are in the drawer, depends on my interests and purposes.

A functional breakdown of an automobile engine would result in a list of a rather small number of relatively complex parts, such as the fuel injection system, the cooling system, the electrical system, and so on. Some of the components in such a functional breakdown, such as the electrical system, can be quite distributed in space. The functionally identified parts causally interact with one another, and may be entangled with one another in intricate ways. In a more atomistic delineation of physical parts, the entire engine breaks down into a fairly large number of simplex objects, including nuts, bolts, plates, fans, belts, hoses, wires, and electronic sensor components, and a few large simplex objects, such as the engine block and the radiator box. The objects we delineate as parts of the engine depend on the dimension and level of granularity chosen, which in turn depend on our purposes for making the delineation.

Despite the causal relatedness of the parts of a complex object, on either a physical or functional delineation of parts, we would resist an argument that would attempt to erode the notion of a coherent physical object by appeal to its physical and causal proximity to other objects. A gear box can be a coherent part of a complex machine, despite the fact that its input rotor communicates kinetic energy from other parts of the machine, and its output rotor meshes with other gears, communicating kinetic energy to "downstream" parts of the machine, and despite the fact that the machine is part of a larger factory production line, and may be physically interpenetrated with other machines on the line. The gear box is a distinct thing if it has enough physical and functional integrity, and we would resist extending its boundaries outward to include the entire machine, the entire production line, or the entire factory, let alone the entire world. A similar discussion could be advanced for social entities such as towns and cities, departments of large companies, and government departments or ministries.

Following the lead of Lewis (1986a) and Moltmann (1997), a similar approach to events will be adopted here. Lewis (1986a: 259), elaborating an example of Jaegwon Kim, spotlights the example of an event of writing "Larry," which has overlapping subevents of writing "Larr," "rr," and "rry." A traditional dance ball is an event, and we readily regard the dancing of a particular couple at the ball as a subevent interwoven with other such subevents as part of the larger event of the ball. We can readily regard the entire ball as a sum of spatiotemporally coherent subevents of dancing couples, the orchestra playing, the serving of drinks, people conversing, and the like. We could define a subevent of several couples dancing in a corner. And with sufficiently strong motivation, we could identify the activities of a particular set of people interspersed among others on the dance floor as a subevent – in *Pride and Prejudice*, for example, the subevent of the Bennett family's participation in the ball at Netherfield. Or we could define the spatiotemporally scattered subevent of the spread of a rumor through the attendees at the ball. If three children, Kristen, Kieran, and Kevin, stand in a room and hop in unison, we can identify the event of all three of them hopping, and we can identify the subevent of any one of them hopping, and, with sufficient motivation, the subevents of any two of them hopping. But at a lower level, we can identify each hop by a single child as a subevent, and at a lower level still, the impact of the child with the floor upon his or her descent in that hop as a further subevent. If Kristen twists her ankle while hopping, we can identify the children's hopping as the cause of this mishap (they shouldn't have been hopping on a concrete surface), or the subevent of Kristen's participation in the hopping, or her last, wayward, hop, or the impact upon descent from this hop, when the ankle was injured. This leaves open questions about the resolution of subevents in a descending sequence ($e_i$) where $e_i$ is a subevent of $e_j$, for each $i > j$, which we acknowledge here just as we acknowledge analogous questions about the smallest components of matter. These questions about subevent resolution have significance for semantic theory; see Eckhardt (2005, 2008) for relevant discussion. But these questions don't have to be fully settled in order for us to appeal to a domain of events in a Davidsonian semantic theory, any more than physics has to settle comparable questions about the composition of physical objects before we can define a domain of objects for semantic theory.

In standard neo-Davidsonian semantic theory, the truth conditions of (1), pertaining to the situation in which Kristen twisted her ankle when the three children were hopping on a cement surface, would be represented as in (2), where Kristen is the individual $x_1$ of the domain; this allows for some equivocation over whether Kristen is an agent or experiencer of that event.

(1) Kristen twisted her ankle.

(2) $\exists e_1\ \exists x_2$ [ankle-of $(x_1, x_2) \wedge$ twist $(e_1) \wedge$ AGENT/EXP $(x_1, e_1) \wedge$ PATIENT $(x_2, e_1)$]

For Davidson and Higginbotham, the choice of $e_1$ as the event being described is quite determinate, since all events with the same causal efficacy as $e_1$, which might be described instead by (1), are in fact identified with $e_1$. But previous considerations suggest that various events at different levels of granularity could be described by (1). On this view, the lexical and compositional semantic structure of (1) essentially provides just a template of the type given in (3), which is used by the speaker to characterize an event at a level of granularity chosen by the speaker, as appropriate to the context. This template is not a representation in a logical language or a model; it is a template for an objectual representation of truth conditions, in which the underline can be replaced by the metalanguage label of an appropriate event.

(3) $\exists\_\ \exists x_2$ [ankle-of $(x_1, x_2) \wedge$ twist $(\_) \wedge$ AGENT $(x_1, \_) \wedge$ PATIENT $(x_2, \_)$]

Variables for events at several different levels of granularity could be substituted in the position of the underline in (3). Reasons for adopting this multiple granularity property of events in a Davidsonian analysis will be reviewed next.

First, consider the granularity of events in the context of the relationship of events to the verbs used to describe them in a neo-Davidsonian framework. If we were to posit a fixed level of granularity for simple, indivisible events corresponding to the Davidsonian variables of action predicates, well-known issues arise in identifying the holders of thematic roles associated with these events in examples such as those in (4); see Parsons (1990), Carlson (1998), and Landman (2000).

(4) a. Max sold the boat to Bill.
b. Bill bought the boat from Max.

The neo-Davidsonian representations of the truth conditions of (4a,b) are given in (5a,b), respectively.

(5) a. $\exists e_1$ [sell $(e_1) \wedge$ AGENT $(x_1, e_1) \wedge$ THEME $(x_2, e_1) \wedge$ GOAL $(x_3, e_1)$]
b. $\exists e_2$ [buy $(e_2) \wedge$ AGENT $(x_3, e_2) \wedge$ THEME $(x_2, e_2) \wedge$ SOURCE $(x_1, e_2)$]

The difficult issues concern the identification of the events $e_1$ and $e_2$. If the purchasing transaction described by (4a,b) is analyzed as a single, indivisible event, then (4a) and (4b) describe the same event, so $e_1 = e_2$, and it follows that the thematic role relations AGENT $(x_1, e_1)$ and AGENT $(x_3, e_1)$ (equivalent to AGENT $(x_3, e_2)$) hold jointly. Now this, by itself, doesn't look as bad as it does on the formulation of Landman (2000: 31), who writes it as AGENT $(e_1) = x_1$, and AGENT $(e_1) = x_3$, where the functional notation tends to enforce the expectation of a unique occupier of a thematic role such as the agent of an action (as Landman, indeed, assumes). The relational notation for thematic role identification used here (and by Parsons [1990] and Carlson [1998]) does not enforce that notion, and with the representations in (5a,b), we are admitting that Max and Bill are both agents of the event described, $e_1$ $(= e_2)$.

In fact, there is reason to embrace this result as one interpretation of (4a,b), as we will do here, even if, as argued in the pages that follow, it is not the only interpretation. The sentences in (4a,b) describe an interaction in which Max and Bill both participate agentively. If we take the whole exchange as one event, then there is nothing wrong with saying that Max and Bill are both agents in this event and that Max is both an agent and a source, while Bill is both an agent and a goal. However, the representations in (5) don't require that the whole interaction of buying and selling is the only level of granularity at which we can characterize what transpired between Max and Bill. We can instead identify $e_1$ as a selling subcomponent of the whole interaction, and $e_2$ as a distinct buying subcomponent of this transaction, so that (4a) and (4b) describe distinct subcomponents of the whole transaction.[1] Then $e_1 \neq e_2$, but $e_1$ and $e_2$ are subevents of a larger event $e_3$, for the whole transaction, and $e_3$ then has the property that Max and Bill both participate as agents in it. But now we can be more specific: Max participates as agent of the subevent $e_1$, but not as agent of $e_2$, and Bill participates as agent of the subevent $e_2$, but not as agent of $e_1$. With these identifications of $e_1$ and $e_2$, we can characterize the

[1] For development of this view, and further rationale for it, see Lombard (1985), Lewis (1986a), Carlson (1998), and Landman (2000).

buying and selling transaction at a fine-grained level with (5a,b), and equally at a coarse-grained level by replacing both $e_1$ and $e_2$ in (5a,b) by $e_3$. On this approach, the lexical semantic representations of the verbs do not fix the level of granularity; the granularity of events characterized is determined by speaker and addressee(s) in the context of use. This raises prospects of ambiguity or grain-level mismatches, but these are exactly akin to those that arise in the talk of complex physical objects, as when Max says that he had to order a part for his car engine. The "part" could be a new fuel injection system, a sparkplug, or a very specialized plate or screw.

With the identifications of $e_1$, $e_2$, and $e_3$ above, $e_1$ and $e_2$ are functionally complementary subevents of $e_3$, and they plausibly sum to $e_3$. But they are to be regarded as subevents of $e_3$, not as different facets or presentations of $e_3$. The sentences (4a,b) are different descriptions of $e_3$, but on the present view we should resist the temptation to say that $e_1$ is $e_3$ from the perspective of Max, and that $e_2$ is $e_3$ from the perspective of Bill. But we can say that $e_1$ is a Max-centered subevent of $e_3$, consisting of Max's participation in $e_3$, and that $e_2$ is a Bill-centered subevent of $e_3$, consisting of Bill's participation in $e_3$, so that $e_1$ and $e_2$ are functionally complementary parts of $e_3$. Each of (4a,b) can serve as a description of the entire purchasing transaction, as well as of the buying and selling subevents. To interpret (4a,b) as descriptions of the larger event, we can replace both of $e_1$ and $e_2$ with $e_3$ in the conditions in (5a,b), and interpret both as descriptions of $e_3$. The resulting representations yield alternate, equally valid, interpretations of (4a,b). There is no single level of event granularity to which (4a) and (4b) naturally or necessarily apply.

The difference between $e_1$ and $e_3$, or similarly between $e_2$ and $e_3$, is not easy to render in terms of different possibilities of adverbial modification. For example, if Max sold the boat to Bill hesitantly, this could be captured by adding to (5a) the condition that Max is hesitant in $e_1$, but it could be captured equally well by adding the condition that Max is hesitant in $e_3$ to a representation of the truth conditions of (4a) consisting of (5a) with the coarser-grained event $e_3$ in place of $e_1$. But consider adverbial modification of (4a) with *expediently*, as in, *Max expediently sold the boat to Bill.* Adding to (5a) a condition that expresses the fact that $e_1$ was expedient for Max permits the representation of a sharper focus on the fact that it was Max's participation in this event, the selling of the boat, that was expedient for Max.

Functional subevents are distinguished by the functional relatedness and lack of arbitrariness of the component events. Subevents delineated on other grounds may or may not seem so natural. When Kristen, Kieran, and Kevin stand in a room and hop in unison, we can identify the event of all three of them hopping, and we can identify the subevents of any one of them, or any two of them, hopping. Any choice of a nonempty subset of the three children yields a (proper or improper) subevent. The subevent consisting of just Kieran and

Kevin hopping may seem arbitrary, if there are no obvious grounds for excluding Kristen's hopping from it, but it is no less a subevent. It suffers only in lacking salience and a simple description. If, on the contrary, Kieran and Kevin are both from the Jones family, and Kristen isn't, then the subevent of Kieran and Kevin hopping has a description that makes it feel less arbitrary: it is just the subevent of hopping confined to members of the Jones family.

We could define different sums according to different principles: functional sum for functionally complementary subevents; spatiotemporal sum for spatiotemporally contiguous subevents; and so on. Although we won't assume this differentiation here, we assume that it is possible, even if it is sometimes difficult to explicitly specify what counts as events at the same level of granularity, just as it seems fitting to classify physical objects according to appropriate typologies even if it can sometimes be difficult to explicitly identify criteria for choosing a given level in a typology. We can identify subevents according to any principle that fits our purposes and interests on the occasion of utterance. When we individuate events according to our interests and purposes, we also sum events according to our interests and purposes. If we unite a collection of disparate car engine parts at different levels of granularity by putting them into a tub labeled "engine parts," we just violate our sense of the coherence of different levels of granularity, and we could instantiate this as a formal error in the application of a formal typologically based summing operation. Likewise, if we unite a collection of disparate "subevents" of an event, chosen at different levels of granularity, we just do violence to our notion of subevent relatedness, which could be captured as a formal error in the application of formally defined event summing operations.

A fundamental question for this approach is how the upper boundary of an event is determined, and whether each event can be considered to be part of a larger, more all-encompassing, event. More generally, what constitutes the boundary of an event? The formal operation of summing may not prevent us from summing an event with nearby events, and there may be no formal constraint against performing this process without bound, summing in ever more distant and unrelated events. On a functional delineation, if Max's sale of the boat to Bill, $e_1$, and Bill's purchase of the boat from Max, $e_2$, are functionally integrated parts of a more complex event $e_3$, then what about Bill's walk down the dock to make the purchase? Or his bus ride to the seashore, where he disembarked onto the dock? Or the breakfast he ate just before he boarded the bus? Or his night of sleep the night before? Does the swaying of the boat as Max and Bill stood on it when they conducted the transaction count as a subevent of the transaction? If so, would the swaying of the boat count equally if they were standing next to the boat, rather than on it? If so, would it count if they made the purchase ten miles away in a lawyer's office? If we admit any of these causally related events as subevents of the transaction of the purchasing and selling of the boat, does it follow that we have to admit all of them, and by extension, everything that happens in the world, so that all the world becomes one big event?

But parallel concerns arise in the delineation of ordinary physical objects, and there, the lack of bounds imposed by formal or mereological summation does not prevent the delineation of naturalistic boundaries, decided by the interaction between the intrinsic structure and causal relations among parts, in ways that depend on our interests and purposes. Similar interactions will suffice to resolve the parallel difficulties with the delineation of events. Events, on the present conception, like other entities, have naturalistic boundaries that provide grounds for resisting a slippery-slope argument that they must embrace everything in the world. When the three children hop together, we can physically delineate the event of them hopping as what goes on in the room in a particular time interval, or as what happens involving the three of them at a particular time. During the boat purchase, the swaying of the boat would be naturally considered a part of the event, if (e.g.) Max and Bill had to struggle to steady themselves as they stood on the boat signing forms or exchanging money for a deed. Or the swaying of the boat could be deemed irrelevant if it played no significant functional role. If we focus on the financial transaction, the swaying of the boat is likely to be irrelevant, while if we focus on the purchasing transaction as a physical undertaking, then the swaying of the boat and the brightness of the sun (or, alternatively, a light rain that was coming down) could be regarded as natural parts of this event.

It might be feared that if we admit fine granularity of events, then (4a), reporting on Max's sale of the boat to Bill, would not entail (4b), reporting Bill's purchase of the boat from Max, since the existence of the event that makes (4a) true does not make (4b) true. But the entailment does go through, whether we choose fine-grained events or coarse-grained events in the interpretation of (4a,b). To begin with, if we use (4a) and (4b) to directly characterize the larger, more complex event of sale and purchase, then the same event is involved in both (4a,b), and then (4a) does entail (4b). If we are going to interpret (4a) and (4b) using the finer-grained events, then the larger, more complex event is nevertheless implicit, to ensure that the selling and the buying subevents are in their proper reciprocal relationship: that a selling of the boat to Bill by Max is accompanied by a buying of the boat from Max by Bill. By appeal to this larger event (or a meaning postulate capturing the relationship of the two subevents within the larger event), the entailment goes through even with fine-grained events.

Some sustained challenges to the view of events adopted here, and their role in semantic theory, will be considered in the next two sections. Before taking them up, a couple of other issues will be addressed briefly, to ascertain that they should not derail the project. Lewis (1986a: 241) notes that "[t]here is no guarantee that events made for semantics are the same as the events that are causes and effects. It seems unlikely, in some cases at least. A mathematical sequence converges." Lewis proceeds to note that we can nominalize this, talking of the converging of the sequence, and talk of it converging rapidly, or not, but that there is no event, in his sense, described. This use of *converge* in a nonaction sentence can be explained if we recognize polysemy and metaphorical extension. The mathematical use of this verb can be shorn of eventive content or significance in the semantics, even if a mathematical constructivist might try to sneak it back in.

For Lewis, an event is a property of spatiotemporal regions, and thus can be identified (extensionally) as a class of regions, at most one per possible world ("one region each from some worlds, none from others, and never more than one from the same world," p.245). Lewis notes the difficulty in trying to identify an event uniquely through a description of the event. But such difficulties shouldn't surprise or disturb us. Individuals have properties, but it is typically hard to specify individuals uniquely by description, except by using relational properties that identify the individual in terms of its relation to other individuals (e.g., *a is three feet due north of b*). The difficulties involved in uniquely describing an event are parallel: it is easy to uniquely specify an event using relational properties to delimit an event, fix its granularity, and place it in space and time. In this, events simply share in the behavior of other complex individuals.

### *5.1.2 A challenge: events in the analysis of perception sentences*

On the present view, events have a richer range of possibilities of part-whole structure than a strictly physical mereological structure. It will be argued here that these additional resources are sufficient to overcome some objections to Higginbotham's (1983) account of the interpretation of perception reports such as (6), without going so far as to embrace the alternative situation semantic account, and while nevertheless preserving the central idea of Higginbotham's approach, that these reports are extensional in a way in which standard propositional attitude reports are not.

(6) Maria saw/heard Max leave at noon today.

In the relevant literature (including both Higginbotham and Situation Semantic accounts), [Max leave] is assumed to be a constituent in (6), an uninflected Naked Infinitive (NI) occurring as the complement of the perception verb.[2]

[2] This syntactic structure is supported by standard constituency tests that do not interfere with the assignment of accusative case to the subject of the NI complement. In Principles and Parameters syntax, where expletive *it* can only be a subject, the fact that the postverbal nominal is a subject of the NI complement, and not an object of the perception verb, is critical for explaining how sentences such as *I heard it raining* are possible.

In Situation Semantics, a situation is a complex of objects and spatio-temporal locations in the world, and relations among objects and spatio-temporal locations; these include static states of affairs, and more eventful courses of events. (See Barwise and Perry [1981a, b], [1983].) A sentence is an indefinite description of a situation; if S = *Max left at noon today*, then a situation σ supports S iff Max left at noon today in σ. Let [[S]] be the set of situations that support S.[3] Situation Semantics takes a very different approach to the problem of upper bounds addressed earlier. Instead of relying on naturalistic boundaries to define the situation being described, Situation Semantics regards any complex of objects and spatiotemporal locations in which Max left at noon today as a situation in which Max left at noon today. The state of affairs consisting of a snapshot of the entire universe at noon today is one such situation, as is the course of events of everything that has ever happened, and ever will happen, in the universe. But the set [[S]], which contains all situations in which Max left at noon today, at all levels of granularity, will also include very fine-grained local ones, and these serve to distinguish the meaning of S from the meaning of *Max left at one minute past noon today*, and *Max left at noon yesterday*, as well as from *Max didn't leave at noon today*, or *Max died at noon today*.

The sentence in (6) (with *see*, for this discussion) is supported by any situation in which Maria is in the "see" relation to a scene in [[S]], for S = *Max left at noon today*, where a scene in [[S]] is a perceptually accessible situation (specifically, a course of events) in [[S]]. The situation of Max leaving at noon today will have many scenes associated with it, and many of these may be perceptually accessible to Maria; (6) asserts that Maria is in the "see" relation to one of those scenes. This is stated more succinctly in (7).

(7) $\exists s \subseteq \sigma\ (\text{see}\ (x_1, s) \wedge \sigma \in [[\text{S}]])$

The scene in (7) is a complex of objects, relations, and spatio-temporal locations, and not itself a singular individual. This provides rich descriptive resources in specifying what Maria saw, and thereby readily allows not only for perception reports such as (6) but also ones in which the NI complement is negated, or involves a quantified nominal. For this reason, Barwise and Perry (1983) regard the analysis of sentences like (6) as a major step toward the analysis of general attitude reports such as *Maria saw/knows/believes that* S.

This is in contrast to the proposal of Higginbotham (1983), on which (6) is true by virtue of Maria being in the "see" relation to the event of Max leaving at noon today, where this event is a singular individual. Specifically, Higginbotham proposed that (6) is interpreted as in (8), where "see" is a three-place predicate in the logical metalanguage that relates the event seen,

[3] In this section, I adopt the practice of Barwise and Perry (1983) in letting [[S]] be the semantic value of (an utterance of) a natural language sentence S, rather than of a syntactic phrase marker.

$e_2$, of Max leaving, to the event $e_1$ of Maria seeing $e_2$, and to the experiencer, $x_1$ (denoted by *Maria*), of $e_1$. The temporal modifiers are assumed here to be in the main clause, where they modify the seeing event $e_1$, but due to the lack of tense in the NI complement, it follows that $e_2$ was also at noon today.

(8) $\exists e_1 \exists e_2$ [see $(x_1, e_2, e_1) \wedge$ leave $(x_2, e_2) \wedge$ today $(e_1) \wedge$ at-noon $(e_1)$]

In neo-Davidsonian formalism, this can be expressed as in (9).

(9) $\exists e_1 \exists e_2$ [see $(e_1) \wedge$ EXP $(x_1, e_1) \wedge$ THEME $(e_2, e_1) \wedge$ leave $(e_2) \wedge$ AGENT $(x_2, e_2)$]

This representation relates the experiencer Maria to a singular individual event, rather than to a complex composed of individuals, relations, and spatio-temporal locations. This makes the analysis of (6) entirely extensional and first-order, rather than a major first step toward the analysis of propositional attitude ascriptions. Higginbotham regards this as a virtue of the account. But others have argued that it lacks the resources to make fine-grained distinctions that are observable, or to account for perception sentences with negated NI complements, and ones containing quantified nominals.

The Davidsonian conception of an event plays an important role in the application of the analysis to the interpretation of perception sentences with NI complements. As discussed in Section 5.1.1, Davidson (1967: 163–180) identifies two events just in case they have the same causes and effects. This ensures that events for semantic theory will not be multiplied beyond a minimal set of causally efficacious events. But it also establishes a fixed and quite coarse granularity of events for semantic theory, and the objections to the account hinge on this property of it.

To take up an objection raised by Asher and Bonevac (1985), suppose that Buck shot Ed, and that, as he did so, Maria saw Buck raise the rifle and pull the trigger to fire, but did not see Ed get hit since the target of Buck's shooting was in shadows, or otherwise obscured for her. In this case, a Davidsonian approach with the relatively coarse-grained events assumed by Davidson (and Higginbotham) would identify the event of Buck pulling the trigger with the event of Buck shooting Ed. The sentences in (10) and (11) would then have identical representations as asserting that Maria is in the "see" relation to this event, and would therefore have to have the same truth value. Since (10) is true, it would follow that (11) is true as well, a result that seems intuitively to be incorrect in the case at hand.

(10) Maria saw Buck pull the trigger.

(11) Maria saw Buck shoot Ed.

Furthermore, since the events described by the NI complements are the same, (11) would entail (10) as well, and this also seems incorrect: if Maria was far away, she could see Buck raise the rifle and fire, and see Ed fall, so that (11) is true, but be too far away to see Buck pull the trigger.

In Situation Semantics, the NI complements in (10) and (11) have distinct meanings: (10) is true by virtue of Maria being in the "see" relation to a scene that is part of a situation supporting $S_1$ = *Buck pull the trigger*, while (11) is true by virtue of Maria being in the "see" relation to a scene that is part of a situation supporting $S_2$ = *Buck shot Ed*. The situation supporting $S_1$ need not support $S_2$ (Buck could pull the trigger without shooting Ed), and vice versa (Buck could have shot Ed if the gun went off accidentally, without Buck pulling the trigger).

The present approach, a Davidsonian analysis founded on an event ontology that admits events with multiple levels of granularity, has some features of each account. It appeals to events as individuals, as objects of "see," for example, not proposition-like complexes of objects, relations, and locations. But the events appealed to here admit multidimensional part-whole structures. On this approach, Buck's shooting of Ed is a complex event-system, with various levels of granularity. At the finest level of granularity, the focus of the event is the pulling of the trigger. At a more comprehensive level, the event includes Buck's immediate preparations and execution of the action: cocking and aiming the rifle, pulling the trigger, and then lowering the rifle. At a much more comprehensive level, it includes the long-term preparations and long-term ramifications of the shooting. Larger-scale events would include one centered on Buck, including Buck making preparations, perhaps as long as weeks beforehand, leading up to the shooting, and including the shooting itself, and Buck's escape afterward. And it could include event scales in between these two extremes. In a report by a security service, (12) could report on the "achievements" of the agency during the month of March, including one such fairly elaborate operation involving weeks of planning, leading up to Buck shooting, and thereby assassinating, Ed. Or in (13), Sam could remind Bill of a party they attended, at which Buck shot and wounded Ed, a commotion ensued, and it took two hours after the police and ambulance arrived for things to settle down to the point that Sam and Bill could leave; in this report, Sam's use of the demonstrative *that* refers to a rather larger event consisting of the shooting and its two-hour-long aftermath.

(12) [Successes for the month of March] Item #16: Buck shot Ed.

(13) Remember, we had a drink in the bar downstairs and talked to Maria for a while.
Then Buck shot Ed. After *that*, we left and went to a diner for a late snack.

On the present approach, in the context where Maria sees the aiming and firing of the rifle but is not positioned to see Ed getting hit, (10) does not entail (11), since (10) is true by virtue of Maria being in the "see" relation to a fine-grained event of Buck pulling the trigger, while (11) would be true only if Maria were in the "see" relation to a larger event, including Ed getting hit, which she is not in this case. (11) would normally be understood as asserting that Maria was in the "see" relation to the most salient event identified as what one would see upon seeing somebody shoot somebody else, namely, the event $e_2$ in (14). This event is salient in its mid-level granularity since it includes everything that is most likely to be important about the event, the culmination of Buck's preparations and the physical genesis of Ed's woes, in a package that could be immediately observed by a well-placed observer in an accessible region of space-time (the domain of mid-sized objects extended to mid-sized events).

(14) $\exists e_1\, \exists e_2\, [\text{see}\, (e_1) \wedge \text{EXP}\, (x_1, e_1) \wedge \text{THEME}\, (e_2, e_1) \wedge \text{shoot}\, (e_2) \wedge \text{PATIENT}\, (x_3, e_2)]$

But (10), meanwhile, could be true by virtue of reporting that Maria is in the "see" relation to the subevent of Buck pulling the trigger. Also note that (15) is true in all of these cases, by virtue of Maria being in the "see" relation to a suitable event of Buck shooting, either all of $e_2$, or a Buck-centered subevent $e_3$. That is, (15) can be interpreted as in (16), with $j = 2$ or 3.

(15) Maria saw Buck shoot.

(16) $\exists e_1\, \exists e_j\, [\text{see}\, (e_1) \wedge \text{EXP}\, (x_1, e_1) \wedge \text{THEME}\, (e_j, e_1) \wedge \text{shoot}\, (e_j) \wedge \text{AGENT}\, (x_2, e_j)]$

Both of $e_2$ and $e_3$ are in the THEME relation to the seeing event $e_1$, as is appropriate since, in the last scenario sketched, both of these Buck-centered events, the shooting of the gun and the trigger pulling, are seen by Maria.

Adopting events at various levels of granularity, different levels of event granularity can be involved in (10) and (11), and (10) does not entail (11) because Maria can see a part of something without seeing a larger whole of which it is a part. That has an analog with concrete objects: I can see the trunk of an elephant, and not be said to see the elephant, at least not in a strict sense analogous to that in which (11) is false when Maria does not see Ed get hit. Whether or not we adopt a sense of "see" whereby seeing a part of something does not entail seeing the whole thing depends on our purposes. For many purposes, if I see the trunk of an elephant sticking out from behind a wall, I could be said to see the elephant. For example, this would suffice if I were looking for a particular lost elephant, and it was the only elephant around. But if the point of looking at elephants in a given context is to see what one looks like, because I've never seen one before, then, in this context, I could be said to see the elephant trunk without being said to see the elephant. There are, similarly, contexts in which the interpretation of "see" as applied to events would be restricted so that (10) does not entail (11). Conversely, if Maria sees Buck shoot Ed from a distance, it doesn't follow that she sees Buck pull the trigger, anymore than it follows that when Maria sees Ed from a distance, she necessarily sees the freckles on Ed's nose.

Perception sentences with a quantified nominal in the lower clause have been argued to be similarly problematic for Higginbotham's account, but adequately explained within Situation Semantics. The perception sentence in (17) can be interpreted as asserting that, for each person in the domain of the quantified subject of the lower clause, there is an event of that person dropping to the ground, which was seen by Maria; this would be appropriately used to describe a situation in which Maria wandered through an extended stage rehearsal and happened to view separate occasions when different cast members dropped to the ground.

(17) Maria saw everyone drop to the ground.

Or, (17) can be used to assert that Maria saw a complex event in which everyone in the domain of the quantifier dropped to the ground (the "group drop" interpretation) at once, or in some coherent, organized fashion. It can also be interpreted as asserting that there is a coherent event, such as an audition for an action part, in which Maria was the casting director, and watched as each auditioner in turn dropped to the ground in separate dropping events that could be scattered in time and space, but still united as being part of the audition process.

If $e_1$ is the (neo-)Davidsonian event variable introduced by *see*, and $e_2$ the variable introduced by *drop*, and $x_2$ is the variable bound by the quantifier, then these three interpretations can be summarized as: $\forall x_2\ \exists e_1\ \exists e_2$, $\exists e_1\ \exists e_2\ \forall x_2$, $\exists e_1\ \forall x_2\ \exists e_2$. These interpretations can all be captured in Situation Semantics by giving the quantifier appropriate scope, and then structuring the situation that supports the complement clause appropriately to the scope of the quantifier. For Higginbotham, the complement clause must denote a singular event that is seen by Maria; this requires the quantified subject of the complement to take wide scope, yielding just the first interpretation.

On the present approach, the quantified subject of the lower clause can adjoin to its local clause, taking scope over the introduction of $e_2$, as represented in (18). This yields the audition interpretation ($\exists e_1\ \forall x_2\ \exists e_2$): $e_1$ is the auditioning event in which each auditioner $x_2$ has an audition $e_2$, viewed by Maria.

(18) $\exists e_1$ [see $(e_1) \wedge$ EXP $(x_1, e_1) \wedge (\forall x_2$: auditioner $(x_2))$:
$\exists e_2$ [drop $(e_2) \wedge$ AGENT $(x_2, e_2) \wedge$ THEME $(e_2, e_1)$]]

In line with the "group reading" of *Everyone dropped to the ground*, where they all dropped at once or in an organized fashion, it must also be possible for the quantifier to scope under the introduction of the lower event ($\exists e_1\ \exists e_2\ \forall x_2$); this is represented in (19), and yields the "group drop" reading.

(19) $\exists e_1\ \exists e_2$ [see $(e_1) \wedge$ EXP $(x_1, e_1) \wedge$ THEME $(e_2, e_1) \wedge$ drop $(e_2) \wedge$ $(\forall x_2$: member-of-group $(x_2))$ AGENT $(x_2, e_2)$]]

Finally, the quantified subject can undergo Quantifier Raising out of the NI complement, and take scope over the matrix clause ($\forall x_2\ \exists e_1\ \exists e_2$); this yields the interpretation represented in (20), in which there are potentially distinct (and potentially scattered) seeing events by Maria for each person who dropped to the ground, as would be the case if Maria just happened to observe each cast member drop to the ground, but not as part of any coherent process.

(20) $(\forall x_2$: on-studio-lot $(x_2))$ $\exists e_1\ \exists e_2$ [see $(e_1) \wedge$ EXP $(x_1, e_1) \wedge$ drop $(e_2) \wedge$ AGENT $(x_2, e_2) \wedge$ THEME $(e_2, e_1)$]

Thus, the functional part-whole structure permits an extensional account of all three interpretations.

Functional part-whole relations can be used likewise to account for why there is an interpretation of (21a) on which it entails (21b), but not vice versa, and to explain why (22a) does not entail (22b), facts that were invoked by Asher and Bonevac (1985) to argue the inadequacy of Higginbotham's account of perception sentences as purely extensional relations between events of perception and events perceived.

(21) a. Maria saw nobody in the audience applaud.
b. Maria didn't see anybody in the audience applaud.
(22) a. Maria sees no student cheat on the exam.
b. No student cheats on the exam.

The sentence in (21a) can be used to report on a composite event of an audience responding in silence to a performance, or the object of *see* in (21a) can introduce such an event into the discourse. With this interpretation, it entails (21b), that Maria didn't see any component event of an audience member applauding. But if Maria missed the performance altogether, (21b) might be true – Maria didn't see any event of an audience member applauding, while (21a) is false – it is not true that Maria saw the event of nobody applauding. Of course, (21a) can also be interpreted as asserting that Maria missed all events of somebody in the audience applauding, obtained when the negative quantifier takes wide scope. (22a) has the latter type of interpretation as well, and under that interpretation does not entail (22b).

### *5.1.3 Conclusion*

Higginbotham (1983) sought to account for perception sentences such as (6) in terms of purely extensional perceptual relations to events, with only first-order quantification over individuals (including events). Asher and Bonevac (1985) argued that Higginbotham's event semantics is inadequate to handle the issues posed by the nonequivalence of (10) and (11), and by the interactions with quantifiers in (17), which are addressed adequately by Situation Semantics. Asher and Bonevac admit that situations in Situation Semantics are intensional, but they suggest that there is no real price in admitting intensionality in our account of perception sentences, since we will need it anyway in the analysis of propositional attitude ascriptions. And, as mentioned earlier, Barwise and Perry (1983) regard their analysis of NI perception complements as a major first step in their account of propositional attitude ascriptions.

The discussion in this section indicates that Higginbotham's original intuition has merit – that perception sentences are different from attitude ascriptions in ways that favor an analysis of them as extensional, and that we should not give up this intuition lightly. The present approach to perception sentences is similar to Higginbotham's in its extensionality, but it appeals to multidimensional part-whole structures on events at various levels of granularity, including very fine-grained, in a way that permits it to match the discriminatory powers of the Situation Semantic accounts of the phenomena discussed earlier. Propositional attitude ascriptions, on this approach, are more fundamentally intensional than perception reports with NI complements, as borne out by the interpretations of attitude ascriptions examined in Chapters 3 and 4. In Chapter 6, it will be argued that propositional attitude ascriptions are characterized by conditions on an attributed context, an information state attributed to the attitude experiencer (the denotation of the subject of the ascription), where this condition on the update of the attributed context may or may not characterize a mental state of the experiencer. It will be argued further that this semantic dimension is absent in perception sentences.

## 5.2 Individuating events

Recall Davidson's (1967 [1980]: 179) criterion for the individuation of events: "events are identical if and only if they have exactly the same causes and effects." Consider again the case of Buck shooting Ed, where the event $e_3$ of Buck pulling the trigger is part of the larger event $e_2$ of Buck shooting Ed. All causes of the larger event, $e_2$, such as Ed's earlier betrayal of Buck, or Buck's resentment of Ed, are contributory causes to the event of Buck pulling the trigger. All causes of Buck pulling the trigger, such as the contraction of the muscles of his finger, are contributory causes to Buck shooting Ed: if his finger didn't contract, he would not have shot Ed. (Or if his finger froze up, whatever Buck eventually did to press the trigger of the gun and cause it to fire and shoot Ed would be a contributory cause to Buck shooting Ed, because without it, the shooting would not have happened.) All effects of Buck's shooting of Ed, such as the penury of Ed's dependents, and Buck's arrest and incarceration, have the pulling of the trigger as a critical, contributory cause. And all effects that we would primarily identify as effects of (events caused by) Buck pulling the trigger, say a soreness in his trigger finger, are results of Buck shooting Ed, however minor, and however overshadowed by other results.

Therefore, Davidson's criterion of event identity would identify the event of Buck shooting Ed with the event of Buck pulling the trigger. But it was this view of events that made Higginbotham's analysis of NI perception reports vulnerable to the objections of Asher and Bonevac, as discussed in Section 5.1.2. And this view of events is incompatible with the delineation of events at various levels of granularity, and the part-whole structures defined on events at different levels of granularity, argued for in Section 5.1.1. Davidson's criterion, interpreted extensionally, is incompatible with the existence of fine-grained events participating in part-whole structures.

If we apply Davidson's criterion of event identity to events across possible worlds, the event of Buck shooting Ed would not be identified with the event of Buck pulling the trigger. In another world, the trigger of the gun might have been disconnected from the firing mechanism, and Buck might have found a way to shoot the gun by releasing the firing pin, without pulling the trigger. However, if we formulate Davidson's criterion as applying to events across possible worlds, it is so restrictive that a given event would have few, if any, counterparts across possible worlds. Continue consideration of the event of Buck shooting Ed. There are many possible worlds in which Buck's preparations deviate ever so slightly from those in the actual world, and many possible worlds in which the consequences of Buck shooting Ed diverge ever so slightly from those in the actual world. (For instance, Buck might have purchased the gun thirty-one days and six hours before the shooting, instead of thirty-one days and five hours; the bullet might have lodged in the upper left-hand corner of a wall portrait after passing through Ed in one world, and the upper right-hand corner in another.) For most purposes, we would want to say that these events should be identified across possible worlds; Davidson's criterion, interpreted strictly, prevents that.

In fact, once we establish events as individuals in the world, we have to address questions about their existence, or the existence of their counterparts, across possible worlds in the same way we do for physical objects and other individuals. On the view that events are just a kind of individual, this should follow as much as possible from a counterpart theory for individuals.

In line with the view of individuation of objects outlined in Section 5.1, we individuate objects across possible worlds, and identify counterparts, in line with our interests and purposes, in ways that may vary depending on those interests and purposes. In the classical example of the ship of Theseus, the planks of the ship are replaced one by one, as the ship continues to sail and put to port, until all the planks have been replaced, at which point the discarded planks are reassembled, reconstituting the original ship. In this example, we can identify the physically reconstituted ship assembled from discarded planks as the "same ship" as the original, if sameness of material constitution is most important to us, or, alternatively, we can identify the result of plank-by-plank substitution as the "same ship," if spatiotemporal continuity is most important. We might have different criteria of identification, yielding different results, depending on our interests and purposes. For tax purposes, with concern for the operation of the ship as it sails the sea and puts into port, spatiotemporal continuity might be paramount; for purposes of historical preservation, with concern for putting on museum display the physical ship that was sailed during key historical events, the ship assembled from discarded planks might be paramount.

In a variant of the ship of Theseus example, we can imagine the original ship persisting untouched in the actual world, $w_o$, while the process of substituting planks one by one, and then physically reassembling the discarded planks, all happens in a world $w'$, distinct from $w_o$. At a given point in time after the reassembly of the planks is complete in $w'$, we can ask which ship in $w'$ is the counterpart of the ship in the actual world: the one that has maintained spatiotemporal continuity with what was previously the unambiguous counterpart of the actual ship, and which has (we will assume) continued to sail missions and put into port in $w'$, playing the role of the actual ship, as its planks were replaced one by one, or the museum ship reassembled from the discarded planks? The answer depends on our criteria of identification, which in turn, depend on our interests and purposes.

We can pose similar problems for the individuation and identification of events across possible worlds. Suppose that Max, who works for the CIA, types up and prints a list of CIA agents who work abroad, along with their contacts. Then he deliberately and knowingly arranges to meet with Maria, whom he knows to be the agent of a hostile power, on the pier, one night at midnight, to give her the list, thus betraying his country. In another world, $w'$, Max types up the same list, and meets the same person Maria at the same spot on the same pier at the same time, and gives her the list, but in $w'$, Max knows that Maria is a friendly double-agent, and by giving Maria the list, he is making it possible for a warning to be sent to the agents abroad and their contacts, thus preventing them from being compromised and arrested. Ten minutes later, sitting in his office, Max changes his mind and with the press of a key on his computer keyboard, sends the list to the agent of a hostile power. Which of the two events in $w'$ is the counterpart of the event in the actual world? If the physical execution of the action is designated as criterial, then the counterpart event will be that of Max meeting Maria on the pier at midnight, and handing her the paper with the list of names printed out on it. That might be identified as the "same event" if we are tracking how Max caught a chest cold that aggravated his acute bronchitis and caused him to die two days later (in both $w_o$ and $w'$). If the functional significance of the event and its ramifications in the spy world are designated as criterial, then the physically quite different event of Max pressing the button on his computer keyboard in his warm, cozy office is the counterpart event.

Thus, identifying counterparts of individuals or events across possible worlds involves specification of criteria of identity, dependent on our interests and purposes.

## 5.3 Stative attitude ascriptions and neo-Davidsonian representations

The neo-Davidsonian representation of the eventive sentences in Section 5.1 can be extended to the analysis of stative sentences such as (23a) and (24a) given in (23b) and (24b), as modifying a state introduced by the state predicate; see Parsons (1990), Higginbotham (2000), Landman (2000), Mittwoch (2005).

(23) a. The statue stands in the park.
b. $\exists s_1$ [stand $(s_1) \wedge$ THEME $(x_5, s_1) \wedge$ in $(x_5, x_{17}, s_1)$]

(24) a. Alex loves New Orleans passionately.
b. $\exists s_2$ [love $(s_2) \wedge$ EXP $(x_3, s_2) \wedge$ THEME $(x_7, s_2) \wedge$ passionate $(s_2)$]

These representations capture basic properties of adverbial modification, most prominently those identified by Parsons (1990) which Katz (2008) calls the properties of Permutation, Drop, and Non-Entailment. These are illustrated for action sentences in (25): they are the properties that (25a) is equivalent to its counterpart with the modifiers reversed (Permutation); that (25a) entails each of (25b) and (25c) (Drop); and that (25b,c) do not jointly entail (25a) (Non-Entailment), since (25b,c) might be true by virtue of the existence of two distinct events, one in which Bill stabbed the tabletop violently, but with an axe, and one in which he stabbed the tabletop gently with a knife.

(25) a. Bill stabbed the tabletop violently with a knife.
b. Bill stabbed the tabletop violently.
c. Bill stabbed the tabletop with a knife.

Likewise for states, in (26) (see Parsons [1990: 195], Landman [2000: 15–17], Mittwoch [2005]), where the two modifiers in (26a) can be permuted with no change in truth conditions (*with reservations, but passionately*); (26a) entails (26b) and (26c); and Non-Entailment is exhibited in application to a situation in which (26b) was true only before Hurricane Katrina, and (26c) was true only after Katrina.

(26) a. Sam loved New Orleans passionately, but with reservations.
b. Sam loved New Orleans passionately.
c. Sam loved New Orleans, but with reservations.

Katz (2000, 2003, 2008) argues that the extension of the Davidsonian analysis to sentences with state predicates is flawed since it predicts a symmetry between eventive and stative predicates for adverbial modification, nominalization, and a host of other properties, a symmetry that Katz claims does not obtain. Katz (2000) argues that these symmetries break down quite generally; Katz (2003, 2008) argues that each class of stative predicate, and each type of VP-modifier, must be considered in its own right, and that when this is done, a variety of alternative, more satisfactory, analyses of the interpretations of stative predicates and their modifiers are obtained, undercutting the rationale for the extension of the Davidsonian analysis to stative predicates.

One of the general arguments presented by Katz (2000) concerns anaphoric reference. The eventive sentence *Smith stabbed Jones* sets up immediate anaphoric reference to the associated fact in (27a), or to the event in (27b). [Katz's ex. (6)]

(27) a. Smith stabbed Jones. That bothers me.
b. Smith stabbed Jones. It happened at noon.

But there is no symmetric and uncontroversial anaphoric reference to the state introduced by a stative; Katz claims that *it* is a fact anaphor in both (28a,b). [Katz's ex. (8)]

(28) a. Peter is sick. It is worrying his mother.
b. Danny owns a car. It makes it easier for him to get around.

To begin with, it has been established in the literature on reference that facts have relatively low accessibility to anaphoric reference with *it*, as opposed to with *that*, and as opposed to reference to events with *it*.[4] This suggests that *it* refers to something other than a fact in (28a,b). The personal pronoun *it* refers most felicitously to Peter's sickness in (28a), and to the car itself in (28b). (Replacing *it* with *that*, we have felicitous reference to the fact in both examples.) But returning to the point of the argument, Katz is right that there seems to be no reference to the state introduced by the first sentence in (28b). But this is less clear in (28a): can Peter's sickness, which worries his mother, not be construed as his state of sickness?

[4] See Gundel, Hedberg, and Zacharski (1993), Gundel, Borthen and Fretheim (1999), Hegarty, Gundel, and Borthen (2001), Gundel, Hegarty, Borthen (2003), and Hegarty (2003, 2006a, 2007).

Another of the general or global asymmetries presented by Katz (2000) concerns nominalization. Event predicates such as *perform* in (29) [Katz's ex. (12)] have nominalizations that are ambiguous between what Katz calls a "manner" reading, and a fact reading; such nominalizations include the derived nominal subject in (29a) and the "ing-of" gerund in (29b). But the "POSS-ing" gerund (in Abney's [1987] terms) in (29c) fails to exhibit this ambiguity – it has only the "fact" reading, and lacks the "manner" reading.

(29) a. John's performance of the song bothered Mary.
b. John's performing of the song bothered Mary.
c. John's performing the song bothered Mary.

In contrast, stative predicates, as in (30), do not have nominalizations that support a "manner" reading. [Katz's ex. (15)–(16).]

(30) a. Tina's ownership of the house bothered Jeff.
b. John's belief in free-market capitalism bothered Mary.
c. Richard's knowledge of French delighted his girlfriend.
d. Richard's knowledge of French disappointed his teacher.

We get degree readings in (30c,d), but nothing parallel to the "manner" readings of (29a,b).

First, a quibble: the "manner" readings of (29a,b), with event predicates, really seem to be event readings: it is some aspect of the event itself, something about how it transpired, which bothers Mary. So the nominalized subjects in (29a,b) exhibit an ambiguity between describing or denoting an event (or an aspect of an event), and describing or denoting a fact. Katz claims that the nominalized subjects in (30) have only fact-describing or fact-denoting readings. This observation seems to be correct, and supports Katz's claim that stative predicates such as *own* do not have associated neo-Davidsonian state variables. Across a wide range of stative predicates, we are in accord with Katz on this point.

Turn now to the stative predicates that are of interest for the present work, stative ascriptions of belief and other propositional attitudes with *that*-clause complements. The discussion of Chapter 3 showed how these can, in fact, be interpreted without a Davidsonian eventuality variable for a state of belief or other attitude state. However, this will be made clearer in the next chapter, through contrast to uses in which the same ascriptions, or similar ones, are used to characterize a state of belief or other attitude state. Thus, for these predicates, we will conclude that Katz is half right. Most of the remainder of the present chapter is devoted to examining Katz's arguments as they apply to belief ascriptions, in contrast to how they apply to other stative predicates, in order to get a sense of whether there is a possibility that predicates of belief and other stative propositional attitudes can, in fact, be used to characterize a state of the experiencer subject. Admitting from Chapter 3 that such ascriptions can be interpreted without a Davidsonian eventuality variable, the question is whether they necessarily lack, or are excluded from, an interpretation as characterizing an eventuality.

There is a candidate for a state-denoting reading of the nominalized subject in (31), analogous to the event-denoting interpretations (not the fact-denoting interpretations) of the subjects in (29a,b).

(31) John's belief that Mary is a murderer is eating away at him.

Whether or not we can say that the fact that John has this belief is what is eating away at him (we would probably really be saying that it is the belief itself which is doing the eating), the most natural interpretation of (31) is that it is the manner in which John indulges in his belief, or his emotional response to the belief content, which is eating away at him; just as on the event-denoting interpretation of (29a,b), it is some aspect of the event itself (not the fact that it happens) that bothers Mary.

So why is a corresponding state-denoting interpretation not more generally available in examples such as (30)? The fact that the effect of the eventuality denoted by the complex nominal in (31) is on John, the expressed experiencer of that state, and not on a third party, seems to be relevant.[5] But even if we make *Tina* the object of *bothered* in (30a), *Tina's ownership of the house bothered her*, it seems that it is the fact of Tina's ownership of the house that bothered Tina, and other events associated with her obligations as owner, not the state or eventuality of her ownership itself. If we follow Davidson in assuming that causal efficacy is a major motivation for assuming that events (and by extension, other eventualities) exist, and extend this principle to states, then the lack of causal efficacy for the states described in (30) would be grounds to assume that there is no neo-Davidsonian variable associated with the sentences in (30), though there may be one for the sentence in (31). This suggests that state-denoting interpretations of nominalizations such as in (30) and (31) are not absent in general or in principle, though they may be absent with particular verbs, in particular examples. This would also explain the failure of state predicates to occur as the main predicate in perception complements, the next general asymmetry noted by Katz (2000), as in (32) versus (33). [Katz's exx. (31) versus (44)].

(32) John saw Maria perform the song.

(33) a. * John saw Mary own a car.
b. * John saw the socks sit on the floor.

[5] I am indebted to Jerry Weltman for discussion of this point.

Katz (2000) argued that it is dubious to posit stative eventuality variables by symmetry with Davidsonian events in active sentences, since there are significant asymmetries in semantic behavior. Katz (2008) proceeds by examining arguments for the existence of stative eventuality variables given in the literature. He argues that temporal modifiers analyzed along the lines of **for-a-long-time** (*e*) and **at-midnight** (*e*) on the neo-Davidsonian approach can be analyzed just as well, if not better, as predicates of times. And he argues that state predicate modifiers such as *completely* (e.g., in *John loves Mary completely*) or *well* (e.g., in *John knows French well*) are better analyzed as degree modifiers. Furthermore, Katz notes that stative predicates exhibit a high degree of lexical selectivity for particular modifiers, with variation in this selectivity even among stative predicates of the same semantic type, implicating a lexical (or distributed morphological) analysis of them. (E.g., *John knows French well* and *John loves Mary completely*, versus # *John knows French completely* and # *John loves Mary well*). With this in mind, we can understand why few, if any, modifiers of state predicates exhibit the three properties of Permutation, Drop, and Non-Entailment that are characteristic of sentences with event predicates. Typical of this failure is (34) [Katz's ex. (28)], where the two modifiers in (34a) can indeed be interchanged without change of truth value (Permutation), and (34a) does entail both of (34b) and (34c) (Drop), but (34b) and (34c) do jointly entail (34a), so that Non-Entailment fails.

(34) a. John loves Mary quietly with all his heart.
b. John loves Mary quietly.
c. John loves Mary with all his heart.

Katz argues persuasively that temporal and aspectual modifiers do not provide good evidence of a neo-Davidsonian variable for any predicate, if tense and aspect are interpreted in terms of times and time intervals. His argument for manner modifiers is less straightforward. He analyzes (35) as making an implicit generic statement about events that mark John's behavior toward Mary, asserting that they are characterized by passion.

(35) John loves Mary passionately.

Whether or not this is adequate for the analysis of the contribution of the adverb *passionately* to the interpretation of (35), it does not seem adequate for the analysis of the contribution of this adverbial, and others like it, to the interpretation of belief ascriptions such as (36), the object of interest here, or more generally, ascriptions of the form given in (37), for S a subordinate clause.

(36) Alex passionately believes that Mary is innocent of Bill's murder.

(37) Alex believes that S.

Adverbial manner modifiers such as *passionately* still have some claim to provide evidence for a neo-Davidsonian variable when they are used to modify (37) since their interpretive contribution seems to be best rendered as a relation of the form passionately ($x_3$, $e_4$), between an experiencer subject and an event or state.

To probe this idea, consider the wider range of adverbial modifiers in (38) and how they would be interpreted as modifiers of the main clause VP in (37) (including *passionately* for completeness); as included in this list, *tentatively* characterizes Alex's manner, not his degree of belief, *shamefully* means that Alex feels shame in believing S (e.g., in *Alex shamefully believes that he is responsible for the team's defeat*), as distinct from the speaker-oriented use to characterize the speaker's judgment of the experiencer subject of the ascription, and likewise, *perversely* is included for its use to assert that Alex himself is perverse in holding the belief that S, not as a speaker comment on the fact that Alex has the belief ascription.

(38) *passionately, wholeheartedly, sincerely, honestly, angrily, happily, sadly, triumphantly, stubbornly, obstinately, obdurately, confidently, arrogantly, humbly, meekly, excitedly, placidly, calmly, coldheartedly, calculatingly, rationally, clearsightedly, confusedly, perversely, crazily, frantically, desperately, indignantly, secretly, furtively, conspicuously, conditionally, guardedly, tentatively, shamelessly, shamefully*

Some uses of some of these in (37) might be argued to modify associated external events (e.g., *conspicuously, furtively*), rather than the internal state of Alex as the experiencer in (37), but most uses of these in (37) seem to modify Alex in relation to his belief state, or the manner of his belief. They all have corresponding adjectives that can head an Adjective Phrase that fits into the paradigm in (39) in a way that paraphrases (37) as modified by the corresponding Adverb from (38); for example, if Alex placidly/angrily/arrogantly/perversely/stubbornly/confidently believes that Mary is innocent of Bill's murder, then Alex is placid/angry/arrogant/perverse/stubborn/confident in his belief that Mary is innocent of Bill's murder.

(39) Alex is AP in his belief that S.

It might be said that the adverbs in (38) merely denote a property of the experiencer, Alex. But then (37) modified by one of the VP-adverbs in (38) would be interpreted as in (40).

(40) believe $(x_1, \Phi)$ $\wedge$ placid/angry/arrogant/perverse/stubborn/confident $(x_1)$

By itself, this is inadequate since it doesn't tie Alex's anger, arrogance, placidity, perversity, stubbornness, or confidence in any way to his belief. We don't just want to say that Alex has a belief that S, and he is passionate (or perverse, or arrogant, etc.); we want to capture the idea that Alex is passionate (or perverse or arrogant) in that belief, which can correlate with degree of emotion (in some cases), or manifest itself as the way in which Alex persists in the belief (in other cases), but which is not reducible simply to degree modification or a characterization of related external events. Consider, for example, *Alex arrogantly believes that he is the greatest novelist in the world, even though he is always humble in demeanor*. This can be used to express the judgment that Alex is arrogant or presumptuous in having the belief in question. We could analyze this as saying that Alex is arrogant *because* he believes he is the greatest novelist in the world, or arrogant *in that* he believes he is the greatest novelist in the world. Either of these serves to underscore the incompleteness (at best) of the interpretation in (40). But this doesn't, by itself, suffice to show that (40) is incomplete through lack of an eventuality modified by the adverbial along Davidsonian lines by addition of a term "arrogantly ($a_5$, $e_4$)" relating Alex to a belief state; this would be one way to explicate the relation between Alex and the content of the complement clause that makes it a belief relation, but it may not be the only, or best, way of doing this.

Evidence that points more specifically toward a neo-Davidsonian variable associated with some instances of stative attitude ascriptions with predicates such as *believe* and *think* (on the stative use of *think*) is the pronominal reference in examples such as (41)–(43), where each (a) example is the utterance of a speaker, and the (b) examples present three options for a follow-up utterance, either by that speaker, or by an addressee of the (a) utterance.

(41) a. Alex believes that Mary murdered Bill.
b. i. That began/started last March.
ii. That has lasted long enough.
iii. The forensic pathologist's report will put a stop to that.

(42) a. Alex thinks that everyone is talking about him.
b. i. That began/started last month.
ii. That has lasted a long time.
iii. His therapist will put a stop to that.

(43) a. Billy wants to go to Disneyland.
b. i. That began/started last week.
ii. That has lasted a long time.
iii. (mean Dad) I'll put a stop to that.

The predicates of initiation, persistence, and cessation that take the demonstrative *that* as their argument in the (b) examples do not sensibly or comfortably apply to facts. And although we might follow Katz in taking temporal adverbials such as *at midnight* to be predicates of time intervals, it does not seem sensible to interpret (41b-iii), for example, as asserting that the pathologist's report will put a stop to a time interval. It is more straightforward for *that* to refer to Alex's belief or belief state in (41b) and (42b), and Billy's volition or volitional state in (43b). When we use responses such as (41–43b) in a context where the antecedent of *that* is to be retrieved from a sentence with an indisputably stative predicate, we obtain examples such as those in (44) and (45).

(44) a. Mary owns the car.
  b. i. That began last March.
    ii. That has lasted a long time.
    iii. My lawyer will put a stop to that.

(45) a. The socks sit on the floor.
  b. i. That began this morning.
    ii. That has lasted long enough.
    iii. Billy's mom will put a stop to that.

The use of *that* to refer to the states described by the (a) sentences is distinctly odd in (44b-i) and (44b-ii), and likewise in (45b-i) and (45b-ii). In (44b-iii), we can readily understand that the lawyer might find a legal means to nullify Mary's ownership of the car, or to transfer the ownership to another person, but "put a stop to that" seems at best a colorful way to put this. If we take it literally, the speaker's lawyer is understood to be not actually stopping Mary's ownership of the car, but something more like putting a stop to an aggressive campaign by Mary to acquire ownership of articles in which the speaker has an interest (e.g., in a messy divorce case). In (45b-iii), we understand that Billy's mom will put a stop to events of Billy leaving socks on the floor, or Billy's habit of leaving socks on the floor, not to a current state in which the socks sit on the floor.

So even if we accept Katz's arguments that the bulk of stative predicates do not modify a Davidsonian state, it seems that stative attitude predicates can introduce an associated state, as an interpretive option. A plausible reason for this is that we can interpret these ascriptions as describing attitude states that can engage in causal interactions in examples such as (41)–(43); such interactions underlie Davidson's conception of an event. The formal concomitant for why these predicates can behave in the ways noted, but (at least most) other stative predicates don't is that these predicates can have an associated Davidsonian state variable in their argument structure, which makes them subject to forms of modification and anaphoric reference in exactly the way that sentences with eventive predicates are.

Thus, for the interpretation involved in (36) as well as (41)–(43), there seems to be reason to posit a belief state $e_4$, which is experienced by Alex (or Billy), such that the relational condition "passionately ($x_3$, $e_4$)" holds in (36), alongside a condition explicating the relation between Alex and the content of the complement clause that makes it a belief relation. This implicates a representation for (36) along the lines sketched in (46); the condition relating $e_4$ to the content of the complement clause is the topic of investigation in Chapter 6.

(46) $\exists e_4$ [EXP ($x_3$, $e_4$) ∧ (a condition relating $e_4$ to content of S) ∧ passionate/perverse/stubborn/confident ($x_3$, $e_4$)]

The characteristics of Alex's belief state that are captured by the adverbials would not necessarily have to manifest themselves in associated acts or events that could be characterized as perverse, stubborn, confident, or arrogant. These adjectives, and the corresponding adverbials in (38), could be used to characterize the state of Alex in holding the belief.

To give due credit to Katz's view, other adverbials in (38), upon modifying the main clause in (37), would most plausibly be interpreted as modifying one or more associated events that mark Alex's behavior. For example, *Alex calculatingly believes that S* might be true by virtue of one event of calculation that brought Alex to his belief. So only a subset of the verbs in (38) are being presented here as ones that would modify (37) (under one interpretation) in a way that is best characterized as describing Alex in relation to his state of belief.

There are also PPs that can serve as modifiers of the main clause VP in (37), including those in (47).

(47) *with mounting horror, with increasing indignation, with unmitigated gall/arrogance.*

When they are not antonyms, or otherwise semantically incompatible, the modifiers in (38) can be combined with one another, or with those in (47), exhibiting the properties of Permutation, Drop, and Non-Entailment, as in (48). Similar combinations are listed in (49).[6]

(48) Alex conditionally but wholeheartedly believed that Bill should be executed.

(49) frantically and perversely, secretly and shamefully, conditionally but confidently, confusedly and desperately, calmly and calculatingly, excitedly and confusedly, guardedly but stubbornly.

[6] As with active predicates, Non-Entailment may hinge on evaluation at different times. Just as (25a,b) do not entail (25a) if the violent stabbing happened on one occasion (with an axe) and the stabbing with a knife happened on another occasion, so too, *Alex wholeheartedly believed that Bill should be executed* could be true in a situation in which his belief is unconditional, until a doubt arises, after which *Alex conditionally believed that Bill should be executed* is true; indeed, in such a case, (48) is never true, so Non-Entailment is obtained. It might be argued that such "time-shifting" should not be permitted in evaluating Non-Entailment, but if so, it should be disallowed for active as well as stative predicates, which would nullify its utility as a criterion for the existence of a Davidsonian variable.

With this background, consider some other VP-modifiers, and how they are interpreted when they modify the main clause VP in (37). Consider the examples in (50).

(50) a. Alex really believes that Bill stole the artifact.
b. Alex strongly believes that Bill stole the artifact.
c. Alex firmly believes that Bill stole the artifact.
d. Alex passionately/stubbornly/perversely/arrogantly/... believes that Bill stole the artifact.

These are all quite distinct. In (50a), following Romero and Han (2004), *really* can be interpreted as a degree modifier, or as a verum operator asserting that it is really true that Alex has the cited belief. See Katz (2008) and Kennedy (1999b, 2001) for an analysis of the degree modifier interpretation. In (50b), *strongly* can also be interpreted as a degree modifier. In (50c), *firmly* is more likely than *strongly* to be a modifier not of the degree of belief, but of its soundness or its evidential basis, as indicating that the belief has what Alex considers to be a firm evidential standing. In (50d), already considered, the adverbial describes the state of Alex in holding the belief. To illustrate these differences: I firmly believe that it is sunny outside today (I can see the sun), but it would not be correct to say that I passionately believe that it is sunny out today. It is even odd to say that I strongly believe it, unless the point is being challenged – the evidential basis for it being sunny is so firm that I hardly need a high degree of engagement or commitment to affirm that it is sunny out now, or for an ascription ascribing that belief to me to be true.

Katz (2003) offers a different sort of argument for the lack of a Davidsonian state variable than that given by Katz (2008). Katz (2003) notes that adverbs such as *quickly* modify active predicates, but not statives, as shown in (51).

(51) a. John kissed Mary quickly.
b. * John loved Mary quickly.

But there are no adverbs that have the complementary distribution, in that they can modify statives, but not actives. That is, there are no adverbs that fit the paradigm in (52).[7]

(52) a. John loved Mary ADVERB.
b. * John kissed Mary ADVERB.

Katz explains this on the grounds that stative predicates lack a Davidsonian event variable that supports the interpretation of active adverbials such as *quickly*. As a result, *quickly* cannot modify a stative predicate such as *love* in (51b). However,

[7] Ann Mulkern (p.c.) notes that the adverb *steadfastly* seems, in fact, to fit this paradigm.

an adverb that can be interpreted as modifying the stative predicate in (52) does so by modifying degrees or other features of the interpretation of the predicate, and all such features can be realized on active predicates as well. There is, therefore, a Stative Adverb Gap, best explained on the grounds that stative predicates are distinguished by the lack of a Davidsonian event variable.

But Sara Monti notes (p.c.) that, in exactly the same way in which these facts manifest a Stative Adverb Gap, there is an activity adverb gap. Any adverbs that felicitously modify activity predicates such as *walk*, *run*, and *work* either modify accomplishment predicates (*climb the mountain quickly/steadily/...*) or they modify statives (*sit on the shelf for an hour/continuously/...*). So there are no adverbials that can only be used to modify activity predicates. These facts do not really show that stative predicates are special; rather, they indicate the need for a more explicit ontologically grounded account of the eventualities in the Vendler classes of states, activities, accomplishments, and achievements, and thus, a corresponding ontologically based account of adverbial modification of predicates that are used to describe such eventualities.

In Chapter 6, a theory of the interpretation of propositional attitude ascriptions will be developed, which will provide a better foundation for the claims made here. In particular, Chapter 6 will present evidence that there are two interpretations of a sentence like *Alex believes that Mary murdered Bill* that are explicable in relation to one another if it is supposed that one of them introduces and/or qualifies a state of belief experienced by the experiencer subject of the ascription, and that the other doesn't.

### *5.3.1 Conclusion*

Katz (2008) makes a good case that stative predicates should not automatically be accorded a neo-Davidsonian event/state variable, and that such a representation is not appropriate for most stative predicates. Katz's discussion also establishes an important methodological point: that different classes of stative predicates, and different types of modifiers of stative predicates, have their own semantic properties, and should be considered carefully in their own right. We can accept these points, even as we advance evidence that some stative attitude predicates are sometimes used to describe a state of the experiencer subject, and thus should be accorded a neo-Davidsonian eventuality variable for the attitude state in the interpretation of sentences with that usage.

On the account developed in the chapters to follow, even for stative attitude predicates for which we have argued here that we should invoke a neo-Davidsonian variable, this interpretation exists alongside ones without this variable, as was developed in Chapters 4 and 5. And this latter interpretation, based solely on quantification over possible worlds, is, if anything, the more natural and common one with some attitude predicates.

# 6 Full attitude ascriptions: individuating mental states through propositional content

## 6.1 Something is missing

Chapter 3 examined the prospects for interpreting a belief ascription such as that in (1) in terms of quantification over a doxastic context, $\cap D_3(w)$, a set of doxastic alternatives for the experiencer subject.

(1) Alex believes [that Bill stole the artifact].

It was argued in that chapter that ascriptions would have to be acknowledged at varying degrees of strength, and that quantification over the doxastic context should be modulated by an ordering source $\Omega_3(w)$ on doxastic alternatives. This led to the adaptation of Kratzer's analysis of modal interpretation to this purpose, representing ascriptions with *be sure* and *be certain* with the quantificational schema of modal necessity, rendering ascriptions with *believe* and *expect* with the quantificational schema of weak necessity (SHOULD), and representing ascriptions with *suspect* and *imagine* as the dual of weak necessity, which is below the midpoint of the scale of doxastic attitudes, but stronger than the dual of *be sure* (which is not lexicalized by any predicate of English). This led to truth conditions such as those in (2) for the top of the scale, *be sure*, where $\mathbf{S}_1$ is the phrase marker of the complement clause in (1) .

(2) $[[\textbf{Alex is sure that } \mathbf{S}_1]]^{c,w} = 1$ iff

a. $\forall u \in \cap D_3(w)$: $(\exists v \in \cap D_3(w)$: $v \leq_{\Omega 3(w)} u)$ $(\forall z \in \cap D_3(w)$: $z \leq_{\Omega 3(w)} v)$ $[[\mathbf{S}_1]]^{z} = 1$,
iff

b. every sequence of worlds in $\cap D_3(w)$ that is weakly ordered by $\Omega_3(w)$ has an $\mathbf{S}_1$–uniform tail, $\text{iff}_{\text{def}}$

c. MUST $(\mathbf{S}_1 \mid D_3(w) \parallel \Omega_3(w))$.

But this was predicated on the assumption that the doxastic base $D_3(w)$ is consistent. If it happens to be inconsistent, then the doxastic context set $\cap D_3(w)$ is empty, and the truth conditions in (2) are not viable. So instead of quantifying over the doxastic context, the doxastic base $D_3(w)$ was intersected with the ordering source to form the derived ordering source, $\Omega{:}D_3(w)$. Quantifying over all possible worlds **W** with this derived ordering source, the import of the interpretation emerges from what happens as possibilities become progressively more highly ranked according to propositions common to Alex's doxastic base and what was formerly the doxastic ordering source. This led to the truth conditions for *be sure* in (3), and for the other points of the scale in (4) and (5).

(3) [[**Alex is sure that** $\mathbf{S_1}$]] $^{c,w}$ = 1 iff

a. $\forall u \in \mathbf{W}$: $(\exists v \in \mathbf{W}: v \leq_{\Omega:D3(w)} u)$ $(\forall z \in \mathbf{W}: z \leq_{\Omega:D3(w)} v)$ $[[\mathbf{S_1}]]^{z} = 1$, iff

b. every sequence of worlds in **W** that is weakly ordered by $\Omega{:}D_3(w)$ has an $\mathbf{S_1}$–uniform tail, $\text{iff}_{\text{def}}$

c. MUST $(\mathbf{S_1} \parallel \Omega{:}D_3(w))$.

(4) [[**Alex believes that** $\mathbf{S_1}$]] $^{c,w}$ = 1 iff SHOULD $(\mathbf{S_1} \parallel \Omega{:}D_3(w))$

(5) [[**Alex suspects that** $\mathbf{S_1}$]] $^{c,w}$ = 1 iff $\text{SHOULD}^{\perp}$ $(\mathbf{S_1} \parallel \Omega{:}D_3(w))$

There is a dimension to the meaning of some ascriptions of belief and other attitudes that is left out by the truth conditions in (3)–(5). It is highlighted by the pair in (6), from Prince (1976: 417), either of which could be appropriately uttered by someone entering a room of the gothic Old State Capitol building in Baton Rouge, Louisiana (reputed to be haunted).[1]

(6) a. I'm imagining that there are ghosts in this room.

b. I imagine that there are ghosts in this room.

The ascription in (6a) describes an event of active imagining of ghosts, involving images or feelings characterized by the propositional content of the complement clause, while (6b) asserts that the speaker harbors a suspicion or expectation that there are ghosts in the room, and thus is entertaining not mental images or feelings, but just the propositional content itself. The contrast obtains as well without the progressive. Compare the active imagining reported in (7a,b) with Sue's response in (8), which is a belief report with weak force.

(7) a. Alex fearfully imagined that he would be mugged on his way home.

b. Alex fearfully imagines his house being foreclosed on.

[1] This distinction was also noted by Lindholm (1969) and by Portner (1992: 88–89). Stephenson (2007: 168) also notes this distinction, but then explicitly puts it aside to focus on paradoxes with the mental image interpretation, particularly in imperatives embedding a modal operator; she also focuses on a different distinction, between the subjective or internal image involved in the interpretation of *I imagined skiing down the hill*, versus the more external or objective imaging involved in the interpretation of *I imagined that I was skiing down the hill*. The present discussion abstracts away from the finer-grained distinction between the latter two sentences discussed by Stephenson.

(8) Max: When do you think your office can finish the environmental impact report?
Sue: I imagine we can get that done by Friday.

In (6a) and (7a,b), used as described, the speaker imputes a mental state or event to the experiencer, and uses the content of the complement clause to characterize that event or state. In (6b) and Sue's response in (8), there are no feelings or mental images reported; the speaker uses the complement clause just to characterize possibilities for the future and their likelihood of being realized in the actual world.

In order to address this distinction, we need to revisit the individuation of objects and events across possible worlds, from Chapter 5, and extend the interpretations of attitude ascriptions developed in Chapter 3 to new cases where the propositional content invoked in the interpretation of the ascription plays a role in the individuation of the mental state of the attitude experiencer across possible worlds.

## 6.2 Individuating and identifying by content

The indefinite nominal in (9a) could be used to introduce an object, a particular piece of silk ribbon, into the discourse. The definite in (9b) could be used to refer to a silk ribbon that is (at least) activated for the addressee, in the sense of Gundel, Hedberg, and Zacharski (1993), prior to its mention in (9b).[2]

(9) a. The rolled up parchment was tied with a silk ribbon.
b. This silk ribbon was tied around the rolled up parchment.

The indefinite in (9a) would be mapped to a new discourse entity $x_7$, restricted by the conditions ribbon ($x_7$) and silk ($x_7$). The definite in (9b) would refer to a familiar entity $x_7$ satisfying those conditions. From this point on, the discussion will address the case of the demonstrative definite.

If the nominal *this silk ribbon* denotes the individual $x_7$ in $w$, we can ask whether it would denote the same individual in a world $v$ distinct from $w$. As discussed in Chapter 5, this depends on the interests and purposes of participants in the discourse. Suppose that there is a ribbon in $v$ that shares the spatio-temporal history of $x_7$ in $w$, and in particular, has been used for the same purposes, except that it is not made of silk, and suppose that there is otherwise no plausible counterpart of $x_7$ in $v$. Is this the same individual? We would

[2] The use of *this silk ribbon* in (9b) is to be distinguished from its use where no prior acquaintance with the ribbon on the part of the address is assumed, as in, *I just went past the firehouse; there was this black silk ribbon draped across the front to memorialize a fallen firefighter*. The latter has the lower cognitive status of "Referential" in the hierarchy of Gundel, Hedberg, and Zacharski (1993).

probably say so, if its history, function, and appearance were more important than the material that composes it. This is reflected in the fact that when we use the nominal in a modal sentence, we can count as accessible worlds ones in which the best counterpart of the ribbon, sharing its spatio-temporal history and uses, is not made of silk. Consider the examples in (10).

(10) a. The parchment must/might have been tied with this silk ribbon.
b. This silk ribbon must/might have tied the parchment roll.

The individuation conditions are reflected in the ordering source for these modal statements. If the material of composition of the ribbon is of paramount importance, then any possibility in which the best counterpart to the ribbon is not composed of silk will be ranked as relatively inaccessible, a poor fit for the modal generalization. If the material of composition is of little importance in the discourse, the ordering source will make little distinction among worlds based on this factor. In the latter case, either of (10a,b) could be true even though they include in the modal quantification some worlds in which the counterpart of the ribbon shared its spatiotemporal history, but was not made of silk. The adjective *silk* would not apply to the ribbon in these worlds, but that would just mean that these are cases of reference with a nominal that is not fully or correctly descriptive, a common phenomenon, discussed extensively in the literature, especially since Donnellan (1966); for current approaches, see Reimer and Bezeuidenhout (2004).

The moral to be drawn from this is that when reference to an object can be obtained across possible worlds, including ones in which the description might not fully apply, then those worlds in which the description would not apply to the counterpart of the object can nevertheless be included in the domain of modal quantification supporting the truth of a modal statement, if the properties denoted by the inapt description are not critical to the discourse.

The main thesis here is that this is what happens when an entity such as a picture, a story, or a belief is characterized in terms of its content. Consider the example in (11).

(11) a. This portrait of Benjamin Franklin must/might have hung between the two windows.
b. They must/might have hung this portrait of Benjamin Franklin between the two windows.

Would the modal quantification in (11) include worlds in which the best counterpart of the portrait was a portrait with the very same spatio-temporal history as the actual one, but one that happened to be a portrait of Lavoisier? That depends on the purposes and interests of the participants in the discourse. If the portrait and other objects are being addressed simply as furnishing, with no regard for the content or historical significance of who is depicted in the portrait, then the ordering source would not strongly favor possibilities in which the portrait was of Franklin, but instead, might favor possibilities in which the portrait counted as the "same item of furniture," by virtue of a shared or similar spatiotemporal history, as furniture. The modal quantification would then include possible worlds in which the portrait was of someone else, and count it as "the same portrait" for the purpose of a statement about what sort of furnishings the room had. In more typical discourse contexts, where the fact that the portrait is of Benjamin Franklin is critical to the purposes and interests of the conversation (e.g., the portrait is to be put up for auction), worlds in which the same canvas was used for a portrait of an obscure contemporary of Franklin's, or even a famous contemporary such as Lavoisier, and subsequently underwent the same spatiotemporal journey, would not be included. In such cases, a world with a portrait of the obscure gentleman in place of the given portrait of Benjamin Franklin could not count as accessible and contribute to the truth of (11); in fact, if otherwise suitable for the modal statement, the existence of such a world could make the necessity modal statements in (11) false. In such a case, the content of the portrait would be the major factor (even the decisive one) in identifying it in other possible worlds than the world of evaluation.

Next, consider (12), said of a particular story with a certain provenance (say, a story about a troll who was captured in the woods during medieval times and held in the dungeon of the castle), to say that the owner of the castle must or might have regaled his visitors with it.

(12) This story must/might have been told to visitors to the castle for years.

Would the modal quantification in (12) include worlds in which a story with the same provenance but very different content was told? (Say, a story about an explorer who discovers an island in the Pacific, with no moral or narrative parallels to the story about the troll?) The answer is clearly no. The alternate story with substantially different content would definitely not count as a counterpart of the story about the troll in a way that would make a world in which it occurs in place of the troll story count for the truth of the modal statement in (12). But a world in which a minor variant of the troll story was told could contribute to the truth of (12), if the variant is close enough to the story in the actual world to count as the "same story" for the purposes of the statement. Again, these distinctions will be reflected in the ordering source.

Examples (11) and (12) show that content can be (and often is) essential to a discourse entity such as a painting or a story. In such cases, a nominal that refers to such an entity in the world of evaluation must refer to an entity with the same content in other worlds, or content that is close enough to count as the same. The material composition of the ribbon in (10) would usually not be essential in this way. As discussed here and in Chapter 5, a counterpart to an inanimate physical object in another world might have the same material composition as the object in the actual world (or any given world), or it might not, and this typically depends on the interests and purposes with which objects are being individuated.

Although we refer to counterparts in this discussion, we don't need to adopt a full counterpart theory of cross-world identification. We can assume that an individual exists as a cross-world entity in those worlds in which we find suitable counterparts, and differences among them are not markers of distinct individuals, but simply manifestations of variability in the properties an individual has in different worlds. This would permit us to avoid a formal theory of counterparts and counterpart relations.

## 6.3 Two types of attitude ascription

Returning to doxastic attitude ascriptions, the interpretations of (13) using Kratzer's quantificational schemas for modality, given in (3), (4), and (5), and repeated in (14), are in accord with Katz's (2000, 2003, 2008) arguments against variables for Davidsonian eventualities in stative sentences.

(13) Alex suspects/believes/is sure that $S_1$

(14) [[**Alex suspects/believes/is sure that $\mathbf{S_1}$**]] $^{c,\,w} = 1$ iff

a. SHOULD$^{\perp}$ ($\mathbf{S_1}$ || $\Omega$:$D_3(w)$)

b. SHOULD ($\mathbf{S_1}$ || $\Omega$:$D_3(w)$)

c. MUST ($\mathbf{S_1}$ || $\Omega$:$D_3(w)$)

These report on what Alex takes to be the case in the world. For example, the point of Sally's reply in (15) is to comment on the prospects of the passage of the bill.

(15) Sam: What do you think about the bill's prospects?
Sally: I believe the legislature will pass it.

Since the ascription is first person, the experiencer is the speaker, with doxastic base $D_1(w)$, part of which is contained in the doxastic ordering source $\Omega$:$D_1(w)$. If $\mathbf{S_1}$ is the phrase marker for the complement clause in Sally's assertion, her reply is interpreted as asserting the condition, SHOULD ($\mathbf{S_1}$ || $\Omega$:$D_1(w)$), stating, in effect, that $\mathbf{S_1}$ improves in truth in accessible possibilities as they satisfy progressively more propositions in Sally's derived ordering source.

But we have seen examples in which these attitude predicates seem to report on a mental event or state of the experiencer subject. Recall that (7a,b), repeated below, can be used to report on images or prospects that frighten Alex.

(7) a. Alex fearfully imagined that he would be mugged on his way home.
b. Alex fearfully imagines his house being foreclosed on.

And consider examples such as (16), similar to ones discussed in Chapter 5.

(16) Alex believes that Mary murdered the victim, and it is tormenting him.

In (16), the pronoun *it* is used to refer to something with the causal efficacy to torment Alex. This wouldn't have to be an unspecified set of attendant events that accompany his belief; it can instead report that the belief itself, Alex's belief state, through its causal consequences, is what is tormenting him. When Sue makes the report in (17), and Sam responds as shown, and then outlines her plans to make Kim unappealing to Ashley and thereby change Ashley's mind, Sam uses the demonstrative pronoun to refer to Ashley's desire or intention, a causally active mental and emotional state.

(17) Sue: Ashley wants to go out with Kim.
Sam: I'll put a stop to that!

Likewise, the examples in (7) and (16) report that Alex has a mental and emotional state, or is experiencing an internal mental or emotional event. And recall the contention of Chapter 5 that an example like (18) can mean (19), which we should interpret as in (20), positing a belief state $e_2$ of which Alex is the experiencer.

(18) Alex passionately/stubbornly/confidently believes that $S_1$.
(19) Alex is passionate/stubborn/confident in his belief that $S_1$.
(20) $\exists e_2$ [EXP $(x_3, e_2)$ $\wedge$ (a condition relating $e_2$ to content of $S_1$) $\wedge$ passionate/perverse/stubborn/confident $(x_3, e_2)$]

The condition relating $e_2$ to the content of $S_1$ should include the quantificational condition in (14b), SHOULD ($\mathbf{S_1}$ || $\Omega$:$D_3(w)$). But the relation between $e_2$ and the modal quantificational condition is not just their co-occurrence in the world of evaluation, *w*. That, after all, could be just a coincidence. In order for $e_2$ to be the state of belief that $S_1$, it must co-occur with the modal quantificational condition in every world in which $e_2$ exists. It is an essential condition, characterizing $e_2$ across possible worlds, as in (21).

(21) $\exists e_2$ [EXP $(e_2, x_3)$ $\wedge$ [($\forall v \in \mathbf{W}$: EXP $(e_2, x_3)$ in $v$): SHOULD ($\mathbf{S_1}$ || $\Omega$:$D_3(v)$)]]

As discussed at the end of the previous section, we eschew a full-blown counterpart theory for the view that individuals and events identified across possible worlds are cross-world individuals that manifest themselves as the counterparts in different worlds. Thus, to write "EXP ($e_2$, $x_3$) in $v$" in (21) presupposes that $x_3$ is individuated in $v$ and identified with Alex, and likewise, that a mental event or state $e_2$ is individuated in $v$ and identified with the mental state of Alex of which the speaker asserts EXP ($e_2$, $x_3$) in the world of evaluation $w$. This also highlights the fact that the ascription of predicates to an individual, such as EXP ($e_2$, $x_3$) in (21), are implicitly in the world of evaluation, unless specified otherwise. In a fully intensional formulation of the semantics, predicates will have a world argument.

The proposal, then, is that the truth conditions in (14) capture the interpretations of doxastic attitude ascriptions in those cases in which the ascription is about the prospects for $\mathbf{S_1}$ being true in the world of evaluation. In Kratzerian fashion, those prospects can be characterized in terms of the truth conditions of $\mathbf{S_1}$ in possibilities that accord ever more fully with the doxastic base and doxastic ordering source of the experiencer subject. But the truth conditions in (21) are required for those cases in which the ascription is about a mental event of image formation or a mental state with causal properties. As a terminological issue, the latter might be argued to be a mental event, rather than a state. If a heavy base sits in the same spot on a table top for several years and gradually creates a depression in the table top, we can say that the presence of the vase on the table all that time is not purely stative, that it is a slow, inexorable event of wearing down the integrity of the wood underneath, causing a depression. So, too, a causally potent mental condition, such as Max's belief that his sister Mary is a murderer, which eats away at him and ultimately causes disease, might be regarded as a slow event, rather than a state. On these grounds, our conclusion need not controvert Katz's claim that truly stative attitude ascriptions have no associated Davidsonian eventuality. What matters for present purposes is that we acknowledge the need to interpret some doxastic attitude ascriptions as in (14), and some as in (21). Further support for maintaining that distinction will be found in Chapters 7, 8, and 9.

The present proposal was developed out of long-standing engagement with the themes of the present work (see Hegarty [2006a]), which originally had roots in work on reference to facts and propositions introduced by clausal complements (see Hegarty [2003, 2006b]. Nevertheless, there are kindred proposals in the literature coming from other directions. The closest is Hacquard (2006), developed in the course of an investigation of interactions between modality and tense and aspect (which we will return to in Chapter 8). Hacquard (2006: 136–139) interprets a belief report as asserting the existence of an event/state with associated propositional content. In Hacquard's approach, if $s$ is the belief state ascribed to Darcy by (22), then its relation to its content is established through the set $D_4(s)$ of all beliefs Darcy has during $s$, given in (23).

(22) Darcy believes that it is raining.

(23) $D_4(s) = \{p\text{: } p$ is a belief of the experiencer of $s$ in the time interval reported on$\}$

The ascription in (22) asserts the existence of this belief state, with Darcy as experiencer, and asserts that it is raining in all possible worlds compatible with every belief in $D_4(s)$. This is rendered in simplified form in (24), with Hacquard's account of tense/aspect interpretation left out for ease of comparison with the account developed here.

(24) $\exists s$ [believe $(s) \wedge$ EXP $(s, x_4) \wedge [(\forall v \in \cap D_4(s)$: it is raining in $v$]]

One difference between this and (21) is that (21) analyzes away the predicate "believe" in terms of a modal quantificational schema using an ordering source incorporating part of the doxastic base. Another difference is that the present approach does not posit the existence of an eventuality in all attitude ascriptions, distinguishing those that seem to be about a mental event or state from those that merely position the experiencer subject with respect to the circumstances or prospects for the truth of the complement clause. This puts it fundamentally at odds with Hacquard's purpose, which is to use the associated eventuality to account for interactions with tense and aspect. The present approach requires that the interpretation of tense and aspect in the bare attitude ascriptions can be handled in ways that don't involve the postulation of an eventuality or mental state. In this respect, the present work embraces Katz's approach to tense/aspect interpretation. These issues will be addressed further in Chapter 8 since the motive for Hacquard's event/state for attitudes, and the use to which she puts them, will be examined more closely there.

## 6.4 Reports of attitudes of emotion

Ascriptions of bouletic attitudes were addressed in Chapter 4, where, in response to observations and discussion by Heim (1992), it was argued that ascriptions with *want* involve quantification over all possible worlds with a quantificational schema using a joint doxastic-bouletic ordering source. The modal schemas MUST, SHOULD, and SHOULD$^{\perp}$ provide three different strengths of such a condition, varying in how dense the possibilities become as they improve with respect to the ordering source. But we assumed that ascriptions of *want* and other bouletic attitudes have the force of a necessity modal, so we didn't have the full set of motives for using these quantificational schemas that we had with doxastic attitude ascriptions. Among predicates of preference and desire, there are few, if any, cases where we have a series of increasing strength similar to attitude ascriptions of belief with *suspect* < *believe* < *be certain*. Granted, there are, at least, some pairs that differ in strength. These include *like* < *love* (ignoring other differences between these) and possibly *happy* < *delighted*. But the members of these pairs actually seem to differ from one another more in qualities that do not pertain to strength than they do in strength. And many of the prototypical bouletic attitudes (e.g., *want*, *wish*, *hope*, *fear*, *regret*) aren't members of any clear strength series. With these predicates, variations in strength are more likely to be expressed adverbially (e.g., *rather want* < *want* < *very much want*). Thus, we don't have the immediate lexical motive for the use of the modal quantificational schemas that we had with predicates of belief and attitude assertion.

Nevertheless, we adopted the modal quantificational schemas for bouletic attitudes since they gave us the results of Heim's dynamic semantic analysis through mechanisms we have used for other attitude ascriptions, without the need to introduce any new mechanisms involving sets of maximally similar worlds. The exact modal quantificational strength doesn't matter. For example, as noted earlier, *want* does not seem to be the member of a strength series, lexically distinguished in strength from other predicates of the same type. We therefore chose the strength of modal necessity on the view that a *want*-ascription describes an unqualified preference, and that strength modifiers, as in *really want*, or *kind of want*, typically modify the degree of emotional engagement, not the set of possibilities that reflect the bouletic preference. Nothing much hinges on this declaration; *want*-ascriptions could just as easily be interpreted with the force of weak necessity.

It was observed that *want*-ascriptions vary in their realism. Some are very realistic, or must conform to the presuppositions of predicates used in the ascription. This was the case for *I want to get well*, which quantifies over (and talks about) possibilities in which the speaker is sick, as presupposed by the predicate *get well*, even though alternative possibilities in which the speaker is not sick would ordinarily be preferred by the speaker. Also, an ascription such as *I want Bill to be in the room* can be used when the speaker knows that Bill is in the room, to communicate the speaker's preference for that state of affairs over alternatives in which Bill is absent. Other *want*-ascriptions are highly counterfactual, quantifying over more favored possibilities that are clearly counter to what the speaker knows. On these grounds, *want*-ascriptions were interpreted in Chapter 4 as intersecting the bouletic ordering source with some, all, or none of the doxastic base, allowing that the bouletic ordering source may be inconsistent with the doxastic base. (This is on top of the fact that the doxastic base might be internally inconsistent.) The inconsistency between the doxastic base and the bouletic ordering source can be resolved by trimming one or the other, making the ascription more realistic or more counterfactual, according to the purposes of the speaker in the context at hand. In contrast, consider the behavior of *hope* and *fear*, in (25).

(25) a. Alex hopes that Bill is in the room.
b. Alex fears that Bill is in the room.
c. Alex hopes to be in the room.

If we presume that Alex knows that Bill is in the room, or knows that Bill is not in the room, then (25a,b) are infelicitous. If Alex knows that he himself cannot be in the room for an event under discussion, then (25c) is infelicitous; compare with *Alex wishes he could be in the room*. Or if Alex is already in the room, with no hint of any factor that would keep him from staying there for an event that happens in two minutes, then (24c) would also be odd. The complements of *hope* and *fear* should be neither certain nor impossible. If the doxastic base is inconsistent with the bouletic ordering source, a *hope*-ascription or *fear*-ascription is infelicitous. In the analysis adopted here, where we quantify over an ordering source consisting of the bouletic ordering source intersected with some part of the doxastic base, it seems to be a lexical property of *hope* and *fear* that the two sets must be consistent with one another. (Of course, *fear* flips the polarity of the ascription; as worlds become more in accord with Alex's bouletic ordering source in the interpretation of (25b), the negation of the complement clause becomes more ensured.)

Simple attributions of an emotional state, such as those in (26), are interpreted along the lines of (27), as introducing a state of hope, fear, or worry of which Alex is the experiencer.

(26) Alex is hopeful/fearful/worried.
(27) $\exists e_4$ [EXP $(e_4, x_3) \wedge$ hope $(e_4)$/fear $(e_4)$/worry $(e_4)$]

Sentences such as those in (25) likewise report the existence of an emotional state of the experiencer, but in this case the emotional state is directed toward, or characterized by, the content of the complement clause. To help pin down the strength of these ascriptions, consider contrary predicates such as *prefer* and *disprefer*. If Alex prefers that Bill be in the room, it follows that Alex disprefers for Bill to not be in the room. It seems clear that *hope* does not stand in this relation to *fear*. If Alex hopes that Bill is in the room, it does not necessarily follow that Alex fears that Bill is not in the room. But if Alex fears that Bill is in the room, it does seem to follow that he at least hopes that Bill is not in the room. If $\mathbf{S_6}$ is the phrase marker for *Bill is in the room*, then evidently *hope* is weaker with respect to $\mathbf{S_6}$ in (25a) than *fear* is with respect to **not-**$\mathbf{S_6}$ in (25b).

This motivates the suggestion that *hope* has the quantificational strength of SHOULD on the content of its complement clause, while *fear* has the quantificational strength of MUST on the negation of its complement.

If Alex fears that Bill is in the room, then Alex has a state of fear. Fear is causally potent enough that it seems justifiable to posit an associated eventuality, $e_4$, as we did in (27) for the simple ascription of an emotional state in

(26). However, to do justice to the ascription in (25b), we have to do more than characterize $e_4$ as a state of fear. We have to assert that $e_4$ responds to the propositional content of the complement $\mathbf{S_6}$ in (25b). That means that, with circumstances varying across possible worlds $v$ in which Alex has the state $e_4$, there is a bouletic imperative for Alex for **not-$\mathbf{S_6}$** to hold in $v$. And that imperative is realized as the truth of **not-$\mathbf{S_6}$** in possible worlds branching out from $v$, as they become more in accord with Alex's bouletic conditions. To the extent that Alex's doxastic context is consistent, and his bouletic ordering source $\Omega$:$B_3$ is consistent with his doxastic context, we can express the responsiveness of $e_4$ to the propositional content of the complement $\mathbf{S_6}$ with the incomplete formula in (28), the expansion of fear ($e_4$) in (27), where *fear* takes the complement $\mathbf{S_6}$.

(28) $(\forall v \in \cap D_3(v)\text{: EXP }(e_4, x_3)\text{ in }v)\text{: MUST }(\textbf{not-S}_\mathbf{6} \parallel \Omega{:}B_3(v))$

Plugging this expansion of fear ($e_4$) into (27), we derive (29) as the expansion of (27) for (25b).

(29) $\exists e_4\ [\text{EXP }(e_4, x_3) \wedge (\forall v \in \cap D_3(v)\text{: EXP }(e_4, x_3)\text{ in }v)\text{: MUST }(\textbf{not-S}_\mathbf{6} \parallel \Omega{:}B_3(v))]$

Finally, if the doxastic context of Alex is internally inconsistent, or inconsistent with his bouletic ordering source, the ordering source is revised by intersection with some or all of the doxastic base, with conflicts resolved in favor of a more realistic ascription, or a more counterfactual one, depending on which is appropriate. Let the resulting doxastic-bouletic ordering source be $\Omega^*{:}B_3(v))$, where the starred ordering source results from the contextually dependent process of resolving inconsistency. Then (29) is revised as in (30).

(30) $\exists e_4\ [\text{EXP }(e_4, x_3) \wedge$
$(\forall v \in \mathbf{W}\text{: EXP }(e_4, x_3)\text{ in }v)\text{: MUST }(\textbf{not-S}_\mathbf{6} \parallel \Omega^*{:}B_3(v))]$

A parallel representation for the *hope*-ascription in (25a) would be that in (31), taking account of the strength difference discussed earlier.

(31) $\exists e_4\ [\text{EXP }(e_4, x_3) \wedge$
$(\forall v \in \mathbf{W}\text{: EXP }(e_4, x_3)\text{ in }v)\text{: SHOULD }(\mathbf{S_6} \parallel \Omega^*{:}B_3(v))]$

To see if this representation of the truth conditions is merited, compare *hope* with *want*, in parallel examples in (32), where (32b) has the truth conditions given in (33), as discussed in Chapter 4.

(32) a. Alex hopes Bill is in the room.
b. Alex wants Bill to be in the room.

(33) $\text{SHOULD }(\mathbf{S_6} \parallel \Omega^*{:}B_3(v))$

Comparing *hope* in (32a) with *want* in (32b), putting aside the difference in strength between SHOULD and MUST on the analyses proposed here, a couple of differences between the two ascriptions suggest that the *want*-ascription can have the "minimal" interpretation in (33), which was argued for in Chapter 4, whereas the *hope*-ascription in (32a) can only have the full interpretation represented in (31).

One difference arises in describing someone who commits minor crimes in an unconscious attempt to gain attention. If Alex were such a person, it would be fully felicitous to say, *Alex wants to be caught*, to ascribe an unconscious desire to Alex. But *Alex hopes to be caught* is not quite as felicitous in this context; it seems to impute at least some consciousness of the desire, instantiated as a mental state. Furthermore, the ascription with *want* in such a case is fully compatible with an accompanying statement denying experiencer awareness of the fact stated by the ascription, as in (34a,b), whereas comparable examples with *hope* in (35a,b) are not very felicitous.

(34) a. Alex unconsciously wants to be caught.
b. Alex wants to be caught, but he's not aware of that, himself.

(35) a. # Alex unconsciously hopes to be caught.
b. # Alex hopes to be caught, but he's not aware of that, himself.

If the ascription with *hope* can only have the full truth conditions in (31), featuring the condition EXP ($x_3$, $e_4$), then it states that Alex is the experiencer of a state of hope that he will be caught; this condition is not satisfied in the example under consideration, explaining the infelicity of *Alex hopes to be caught* in this case. Furthermore, it is not clear how Alex could be unaware of the fact that he has an emotional state that he, himself, experiences, explaining the oddness of (35a,b). In contrast, *want* permits a minimal interpretation in terms of the content conditions given in (33), in which no such state is asserted to be experienced by Alex. In this case, the ascriptions in (34) require only that Alex not be fully aware of the tendencies of his own (to some extent unconscious) preferences. If Alex is not very self-aware, the tendencies of his own preferences may, in fact, not be fully transparent to him. A second difference between these two verbs supporting the supposition that *want* permits the minimal interpretation, but not *hope*, will be discussed in Chapter 7.

Note that the proposal here is not that *hope* has a Davidsonian event variable, and *want* does not. The proposal is that *hope* is always interpreted along the lines of (31), with an event variable, while *want* can have that sort of interpretation or the minimal interpretation in (33). Both can take modification by an adverbial such as *fervently*, but only the full interpretation of a *want*-ascription does so. The primary justification for this is not intuition, but rather empirical correlations, including the contrast in (34)–(35), and a further contrast to be discussed in the next chapter.

It might be supposed that the existence of the bouletic ordering source for interpreting an ascription with *want* entails the existence of a state of desire on the part of the experiencer subject of the ascription. After all, the bouletic ordering source $\Omega^*:B_3(v)$ used in (33) can be said, at least pretheoretically, to contain propositions that truthfully characterize the bouletic state of Alex. But this misses the central point of the present proposal regarding the relationship between the linguistic ontology and what is grammaticized for semantic interpretation. The state might exist, in fact, and that is a fundamental potential resource for semantic ontology. But the interpretation of a bouletic ascription does not necessarily make reference to it, and the point of the present proposal is that the "bare" or minimal interpretation of an ascription with *want* does not make reference to it. Whether a stative attitude ascription has an associated event is largely one of grammatical expression, not just ontology. For some attitude verbs such as *regret* or *hope* in English, that grammaticization is obligatory, and this is a fact about the conventional semantic interpretation of those verbs. But for others, as in our discussion of *believe*, we may interpret with an event variable for this verb, choosing to grammaticize the associated state, or we may interpret it as evidential qualification, choosing not to grammaticize an associated state. The conventional semantic fact in the case of *believe* is the indeterminacy of whether the ascription is interpreted as making reference to an associated state. When the state is causally implicated in a significant way, or modified in an active way, we have more reason to grammaticize the event. The facts about two types of attitude ascription show that there is a difference in whether the event variable is present as a lexically determined semantic-grammatical feature of the ascription, or optionally absent. In the latter case, what is being asserted in context is the determining factor, not facts of linguistic ontology.

## 6.5 Reports of attitude assertion

The ascriptions in (36) attribute an assertion to Alex.

(36) Alex said/claimed/asserted that Bill stole the artifact.

These could be used to report an event in which Alex lied, just as well as to report an event in which Alex believes the content of the complement clause. What seems relevant to the truth conditions of the ascriptions in (36) is not that Alex believes the proposition expressed by the complement clause, but that, in making the reported claim or assertion, Alex was intent on influencing the doxastic state of his addressees in the direction of believing that proposition.

Any of the ascriptions in (36), with any one of the three verbs, reports an event in the world of evaluation, *w*, and this event can be identified with similar events in other possible worlds, within certain parameters. In another possible world, Alex might make the claim a few seconds earlier or later in spoken discourse, or minutes earlier or later in an email exchange, and it could still count as the same event. Alex could speak more loudly in a spoken instance, or write more tersely in a written instance, and it might still be the same event as in the world of evaluation. But an event of claiming that Bill stole the artifact is different from an event of claiming that Bill didn't steal the artifact, or an event of claiming that the artifact is a fraud, even if the claim is made at the same instant, and in the same manner. So (36) reports that Alex is the agent of a communicative act characterized and identified in large part by its relationship to the content of the complement clause, $\mathbf{S_2}$. Let this communicative act be event $e_4$. It remains to articulate the relationship between $e_4$ and the content of $\mathbf{S_2}$.

As a start, $e_4$ can be characterized by the fact that Alex wants his addressee(s) to believe that Bill stole the artifact. Taking the agency of Alex to be an intrinsic property of the event, $e_4$ can be identified across possible worlds, by the fact that, in any possible world in which $e_4$ occurs, Alex is its agent, and wants or intends his addressee(s) to believe that Bill stole the artifact.

This suggests that the interpretation of (36) might involve iteration of a doxastic condition within a bouletic condition – that Alex acts out of a desire for somebody to believe something. Although this seems to be true of a context of use, it seems a bit elaborate for the semantic interpretation of (36). It is nevertheless hard to prove that it is too elaborate. Toward resolving this question, consider the ascriptions in (37), one stronger than (36) and one weaker, to see if there is evidence of separation between a bouletic layer and an embedded doxastic layer of meaning.

(37) a. Alex insisted that Bill stole the artifact.
b. Alex suggested that Bill stole the artifact.

In (37a), Alex is understood to be acting out of a stronger imperative than in (36), but he also seems to be enjoining his addressee(s) to a stronger commitment of belief in the proposition that Bill stole the artifact than in (36). In (37b), Alex would typically be acting out of a weaker imperative or internal driving force to impress his views upon his addressee(s), and he would be understood to be more weakly recommending for consideration the proposition expressed by the subordinate clause, with less commitment that it be adopted by his addressee(s). Thus, the strength of the predicate of assertion is reflected in both the bouletic and the doxastic components of the interpretation. This indicates that the bouletic and doxastic layers of interpretation are represented in a single operator.

On the basis of these considerations, the ascriptions in (36)–(37) should have truth conditions along the lines sketched in (38), where MOD-Q is an appropriate modal quantificational schema, and the ordering source is a combination of bouletic and doxastic.

(38) $\exists e_4$ [AGENT $(e_4, x_3) \wedge$
($\forall v \in \mathbf{W}$: AGENT $(e_4, x_3)$ in $v$): MOD-Q [($\mathbf{S_2}$ || bouletic & doxastic]

On the basis of experience with other ascriptions, the appropriate schemas would be MUST for *insist*, SHOULD for *say* and *claim*, and SHOULD$^{\perp}$ for *suggest*.

This interpretation of (36)–(37) would be equally appropriate whether Alex has one addressee, or multiple addressees. If there are multiple addressees, there is a doxastic base that they share for matters at hand. Let J be the set of indices for Alex's addressees: $j \in J$ iff $a_j$ is an addressee of Alex in the event reported. Let $a_J$ be the set (or group) of Alex's addressees in this event. Then part of the meaning of (36) is that $\mathbf{S_2}$ is true as worlds accord better and better with Alex's preferences for what members of $a_J$ believe.

But that points to a further level of intentionality involved in these ascriptions. Despite the argument given previously that these ascriptions don't embed a doxastic ascription within a bouletic ascription, Alex's preference is realized not in terms of the truth of $\mathbf{S_2}$ in worlds that accord with the actual doxastic base of members of $a_J$, but rather, in terms of the truth of $\mathbf{S_2}$ in worlds that accord with what Alex takes the doxastic base of members of $a_J$ to be. Let that attributed doxastic base be $D_{J[3]}(v)$, the shared doxastic base attributed by Alex to the members of $a_J$ in possible world $v$, marked by the index of Alex in square brackets. The main idea, then, is that the ascriptions of attitude assertion in (36) have the truth conditions that $\mathbf{S_2}$ is true as worlds get more in accord with speaker desires (for what the addressees believe) and more in accord with what the speaker takes the addressees to believe. The point is that the speaker is trying to get addressees to believe $\mathbf{S_2}$ based on what the speaker takes the addressee's doxastic state to be. Interpreted in this way, these ascriptions quantify over the intersection of the speaker bouletic ordering source and the doxastic base that the speaker attributes to the addressee (or the shared doxastic base of multiple addressees). The ordering source for these ascriptions is thus given in (39).

(39) $\Omega\text{:Assertion}_{3\text{-}J}\ (v)) = \Omega\text{:B}_3(v) \cap D_{J[3]}(v)$

Without further ado, then, we can say that (36) is true iff there is an event $e_4$ of which Alex is the agent, which is identified across possible worlds $v$ by Alex being its agent in $v$ and by the modal quantificational schema SHOULD holding for the complement $\mathbf{S_2}$ in $v$, whereby $\mathbf{S_2}$ is true on tails of sequences weakly ordered by $\Omega\text{:Assertion}_{3\text{-}J}\ (v)$. This is given in (40), with counterparts for (37a,b) in (41) and (42).

(40) $\exists e_4$ [AGENT $(e_4, x_3) \wedge$
($\forall v \in \mathbf{W}$: AGENT $(e_4, x_3)$ in $v$): SHOULD ($\mathbf{S_2}$ ||
$\Omega\text{:Assertion}_{3\text{-}J}\ (v)$)

(41) $\exists e_4$ [AGENT $(e_4, x_3) \wedge$
($\forall v \in \mathbf{W}$: AGENT $(e_4, x_3)$ in $v$): MUST ($\mathbf{S_2}$ ||
$\Omega\text{:Assertion}_{3\text{-}J}\ (v)$)

(42) $\exists e_4$ [AGENT $(e_4, x_3) \wedge$
$(\forall v \in \mathbf{W}$: AGENT $(e_4, x_3)$ in $v$): SHOULD$^{\perp}$ ($\mathbf{S_2}$ $\|$ $\Omega$:Assertion$_{3\text{-}J}$ $(v)$)

Because they characterize acts of assertion, the verbs of asserted attitude readily admit adverbial modification by manner adverbials, as in (43).

(43) a. Alex abruptly claimed that he wanted out of the contract.
b. Alex clumsily announced that he would have another drink.
c. Alex carefully asserted that he had never seen the murder weapon before.

As is well known (see Keenan and Faltz [1985]), the adverbials in (43) are ambiguous between a strict manner interpretation and another, which I will call situational. On the strict manner interpretation, they characterize the execution of the action: that Alex articulated his claim abruptly in (43a), and his announcement clumsily (out of drunkenness, perhaps) in (43b), and that he carefully phrased and articulated his statement that he never saw the murder weapon before in (43c). These assert that Alex was abrupt, clumsy, or careful in the associated event, call it $e_4$. Possible situational uses of these ascriptions could be paraphrased by saying that, in (43a), Alex abruptly changed course, from accepting the terms of the contract to rejecting them, that in (43b) he was clumsy to announce that he would have another drink, as doing so would be a social faux pas, and in (43c), that he took care not to neglect to mention that he had never seen the murder weapon before. With the situational interpretation, the adverbial characterizes the agency of Alex in initiating $e_4$, that he was abrupt, clumsy, or careful in initiating $e_4$. Characterized in this way, the difference is aspectual and intentional, with the situational version saying that Alex was abrupt, clumsy, or careful in choosing to initiate $e_4$ when he did, or in choosing to initiate it at all, and the strict manner interpretation saying that Alex was abrupt, clumsy, or careful throughout $e_4$, or at critical points through the course of execution of $e_4$. Both of these interpretations of the manner adverbials in (43) can be built upon neo-Davidsonian representations of the form given in (40)–(42), where the agency of Alex in $e_4$ is explicit, with further aspectual and intentional conditions needed to distinguish the strict manner interpretation from the situational interpretation. Neither would be supported by a bare condition on possibilities. We leave aside here the task of characterizing these two types of manner modification in more detail.

## 6.6 Multidimensional attitude reports

Heim (1992) discussed the fact that ascriptions of bouletic attitudes have a doxastic component. The assertion that Sam wants to get well, for instance, presupposes that Sam is (and, for at least a short time, has been) sick. In the present work, we have emphasized that such bouletic ascriptions are variable, from more fanciful to more realistic – when doxastic and bouletic conditions are inconsistent, the conflict can be resolved more in favor of one component or the other.

There are other multidimensional ascriptions, including most, but not all, factive ascriptions. To sort this out, we first consider strictly one-dimensional factive ascriptions, as in (44).

(44) a. Alex recognizes that Bill caused the accident.
b. Alex realizes that Bill stole the artifact.
c. Alex knows that Bill stole the artifact.

These are commonly analyzed as presupposing the truth of the complement clause, while asserting something that is otherwise along the lines of a belief report.[3] In the past tense (or the narrative present), ascriptions with *recognize* or *realize* can report on an occurrent event in which Alex comes to a particular recognition or realization. In this use, these verbs support adverbial modification with adverbials such as *delightedly*, *unhappily*, or *to his dismay*, implicating a neo-Davidsonian variable in their representation. In the present tense, the ascriptions in (44a,b) have a salient noneventive interpretation, and ascriptions with *know*, such as (44c), are invariably noneventive.

We will first look at the noneventive cases in order to address factivity without an accompanying event, in order to get a better idea of what factivity is, in itself. A classic account of the truth conditions of (44c) is that the proposition expressed by the complement clause is true, Alex believes that proposition, and Alex is justified in believing it. Although this characterization has been critiqued (by Gettier [1963], and many since), the common denominator of current accounts is that the complement $\mathbf{S_4}$ is true, Alex believes it, and some further condition of reliability or justifiability holds. The ascriptions in (44a,b), as factive ascriptions, are subject to these conditions, along with ones that characterize them more particularly as states or acts of recognition or realization.

The ascriptions in (45) are more thoroughly multidimensional.

(45) a. Alex regrets that Bill stole the artifact.
b. Alex is surprised that Bill stole the artifact.

The ascription in (45a) reports a doxastic condition, Alex's apprehension of a fact, with factivity. But it also reports an associated negative bouletic condition, accompanying the factive doxastic ascription. Alex's relation to the

[3] See Kiparsky and Kiparsky (1971), Levinson (1983), Kadmon (2001), Kratzer (2002), and others for discussion of factives.

complement, $\mathbf{S_4}$, has a belief component, MUST ($\mathbf{S_4}$ || $\Omega$:$D_3(w)$), and a bouletic component, MUST (**not-**$\mathbf{S_4}$ || $\Omega$:$B_3(w)$), or at least a comparable negative bouletic condition with the force of SHOULD. Furthermore, (45a) reports an emotional state, so these conditions serve to identify the eventuality of that state across possible worlds in which it occurs. The assertion in (45b) also reports Alex's apprehension of a fact, with factivity. But it also reports that the complement $\mathbf{S_4}$ is counter to Alex's expectations, that he was expecting **not-**$\mathbf{S_4}$. Thus, the complement $\mathbf{S_4}$ has the relation MUST ($\mathbf{S_4}$ || $\Omega$:$D_3(w)$) to Alex's current doxastic ordering source (which is taken here to include the doxastic base, with quantification over all possible worlds, as before). And it also has a relation with the force of MUST or SHOULD to an immediately prior doxastic base of Alex, $\Omega'$:$D_3(w)$.

Thus, factive ascriptions are intrinsically more complex in combining doxastic conditions with the justifiability conditions of factivity, but many of them are also multidimensional. This will be significant in explaining their resistance to Neg-Raising in the next chapter.

## 6.7 Conclusion

The proposals made in this chapter account for differences between minimal attitude ascriptions, in which the ascription just expresses a condition on possibilities based on the content of the complement clause relative to the experiencer's doxastic base, versus full attitude ascriptions, in which such a condition is part of an essential content condition on a mental state or event. It accounts for differences between the two in possibilities of adverbial modification with manner adverbials whose interpretation hinges on the presence of the event/state variable associated with the attitude, and for the contrast between *want* and *hope* illustrated in (32). The present account effects a lexical decomposition of attitude predicates, and other predicates taking clausal arguments, that captures variation in modal quantificational strength, and the role of implicit doxastic or bouletic modal criteria. Further support comes from the application of this account to elucidate the nature and extent of the phenomenon referred to as Neg-Raising. This will be taken up in the next chapter.

# 7 Lower interpretation of negation

## 7.1 Delineation of the phenomenon

### *7.1.1 Basic facts*

A number of structures of clausal complementation, including some with propositional attitude and volitional predicates, permit an interpretation in which a matrix clause negation is interpreted with the complement clause.[1] These exist alongside interpretations in which the matrix clause negation has the wide scope associated with its surface position. Thus, the first sentence in (1a) can be interpreted as denying that the senator is of the opinion that the bill will pass, but it can also be interpreted as the attribution to the senator of a particular negative belief, a belief that the bill won't pass. The latter interpretation provides a plausible basis for the second sentence of (1a). Likewise, the first sentence of (1b) can be interpreted as attributing a negative belief to the subject, and the example in (1c), a negative expectation. The interpretation of negation with the lower clause is reflected in the licensing of the strict Negative Polarity *until*-phrase in (1c,d,e). Likewise for the perception complements to *seem* and *look like* in (2a,b) and the complement of the probability predicate in (2c).

(1) a. The senator doesn't think the bill will pass. That's why she declined to sponsor it.
b. I don't suppose this bill has enough votes to pass. So I'm burying it in committee.
c. I didn't expect that they would find the ring until Tuesday.
d. It isn't likely that he will arrive until midnight.
e. They don't want (him) to find the document until Tuesday.

(2) a. The governor doesn't seem to be in town.
b. It doesn't look like they found the document.
c. The governor isn't likely to sign that bill until Thursday.

[1] See Jespersen (1917), Quine (1960), Fillmore (1963), R. Lakoff (1969), Lindholm (1969), G. Lakoff (1970), Prince (1976), Horn (1978, 1989), Tovena (2000), Gajewski (2007), and Romoli (2013), among others.

This interpretation is often called the Neg-Raising interpretation, after an early transformational analysis, a term that is sometimes used to designate the empirical phenomenon even by those who question or explicitly reject an analysis involving a syntactic operation of Neg-Raising. Here it will be called LIN to indicate that negation is interpreted with the complement of the matrix predicate (or in a way that is typically paraphrased by putting negation with the complement of the matrix predicate), but reflecting no assumption about its nature, or about whether it is a syntactic, semantic, or pragmatic phenomenon. Some additional properties of LIN will be noted in the section that follows in an attempt to delineate the empirical phenomenon at issue, incorporating observations from the literature. Then the availability of LIN will be examined from the perspective of the interpretation of propositional attitude ascriptions and other sentences with clausal arguments developed in Chapters 3–6.

Reports of propositional attitudes of assertion uniformly lack LIN. In (3), negation can have scope only over the verb of assertion, in the matrix clause.

(3) a. The senator doesn't claim she has enough votes for the bill.
b. I didn't assert that the senator had enough votes for the bill.
c. He didn't say that they would find the ring.

In (4a,b), the lack of LIN is reflected in the fact that strict NPIs occurring at the end of the sentence can only be interpreted as pertaining to the action described in the higher clause, the assertion in (4a), and the announcement in (4b). They are not licensed within the complement clause.[2]

(4) a. He didn't assert that they found the document until Tuesday.
b. They won't announce that the plan will be revealed until Friday.

Most propositional attitude ascriptions of emotion also lack LIN, although *want* permits it. As pertaining to the 2000 US presidential election, (5a) can't mean that liberals hoped that Bush wouldn't win, and (5b) can't mean that evangelicals feared that Bush wouldn't win, even though these assertions would be highly plausible. They can only be interpreted as describing a lack of hope, and a lack of fear.

(5) a. Liberals didn't hope that Bush would win.
b. Evangelicals didn't fear that Bush would win.

Furthermore, LIN is not possible with factive ascriptions, as the examples in (6) show.

[2] In this respect, *until*-phrases contrast with less strict Negative Polarity Items, such as *any*; contrast (4a,b) with (i), where the NPI *anything* is licensed in the complement clause to a predicate of assertion.

(i) He doesn't claim that he did anything wrong.

(6) a. He didn't realize/recognize that they found the document until today.
b. I didn't regret that they found the ring until Friday.
c. I wasn't surprised that he found the ring until today (when I found out that he hadn't even been looking for it).

The *until*-phrases in (6) can only be associated with the higher clause; they don't have the sort of association with the complement to the factive predicate that the *until*-phrases have with the complement clauses in (1) and (2).

LIN is not available when the main clause contains an adverbial modifying the main clause event, or modifying the subject in relation to this event. To see this, note that the example in (7a) can be interpreted with LIN, with the *until*-phrase interpreted as modifying the infinitival complement clause; this demonstrates that *expect* is a predicate permitting LIN.

(7) a. Alex doesn't expect to find the document until Tuesday.
b. Alex confidently expects to find the document.
c. Alex doesn't confidently expect to find the document (*until Tuesday).

In (7b), the adverbial in the matrix clause modifies Alex in relation to his experience of this expectation. In (7c), negation can have wide scope, or it can be interpreted as constituent negation on the adverbial (if the adverbial is stressed), but it cannot be interpreted with the complement clause, as asserting that Alex confidently expects to not find the document until Tuesday; the unavailability of LIN is reflected by the failure of the *until*-phrase to be interpreted with the infinitival complement. (In this case, the *until*-phrase does not apply very felicitously to the main clause either, unless the entire ascription is delivered in the historical present, as in modern American television documentaries about history. In feeble protest against that usage, it is marked here with an asterisk.)

Negation in a main clause can usually be associated with a complement clause that is multiply embedded, provided both the main clause and the intermediate clause(s) meet all other restrictions on LIN.[3] The examples in (8), but not those in (9), permit LIN with the lowest clause.[4]

(8) a. I don't suppose the governor thinks the bill will pass. I think that's why she's not spending any political capital promoting it.
b. John doesn't think he wants to leave until Sunday.
c. I don't believe that he wants me to think that he did it.
d. They don't think the governor is likely to sign the bill until Thursday.

[3] See Section 4.2 for discussion of some exceptions.

[4] Example (8b) is from McCawley (1988: 592), and (8c) from Fillmore (1963: 220), cited by Horn (1978: 130).

(9) a. I don't think the governor claimed that the bill will pass.
b. I don't think the governor regrets that the bill will pass.
c. I don't suppose the governor has long believed the bill will pass.

The example in (8a) has a very felicitous interpretation according to which it is roughly paraphrasable as, "I rather expect that the governor thinks the bill won't pass. I think that's why she's not spending any political capital promoting it." The example in (9c) does not likewise have an interpretation paraphrasable as, "I rather expect that the governor has long believed that the bill will not pass."

Finally, consider the example in (10a), interpreted with LIN, and the corresponding example with a lower clause negation, in (10b), both from Horn (1989: 342).

(10) a. It's not likely they'll accept your invitation.
b. It's likely they won't accept your invitation.

In forms such as (10a), interpreted with LIN, the predicate is widely regarded (across examples, predicates, and languages) as slightly weaker than in the corresponding form with lower negation, an observation that Horn (1978, 1989) attributes to Poutsma (1928) and Dwight Bolinger. This observation will be reformulated and reconceptualized in the present work, but we record it for now in the terms ("slight weakening") in which it has been described in the literature.

Classes of predicates that exhibit LIN with their complements include, following Horn (1978), predicates of opinion such as *think* and *believe*; predicates of volition such as *want* and *intend*; predicates of perception, such as *seem*, *appear*, *look like*, *sound like*; predicates of probability such as *be likely*; and predicates of judgment, such as *advise* and *suggest*.

But, as discussed by Horn (1978, 1989) and others, even within classes of predicates that permit it, LIN exhibits great sensitivity to lexical item choice of the matrix predicate, even between predicates that seem syntactically and semantically quite similar. And the availability of LIN with some predicates exhibits considerable variability among dialects and individual speakers. LIN is exhibited with complements of *want*, but not with complements of *hope* or *desire*; with *suppose*, but with *guess* only in some dialects; with *intend* and *plan*, but not with *make plans* (*I didn't plan to go there* exhibits LIN, but *I didn't make plans to go there* doesn't); with *expect* for most speakers, but with *anticipate* only for some. There is considerable cross-linguistic consistency among cognate predicates that permit LIN, but also some stark divergences. For example, there is no LIN with English *hope*, but there is with German *hoffen*, as in (11), where the association of negation with the lower clause is, if anything, more prominent than with the matrix clause.[5]

(11) Peter hofft nicht, daß Hans kommt.
Peter hopes not that Hans comes

### *7.1.2 Central questions*

On the basis of this brief survey of the facts about LIN and the properties of predicates that permit it, the central questions for a theory of LIN may be outlined as follows: (1) What distinguishes the classes of predicates whose members generally exhibit LIN (opinion, volition, perception, probability, judgment) from those whose members generally don't exhibit LIN (e.g., attitude predicates of assertion and emotion, factive predicates)? (2) Within these classes, what distinguishes the individual predicates that permit LIN from those that don't? Can a relevant property (or cluster of properties) be identified to distinguish these, and can it be explained exactly how the availability of LIN hinges on that property? And does the identification of this property provide a foundation for an account of why LIN is so cross-linguistically and dialectally variable among similar or cognate predicates? (3) How does a theory of LIN account for the blocking effects of certain adverbials in the main clause? (4) How does LIN allow for the interpretation of negation with a multiply embedded complement (relative to the position of negation)? And, (5) how does it account for the slight weakening of the matrix predicate in examples with LIN, relative to their counterparts without LIN?

### *7.1.3 Horn's account*

Horn's (1978, 1989) account of LIN considerably refines the standard delineation of the phenomenon. For this reason, Horn's account will be outlined here, before we take stock and introduce a new approach to the phenomenon in Section 7.2. Other accounts from the literature will be addressed in Section 7.4.

Horn (1978) presents an account of LIN as a product of scalar Gricean implicature based on a scalar model of predicate strength. On Horn's account, scalar implicature is supplemented by politeness-based implicature and

[5] An anonymous reviewer notes that German examples with infinitival complements of *hoffen*, as in (i), do not so clearly exhibit LIN since the position of matrix clause negation in such examples permits it to be construed equally well with the matrix clause and the infinitival complement clause. The claim that German *hoffen* permits LIN rests more securely on examples like (11), with finite complements.

(i) Johann hofft nicht bestraft zu werden.
Johann hopes not punished to be
"Johann doesn't hope to be punished."/"Johann hopes to not be punished."

conventionalization. On this view, LIN is an instance of the phenomenon by which the contradictory negation *He is not happy* is interpreted as the contrary, *He is unhappy*. This interpretation comes about not because the two are semantically identical; the former but not the latter can be followed up by *but he is not unhappy either*. Nevertheless, when there is reason to avoid asserting *He is unhappy* outright, the assertion *He is not happy* can implicate that he is unhappy, and this implicature of the contrary form through use of the contradictory form, Horn suggests, has become conventionalized to a degree that the assertion of the contradictory form might be judged by speakers to have as its preferred interpretation the meaning of the contrary form.

Horn claims that this principle can apply as well in the interpretation of the clausal examples in (10). For Horn (1989: 337–361), LIN in examples like (10a) involves a short-circuited inference, a semiconventionalized conversational implicature from the assertion of the contradictory (10a) to an interpretation along the lines of the contrary, (10b), in some cases driven by considerations of politeness.

To explain the nature of this implicature, and the range of predicates with which it occurs, Horn notes that the matrix predicates in the same class (opinion, volition, probability, perception, judgment), and of the same type, can be ordered with respect to their strength, on a scale admitting low, mid, and high degrees of strength. This is illustrated in (12), following Horn (1978: 194); see also Horn (1989: 324–325).[6]

(12) a. {suspect} < {believe, suppose, think, expect} < {know, realize}
b. {consider (doing)} < {want, intend, plan} < {require}
c. {possible} < {likely, seem, appear, look like, figure to} < {certain, sure}
d. {can, could} < {supposed to} < {necessary}
e. {allow, permit, let} < {suggest, advise} < {make, order, demand, force, insist}
f. {allowed} < {advisable, desirable} < {obligatory, required}
g. {be a terrible idea} < {be a good idea} < {be an excellent idea}

Horn claims that a necessary condition for LIN is that the contradictory value of the matrix predicate with its complement be functionally close to the interpretation with lower negation. This functional closeness is obtained only when

[6] More specifically, the position of a predicate in the ordering reflects the lower bound for the strength of the predicate. For something to be likely, for example, means that its probability is bounded below by 0.5 (exclusive of this lowest value), making *likely* a "mid-scale" predicate.

the matrix predicate has a (bottom-of-range) value in the mid-range of its scale. For example, the most salient way for it to be the case that Alex does not believe, want, or intend *p* is for the functionally close condition that Alex believes, wants, or intends not-*p* to be the case; so much so, that the assertion that Alex does not believe, want, or intend *p* can implicate that Alex believes, wants, or intends not-*p*. In contrast, it is not the case that the most salient way for *p* to not be certain is for not-*p* to be certain; if *p* is not certain, it is a fully salient possibility that neither *p* nor not-*p* may be at all certain.

When it is available, the implicature can be strengthened by considerations of politeness. Insofar as it is perceived to be more polite to assert [*not advise that p*] than [*advise that not p*], there is reason for the hearer to assume that an assertion with [*not advise that p*] is uttered in lieu of one with [*advise that not p*], and is intended to communicate the meaning of the latter.

Horn's account clearly explains why predicates ordered outside the mid-range of their associated scales do not participate in LIN, since the interpretation of such a predicate is far from its own negated value. But, as Horn acknowledges, being mid-scalar in strength is not a sufficient condition for LIN since many mid-scalar predicates don't permit LIN: *hope* is as mid-scalar as *want*, and *say* is in the mid-region of its scale, stronger than *suggest* and weaker than *insist*. Although Horn's account explains why LIN is not available with matrix predicates of asserted attitudes and attitudes of emotion, insofar as the most salient way for it to be true that Alex did not say that *p* is not for Alex to have said that not *p*, and the most salient way for it to be true that Alex does not hope that *p* is not for Alex to hope that not *p*, nevertheless, it doesn't say what it is about *believe*, *intend*, or *want*, as opposed to *say*, or *hope*, that determines the salience of different alternatives, and thus makes this conventionalized inference go through. Is there a lexical property of these predicates that favors this inference? At this point, Horn's account falls back on an appeal to the notion of the greater functional load of the matrix clause relative to its complement clause. Verbs such as *say*, *conjecture*, or *hope*, which don't permit LIN, are presumed to confer a greater functional load on their clause than verbs such as *believe*, *intend*, or *want*. The relative prominence of the content of the matrix clause with "higher load" verbs blocks the inference. On this view, dialect variation in the availability of LIN with particular predicates reflects idiosyncratic variation in the functional load of the predicate in different dialects, or in the conventionalization of the short-circuiting inference. The availability of LIN with cognate predicates in related languages is susceptible to the same explanation. This mode of explanation extends to explain why LIN is blocked by the presence of the matrix clause adverbial *confidently* in (7c), assuming that this adverbial enhances the functional load of the matrix clause relative to its complement. But there is a problem with the intrinsic vagueness of the notion of functional load appealed to in these explanations. These issues will be discussed further in Section 7.3.

The next section will challenge Horn's delineation of the phenomenon of LIN as one pertaining specifically to mid-scalar predicates that remain essentially fixed in meaning between the interpretation with LIN and the one without it. It will identify the near identity of meaning of the predicate in the two cases (modulo the slight weakening with LIN) as a special case, not a defining property of LIN. The next section will delineate instead the class of predicates permitting LIN as one that mostly contains the class of traditionally recognized Neg-Raising Predicates (NRPs), and will reinterpret what is special about these predicates that makes LIN possible. The account of the next section can be regarded as providing an explicit construction of the greater "functional load" of predicates that don't permit LIN.

## 7.2 Neg-importation and duals

### *7.2.1 Toward a new account*

Horn's account invokes a partly conventionalized short-circuited implicature available for mid-scalar predicates, where the strength of a predicate on the associated pragmatic scale is a conventional, lexical semantic property of the predicate. This suggests that LIN might be derived from a suitable lexical semantic analysis of the relevant predicates, if the analysis could provide a natural distinction between predicates that exhibit LIN with their complements, and ones that don't.

The distinction between full and minimal interpretations of propositional attitude ascriptions and related predicates developed in Chapters 3–6 of the present work is a candidate to play that role in an account of LIN. To begin with, it is striking that the predicates occurring in standard examples of LIN include most of those, such as, *think*, *believe*, *likely*, and *want*, which were argued in Chapter 6 to permit the minimal interpretation quantifying over possible worlds subject to an ordering source. And predicates that only permit the full interpretation, introducing an eventuality experienced by the subject and characterizing it in terms of an essential content condition, are conspicuous among the matrix predicates in examples that don't permit LIN, including propositional attitude predicates of assertion, and most attitude predicates of emotion.

Furthermore, it is easy to see why LIN would not be available in an attitude ascription with the full interpretation. For example, consider (13a), which is the negation of the ascription of assertion in (13b). Since (13b) has the truth conditions in (13c), its negation has the truth conditions in (13d).

(13) a. Alex didn't assert/claim [that Bill stole the artifact].
b. Alex asserted/claimed that Bill stole the artifact.

c. $\exists e_2$ [AGENT $(e_2, x_3)$ ∧
($\forall v \in$ **W**: AGENT $(e_2, x_3)$ in $v$): SHOULD ($\mathbf{S_1}$ ||
Ω:Assertion$_{3\text{-}J}$ $(v)$)]
d. ~ $\exists e_2$ [AGENT $(e_2, x_3)$ ∧
($\forall v \in$ **W**: AGENT $(e_2, x_3)$ in $v$): SHOULD ($\mathbf{S_1}$ ||
Ω:Assertion$_{3\text{-}J}$ $(v)$)]

This is wide scope negation; the interpretation is semantically indeterminate among denying the existence of the event of assertion, denying that Alex was an agent of that event, and denying that the event is characterized (across possible worlds) by the modalized essential content condition.[7] (Association of negation with focus could, of course, disambiguate it.) A similar interpretation with wide scope negation can be obtained (as one interpretive option) when negation is added to the matrix clause of an ascription with a predicate that permits LIN. *Alex doesn't believe that Bill stole the artifact* has a non-LIN interpretation as (14), which (in the absence of focus attracting negation) is semantically indeterminate among denying the existence of the belief state, denying that Alex is the experiencer of that state, and denying that the state is characterized across possible worlds by Alex's belief in the propositional content of the lower clause.

(14) ~ $\exists e_2$ [EXP $(e_2, x_3)$ ∧ [($\forall v \in$ **W**: EXP $(e_2, x_3)$ in $v$): SHOULD ($\mathbf{S_1}$ || Ω:$D_3(v)$)]]

This suggests that the account of propositional attitude ascriptions and other sentences with clausal arguments given in Chapters 3–6 might provide a basis for a lexical semantic account of variation in LIN. The present section will develop the details of such an account, and discuss how it addresses the central questions formulated in Section 7.1.2. We begin with an example beyond the scope of Neg-Raising or LIN, as traditionally delineated.

### *7.2.2 A nonstandard instance*

Consider the example in (15a), and its counterpart with negation in (15b).

(15) a. The governor is sure the bill has enough votes to pass.
b. The governor isn't sure the bill has enough votes to pass.

On one interpretation, (15b) fully exemplifies wide scope negation; it asserts that the governor is not the experiencer of a state of being sure that the bill has

[7] Of course, if (13a) is uttered in a context in which it is given that Alex either claimed that Bill stole the file, or claimed that he didn't, so that Alex definitely made a claim one way or the other, then the utterance in that context has a fully nonconventional conversational implicature to the effect that Alex claimed that Bill did not steal the file. This is a different phenomenon from LIN, but see Section 7.4 on "excluded middle" accounts of LIN.

enough votes to pass. This would be obtained by negating the full representation of (15a), parallel to the interpretation of (13a) as (13d). On another interpretation, at least equally salient, (15b) could be paraphrased roughly as, "The governor thinks it possible that the bill doesn't have enough votes to pass." The latter is a "lower negation" reading; note that the strict NPI *until*-phrase in (16) can be interpreted with the lower clause on this interpretation.[8]

(16) The governor isn't sure the bill will have enough votes until Tuesday.

Note that the paraphrase of the "lower Neg" reading involves a weaker matrix predicate, "think it possible" instead of "be sure." Context makes this clearer, as in the naturally occurring example in (17), uttered by tennis announcer John McEnroe in commentary before the Wimbledon final round match between Venus Williams and Marion Bartoli on July 7, 2007 (discussed in Chapter 3).

(17) I'm not sure Venus will crumble like Henin did yesterday.

In this utterance, McEnroe put forth the possibility, too qualified or hedged to be an outright prediction, that Venus Williams would not crumble against Marion Bartoli in the way that Justine Henin did in the previous day's semifinal match. The tentativeness accorded to the assertion on this interpretation is appropriate in light of Bartoli's consistency through the Wimbledon fortnight. Since McEnroe had noted that there was some inconsistency in the play of Venus Williams, he would not have wanted to say that he thinks Venus would not crumble as Henin had; that would have been too strong and confident a statement, verging on a prediction that Williams would not crumble. Something weaker was meant.

Note that (15b), (16), and (17) would not be recognized as examples of LIN, or Neg-Raising as ordinarily delineated, since the "lower Neg" readings would not be paraphrased with the matrix predicate *be sure*, but with a weaker predicate, a reflex of the fact that *be sure* is not mid-scalar, but at the top of its scale. The "lower Neg" reading in these examples seems to be the result of a semantic interaction of negation with the matrix predicate, whereby the predicate *sure* is replaced with its dual when it commutes with negation.

To recall how this works out in detail, return to the account of propositional attitude ascriptions in Chapters 3 and 4. The wide scope negation reading of

[8] These are distinct interpretations. The second is not simply a logical reduction of the first, as we can see from the fact that the *until*-phrase in (16) is infelicitous or uninterpretable when (16) is steadfastly interpreted as simply denying a state of certainty on the part of the governor, as "It isn't the case that the governor is in a state of being sure that the bill will have enough votes (*until Tuesday)"; it is impossible in that case to work *until Tuesday* into the description of the state of certainty that the governor lacks.

(15b) is the negation of the full representation of (15a). In contrast, recall the minimal interpretation of (15a) as a bare attitude ascription, rendered in (18).

(18) [[**the governor is sure that $\mathbf{S_9}$**]] $^{c, w}$ = 1 iff
a. $\forall u \in \mathbf{W}$: $(\exists v \in \mathbf{W}: v \leq_{\Omega 5(w)} u)$ $(\forall z \in \mathbf{W}: z \leq_{\Omega 5(w)} v)$ $[[\mathbf{S_9}]]^{z} = 1$, iff
b. every sequence of worlds in **W** which is weakly ordered by $\Omega$:$D_5(w)$ has a $\mathbf{S_9}$–uniform tail, $\text{iff}_{\text{def}}$
c. MUST ($\mathbf{S_9}$ || $\Omega$:$D_5(w)$).

When we negate this, negation is imported through classical negation and quantifier interactions, and we obtain an ascription with negation on the complement clause, and in which the main predicate is replaced by its dual, as in (19), where $x_5$ is the governor and $\mathbf{S_9}$ is the phrase marker of the complement, **the bill has enough votes to pass**.

(19) [[**not-(the governor is sure that $\mathbf{S_9}$)**]] $^{c, w}$ = 1 iff
a. $\exists u \in \mathbf{W}$: $(\forall v \in \mathbf{W}: v \leq_{\Omega:D5(w)} u)$ $(\exists z \in \mathbf{W}: z \leq_{\Omega:D5(w)} v)$ [[**not-$\mathbf{S_9}$**]] $^{z}$ = 1
b. there is a sequence of worlds in **W** that is weakly ordered by $\Omega$:$D_5(w)$, with a **not-$\mathbf{S_9}$**–persistent tail, $\text{iff}_{\text{def}}$
c. MIGHT (**not-$\mathbf{S_9}$** || $\Omega$:$D_5(w)$).

This represents an interpretation of (15b) with LIN, roughly paraphrasable as, "The governor thinks it possible that the bill doesn't have enough votes to pass." Likewise, in uttering (17), McEnroe was cueing his viewers to the possibility that Venus Williams would not crumble before Bartoli as Henin did.[9]

Of course, the ascription in (15a) could be uttered with the full interpretation, as asserting the existence of a mental state of the governor, characterized by certainty that the bill will pass. With this interpretation, (15a) asserts the existence of an eventuality $e_4$ with the governor as experiencer, and the modal schema will provide an essential condition on the identity of this eventuality across possible worlds, as in (20).

(20) ~ $\exists e_4$ [EXP $(e_4, x_5)$ $\wedge$ [$(\forall v \in \mathbf{W}$: EXP $(e_4, x_5)$ in $v)$: MUST ($\mathbf{S_9}$ || $\Omega$:$D_5(v)$)]]

[9] This provides an explanation of the data item brought up by an anonymous reviewer, mentioned in Chapter 3, that the example in (i) does not exhibit the scalar implicature we would expect from the fact that negated *sure* is a weaker predicate than *sure*.

(i) # I'm not sure Venus will win; in fact, I'm sure she will.

The first clause of (i) reduces, with LIN, to the claim that I think it possible that Venus will not win. This is incompatible with the follow-up assertion that I am sure she will win.

In order to adequately place this interpretation, suppose that Jill and George are members of the governor's staff holding a press interview together. George, who is less experienced, had misinterpreted some clues and utters (15a) during the interview. Jill steps forth immediately and says, "The governor is not sure that the bill has enough votes to pass," stressing *not*, and then elaborates. Jill asserts the simple denial represented in (20); the elaboration could explain that the governor hasn't thought about the bill yet, or made up her mind about its prospects, which would be to deny that there is any relevant mental state $e_4$ of the governor to be talked about yet. Or Jill's elaboration could be designed to suggest that the governor has a view on the bill's prospects (a mental state $e_4$ exists), but the mental state is not a state of certainty that the bill will pass – it cannot be characterized by the content condition MUST ($\mathbf{S_9}$ || $\Omega$:$D_5(v)$). We could even imagine that Jill suspects that the governor is sure that the bill will pass, so that $e_4$ exists and is experienced by the governor, but Jill is not sure that what the governor has decided is that the bill has enough votes to pass. In the latter case, Jill would know that the governor has a mental state about the bill, but is not confident that she knows the identity conditions of this mental state across possible worlds – which is to say, she is not sure what mental state it is.

Call the process of importing negation as in (19), Negative Importation with Dualization (NID). NID is proposed here as a general analysis of LIN.

Note that we are not saying that negation of the minimal ascription is ambiguous between an interpretation with imported Neg and a wide scope interpretation of negation. These two representations of the minimal ascription are logically equivalent; they are both representations of the LIN. The only wide scope interpretation of negation in (15b) proposed here is that given in (20), a negation of a complex condition with multiple conjoined parts, where negation cannot be imported, except as a disjunction of negative conditions. The ambiguity is between negating the minimal representation, resulting in a dual quantified modal schema, and negating the full representation, with the result in (20). Of course, negation can be imported in (20) too, producing a disjunction of negated conditions, one of which will be an essential content condition with dual force to the original. Thus, deep within the disjunctions involved in the elaboration of (20), (19c) will appear. The difference between the two interpretations of (15b) therefore, with LIN or without LIN, is whether (19c) appears as the interpretation of the negated ascription (with LIN), or whether (19) occurs embedded deep within the interpretation.

### *7.2.3 Standard Neg-Raising as LIN*

Now turn to a standard example of LIN, as traditionally delineated, in (21b), which is one interpretation of the ascription in (21a) with the NRP *think*.

(21) a. The governor believes the bill will pass.
b. The governor doesn't believe the bill will pass.

The report in (21a), without negation, has the truth conditions in (22).

(22) $[[\textbf{the governor believes that S}_9]]^{c,w} = 1$ iff
a. $(\forall u \in \mathbf{W}: [[\textbf{not-S}_9]]^{c,u} = 1)\ (\exists v \in \mathbf{W}: v \leq_{\Omega:D5(w)} u)$
$[[\mathbf{S}_9]]^{c,v} = 1\ \wedge$
$(\exists u \in \mathbf{W}: [[\mathbf{S}_9]]^{c,u} = 1)\ (\forall v \in \mathbf{W}: v \leq_{\Omega:D5(w)} u)$
$[[\mathbf{S}_9]]^{c,v} = 1$, iff
b. every sequence of worlds in **W** that is weakly ordered with respect to $\Omega$:$D_5(w)$ has a $\mathbf{S}_9$–persistent tail, and there is such a sequence with an $\mathbf{S}_9$–uniform tail, $\text{iff}_{\text{def}}$
c. SHOULD $(\mathbf{S}_9 \parallel \Omega{:}D_5(w))$.

Negating this yields the truth conditions in (23).

(23) $[[\textbf{not-(the governor believes that S}_9\textbf{)}]]^{c,w} = 1$ iff
a. $(\exists u \in \mathbf{W}: [[\textbf{not-S}_9]]^{c,u} = 1)\ (\forall v \in \mathbf{W}: v \leq_{\Omega:D5(w)} u)$
$[[\textbf{not-S}_9]]^{c,v} = 1$
$\vee\ (\forall u \in \mathbf{W}: [[\mathbf{S}_9]]^{c,u} = 1)\ (\exists v \in \mathbf{W}: v \leq_{\Omega:D5(w)} u)$
$[[\textbf{not-S}_9]]^{c,v} = 1$, iff
b. there is a sequence of possible worlds weakly ordered with respect to $\Omega$:$D_5(w)$ that has a **not**-$\mathbf{S}_9$–uniform tail, or every such sequence has a **not**-$\mathbf{S}_9$–persistent tail, $\text{iff}_{\text{def}}$
c. $\text{SHOULD}^{\perp}$ $(\textbf{not-S}_9 \parallel \Omega{:}D_5(w))$.

By virtue of the change from conjunction to disjunction, $\text{SHOULD}^{\perp}$ is somewhat weaker than SHOULD, so (23) is a somewhat weaker condition on **not**-$\mathbf{S}_9$ than (22) is on $\mathbf{S}_9$. This accounts for the slight "weakening" or "softening" noted by Poutsma, Bolinger, and Horn, exhibited when we paraphrase the interpretation of (21b) with LIN as, "The governor rather thinks the bill won't pass," or, "The governor is inclined to think that the bill won't pass."

As a purely notational device, for discussion, the interpretation of (21b) with LIN given in (23) will be indicated in the following shorthand: the governor believes$^-$ ~ (the bill will pass). This is not to say that there is a predicate believe$^-$. It is just a way of indicating that the interpretation with LIN in the model is that given in (23), a modal condition involving the dual of SHOULD.

The import of this analysis is to subsume the standard examples of Neg-Raising under the somewhat broader class of examples of LIN, derived through Negative Importation with Dualization. The process of NID produces all examples of LIN, whether or not they are traditional examples of Neg-Raising. But this does not analyze away the class of traditional examples of Neg-Raising as a natural class. The traditional NRPs are formally distinguished by the fact that they are nearly self-dual, having duals that are just slightly weaker than themselves. And they are distinguished on phenomenological grounds by the fact that, when we do NID to obtain LIN with a traditional NRP, it is striking to our intuitions that the matrix predicate remains (nearly) the same with the LIN. This is in contrast to a matrix predicate such as *sure*, which permits NID and thus admits an interpretation with LIN on the present account, but where the result with LIN is intuitively understood as involving a predicate of quite different strength from the original. Thus, the class of traditional NRPs has a legitimate formal definition and can be identified as a legitimate natural class on phenomenological grounds: they are those predicates whose ascriptions can be interpreted with NID as a modal condition in which the associated modal operator is (nearly) its own dual.

But the interpretive apparatus of LIN is NID, and those interpretive mechanisms equally well produce LIN in examples with predicates like *be sure*, which are not traditional NRPs. From now on, NRPs will refer to the natural class of traditional NRPs. But "predicates that permit LIN" will refer to the wider class of predicates that permit LIN, including predicates such as *be sure*.

Finally, note that *imagine*, though traditionally identified as Neg-Raising, doesn't become slightly weaker under LIN; in fact, it becomes somewhat stronger, as illustrated in (24).

(24) I don't imagine you have a hundred dollars you could lend to me.

The best paraphrase of (24) is, "I rather expect that you don't have a hundred dollars you could lend to me." As a request, (24) is more polite than, "Do you have a hundred dollars you could lend to me?" because it is interpreted as in the "rather expect" paraphrase, asserting that the addressee probably doesn't have the money to lend, thus taking pressure off the addressee to comply with the indirect request. On the present account, *imagine* is somewhat weaker than the midpoint of its scale, so its dual, which we take to have the strength of *believe*, is somewhat stronger. The operation of NID to produce LIN in (24) yields an interpretation with the correct strength, in line with the paraphrase.

## 7.3 Applications and discussion

Even with LIN, negation takes scope over the lower clause. This is not the interpretation of negation when it is overtly expressed in the lower clause. To see this, note that negation scopes under modals such as *should* and *must* in monoclausal examples, as in (25).

(25) a. He must not leave. [$\rightarrow$ must (not (he leave))]
b. He shouldn't leave. [$\rightarrow$ should (not (he leave))]

In contrast, with LIN in (26), we get negation scoping over the modals *must* and *should* in the lower clause, and the analysis of LIN as NID successfully accounts for this.[10]

(26) a. I don't think he must leave.
[→ I think⁻ (not (must (he leave)) ⇒ I think⁻ he could stay.
Sketch of derivation:
I don't think he must leave → I think⁻ ~(he must leave)
⇒ I think⁻ (must$^{\perp}$ ~(he leave))
⇒ I think⁻ he could stay.]
b. I don't think he should leave. [→ I think⁻ he should$^{\perp}$ stay. (Similarly.)]

Return now to the basic facts about LIN outlined in Section 7.1.1, and the central questions about LIN formulated in Section 7.1.2. On the analysis developed here, many properties of LIN follow directly from the implementation of LIN as NID. The core fact is that NID is available only when the matrix clause is interpreted as a bare attitude ascription, a modal condition on possibilities relative to the doxastic base of the experiencer, and is not available when this clause introduces or describes a state of the experiencer subject, or an event of which the subject is an agent.

Distinctions among broad classes of predicates, as to whether most of their members permit LIN, or most do not, follow directly from this core fact. Most ascriptions of propositional attitudes of assertion have only the full interpretation, and thus do not permit LIN; likewise for most attitude ascriptions of emotion. Predicates of perception (such as *seem*, *appear*, *look like*, *sound like*) also can be used to describe a state or event experienced by the subject, characterized by the content of the complement clause; this use is salient when the predicate is modified, as in, *It has long seemed to me that Max is dishonest*, used as a statement about the speaker, to express that the speaker's suspicions about Max have been a long-standing experience of the speaker. But these predicates are primarily used just to express a doxastic condition: *It seems to me that Max is dishonest* is commonly interpreted as just asserting that doxastic criteria of the speaker favor adoption of the view that Max is dishonest. Thus, the presence of this class on the list of classes of predicates that permit LIN is, on the present account, entirely expected. Likewise, predicates of volition (such as *want*, *intend*, *plan*) can be used to ascribe a bouletic state to their experiencer subject, an interpretation that is more salient when they

[10] This shows that a transformational analysis of LIN is actually not tenable, since it would raise Neg from its position in the subordinate clause to its surface position in the higher clause, but interpret it in its lower position. But, as seen in (26), negation with LIN is not interpreted as it would be if it occurred in its normal position in the lower clause.

are modified, as in, *Alex has long and fervently wanted to find the document*. But they can also be used to express a minimal content condition according to bouletic criteria of the subject, as in a casual, "just to let you know what's happening" use of, *I plan to leave in a few minutes*. Predicates of probability are just predicates of modality, so it is not surprising that they include ones that permit LIN. This includes statements with traditional NRPs such as *likely*, whose dual under NID is a predicate with slightly weaker modal force, since *likely* is "mid-scalar," that is, it has strength just above the midpoint of its scale. It also includes statements with *possible* and *probable*, as in, *It isn't possible/probable that we will win*; with NID, these convert to their respective standard semantic duals, *It is certain/possible that we won't win*. These are not traditional instances of Neg-Raising, since these two predicates differ significantly in strength from their duals. But they are instances of LIN, derived by NID.

Some factive predicates are used to report on an event of which the subject is an agent, as with *point out* and *notice*. Other factive predicates are used to report on a state experienced by the subject, as with *surprise* and *regret*. Ascriptions with these two types of factive predicates can only have the full interpretation, and therefore do not permit LIN. Factive predicates such as *realize* and *recognize* are not so easily pigeonholed. A statement such as *Suddenly, Alex realized that Bill stole the artifact* reports on an event of realization or recognition experienced by Alex, so the ascription would have the full interpretation with this use of *realize* or *recognize*, and LIN would not be available. On the contrary, with an ascription such as *Alex has long realized that Bill stole the artifact*, it is not clear that it would have to be a report on a state experienced by Alex, and could not be a more minimal report on Alex's doxastic situation, that Alex's doxastic alternatives are all compatible with the truth of the complement clause (and have been so for a long time). In the latter case, an ascription with *realize* or *recognize* should be interpretable as a bare condition on possibilities relative to Alex's doxastic ordering source, as one available interpretive option, and should therefore be expected to permit LIN. Since it doesn't, this indicates the need for a further explanation as to why LIN is not available in ascriptions such as those given in (6a), or in (27b).

(27) a. Alex realizes that Bill stole the artifact.
b. Alex doesn't realize that Bill stole the artifact.

As discussed in Chapter 6, Section 6.6, factive reports, even at their most minimal, don't just report that the complement clause comports with the doxastic base and ordering source of the experiencer. They also say something about the reliability of the experiencer's evidential ground relative to the speaker context (presumptively shared as common ground with the addressee), thus implementing conditions of justified true belief, or whatever one's analysis of factivity might be. As a result, factive ascriptions are too complex to permit

LIN, even when they do not characterize a mental state of the experiencer subject.

Now consider the capacity of certain adverbials in the matrix clause to block LIN. Consider first the example in (28), a constructed dialog between an art seller A and a middleman B, concerning the prospective sale of a collection of paintings to a wealthy collector, a few days after the collector has had an arranged viewing of the paintings.

(28) A: [eagerly] I bet she fervently wishes to see them again.
B: I just heard from her. She doesn't fervently wish to see them again.
In fact, she doesn't wish to see them again at all.

This example shows that negation in the main clause of B's second sentence, the one with *fervently* (call this sentence B2), can have wide scope, and is therefore not confined to be interpreted as constituent negation of the adverb *fervently*. Given that negation in B2 is not confined to be constituent negation, and can, in particular, be interpreted with wide scope over the matrix clause, it is then significant that it cannot be interpreted with the lower clause, to mean that she fervently wishes to not see the paintings again. Finally, B's last sentence, which we will call B3, can be interpreted with LIN, to mean that she wishes to not see the paintings again at all, indicating that *wish* is a predicate that (on its own) permits LIN. In B's second sentence, B2, the adverb *fervently* modifies the collector as experiencer of an emotional state $e_6$ of wishing. Thus, this ascription can be interpreted only as the denial of a complex set of truth conditions asserting EXP ($x_9$, $e_6$), and identifying $e_6$ across possible worlds by stating that a modal quantificational condition with the force of SHOULD, defined on the collector's joint doxastic and bouletic ordering source, holds in each world in which $e_6$ occurs, as in (29).[11]

(29) $\sim \exists e_6$ [EXP ($e_6$, $x_9$) $\wedge$ [($\forall v \in \mathbf{W}$: EXP ($e_6$, $x_9$) in $v$) SHOULD ($\mathbf{S_8} \parallel \Omega^*$:B$_6(v)$)]

The ascription in B3, in contrast, can be interpreted with LIN, as in (30), at the world of evaluation $w$.

(30) SHOULD$^{\perp}$ (**not-S$_8$** $\parallel \Omega^*$:B$_6(w)$)

[11] Recall from Chapter 6, Section 6.4, that the doxastic-bouletic ordering source $\Omega^*$:B$_3(v)$) is the result of squaring the bouletic ordering source with a doxastic context that could be internally inconsistent or incompatible with elements of the bouletic ordering source. It is obtained by intersecting the bouletic ordering source $\Omega$:B$_3(v)$) with those parts of the doxastic context that are necessary to produce a result that is more realistic or more counterfactual, whichever is most appropriate for the interpretation of a bouletic ascription in a context.

Having seen that *wish* permits LIN, but that LIN is unavailable when the adverb *fervently* is present in the main clause, it is then instructive to consider (31).

(31) a. She really wishes to see them again.
b. She doesn't really wish to see them again (until Tuesday).

Following Romero and Han (2004), the adverbial *really* in (31a) can be interpreted as a VERUM operator, pertaining to the veridicality or epistemic status of the assertion,[12] or as an intensifier of the matrix predicate, assigning a high degree of desire to the subject; to use one of their examples, compare this to the contrast between *Sandra really is clever* (VERUM) and *Sandra is really clever* (intensifier/degree modifier).

One interpretation of (31b), which seems to be the most readily available, can be paraphrased as, "In truth, she prefers not to see them again (until Tuesday)." This could be uttered by the middleman, with a slightly sour face, directed to the art seller, in reference to items that the prospective buyer has already seen. (A slightly drawn out pronunciation of *really*, lightly stressing it, favors this.) In this interpretation, the VERUM operator ("in truth") takes wide scope over the result of NID, thus, VERUM [she wishes$^{\perp}$ ~ (to see them again)]. Since *wish* has its lowest point just above the midpoint of its scale, wish$^{\perp}$ is slightly weaker than *wish*, so this interpretation with LIN involves only a slight weakening of the matrix predicate.

Another interpretation of (31b) is obtained when *really* is unstressed and generally quite phonologically reduced, and there is nuclear stress on *wish*. This interpretation is less negative than the other; it comports best with a situation in which the prospective buyer has no strong feelings one way or the other, but reports that she would be just as happy not to see them again (until Tuesday). On this interpretation, *really* strengthens the mid-scalar predicate *wish* to a modified predicate with strength near the top of its scale. The dual of *really wish* is therefore a predicate of weak desire or preference, in the same way that the dual of *be sure* is paraphrasable as "be inclined to think." The derivation can be summarized as follows: ~ [she (really wishes) to see them again] ⇒ [she (really wishes)$^{\perp}$ ~ (to see them again)] ⇒ [she would just as soon (not see them again)]. This interpretation follows directly through NID when *really* is interpreted as an intensifier or degree modifier of the matrix predicate.

But the main point for present purposes is that the unavailability of LIN with an adverbial such as *fervently* in the matrix clause, as opposed to *really*, is purely a function of the fact that the former adverb is interpretable only as a modifier of an event or state introduced or described by the matrix clause, and

[12] Romero and Han (2004: 624–627) analyze the VERUM operator as a "conversational epistemic operator," which doesn't simply assert a high doxastic or epistemic status for the proposition at issue; rather, it asserts that this proposition should be added to the doxastic common ground.

therefore requires the full representation of the ascription, while the latter, on at least one interpretation, and probably more than one, is compatible with an interpretation of the ascription as a bare content condition expressed in terms of one of the modal schemas, permitting LIN. Furthermore, it discredits appeal to an unanalyzed notion of functional load. It is not clear that there is any such notion that would attribute different degrees of functional load to the adverb *really*, interpreted as a VERUM operator or an intensifier of *wish* in (31b), where it permits LIN, and the adverb *fervently* in B2 in (28). What matters to the application of LIN is the sort of interpretation required of the ascription to support modification by *fervently* versus *really*.

Similar effects are observed with adverbials in the matrix clauses of ascriptions of judgment, as in (32).

(32) a. I don't advise you to go there (until Tuesday).
b. I don't wisely/happily advise you to go there (*until Tuesday).
c. I don't really advise you to go there (until Tuesday).

LIN is not available in the presence of the adverbials *wisely* and *happily* in (32b), but it is available with *really* in the matrix clause, in (32c), where *really* is interpreted as a VERUM operator, or as an intensifier of the predicate, and the matrix predicate, whatever its resulting strength, gets weakened to its dual with LIN.

Next, consider an example of LIN with multiple clausal embedding, as in (33b), where negation can be interpreted in the lowest clause.

(33) a. I think John wants to leave.
b. I don't think John wants to leave (until Sunday).

The availability of LIN depends on both clauses of (33a) being interpreted as bare modalized content conditions. Following the discussion of the interpretation of iterated attitude ascriptions in Chapter 3, Section 3.3, the interpretation of (33b) is given in (34), where *John* denotes the individual $x_7$ and $\mathbf{S}_{10}$ is the complement clause.

(34) $\text{SHOULD}^{\perp}\,(\text{SHOULD}^{\perp}\,(\textbf{not-S}_{10} \parallel \Omega^{*}{:}\text{B}_7(w)) \parallel \Omega{:}\text{D}_1(w))$

This is roughly paraphrasable as, "I'm inclined to think that John would rather not leave until Sunday."

We could indicate the effects of NID in (33b) schematically as in (35).

(35) a. ~ (I think John wants to leave until Sunday)
b. I think$^{\perp}$ ~ (John wants to leave until Sunday)
c. I think$^{-}$ [John wants$^{\perp}$ ~ (PRO to leave until Sunday)]
d. I think$^{-}$ [John wants$^{-}$ [~ (PRO to leave until Sunday)]]

LIN can be difficult to process with multiply embedded clauses, even using predicates that readily permit LIN with one level of embedding. In general, first person subjects and infinitival complements facilitate processing of LIN, so examples with LIN into multiply embedded clauses are easiest to obtain when these conditions prevail in either or both of the higher clauses.

On this account, an interpretation with LIN in a multiply embedded clause is derived whenever the higher, "intervening" clauses are interpreted as bare, modalized content conditions; thus, LIN should always be available when the intervening predicates permit LIN and no material in the higher clauses would "block" it (by requiring that the clause be interpreted as a complex condition modifying an eventuality). There are examples where LIN with a multiply embedded clause is not available even when the above conditions are satisfied. But there don't seem to be any systematic factors underlying the exceptional cases, or underlying any significant class of exceptions, in a way that would challenge the idea that the basic mechanism of LIN is NID, as proposed here. These examples will be discussed further in Sections 7.4.2 and 7.4.3, in a discussion of Gajewski's (2007) and Romoli's (2013) proposals that volitional predicates in the higher clause systematically preclude LIN.

All of the properties of LIN discussed so far in this section follow from the core of the present analysis, that LIN is effected through NID, and is therefore dependent on the ascription being interpreted as a bare modalized content condition. This doesn't, by itself, suffice to explain fine-grained lexical, dialectal, and cross-linguistic variation in LIN. For example, it doesn't explain why LIN is available with *wish* but not *hope* in standard varieties of English; why LIN is available with *expect* in most dialects of English, but with *guess* only in some dialects; or why German *hoffen* permits LIN, but not the English cognate *hope*.

But this fine-grained variation might be explicable based not on the mechanics of LIN using NID, but through the need of a language, faced with some vagueness in what counts as a predicate suited to the minimal interpretation, to effectively make a decision about how to classify a borderline predicate as having the minimal interpretation or not, and thus one to which the all-or-nothing interpretive process of NID should apply, or not. This is the sort of decision that could be handled by a suitable notion of markedness among predicates in the same broad lexical class (e.g., opinion predicates, or volitional predicates), whereby those which permit LIN tend to be less marked. First, given predicates *P* and *Q* in the same class, and of the "same type" (what is meant by this will be illustrated in the following section), *P* is more marked than *Q* if a statement using *P* uni-directionally entails one using *Q* in the same position. Thus, *long for* is more marked than *want*, when both refer to self-conscious desire (as opposed, say, to unacknowledged or repressed desire), since *Alex longs for Sue to be here* entails *Alex wants Sue to be here*, but not vice versa; in effect, the set of instances of longing is a proper subset of the set of instances of wanting.

Another component of markedness could be the frequency of a predicate: verbal *want* seems to appear more frequently in corpora than verbal *hope* or *desire*, and *suppose* appears more frequently than verbal *guess*; likewise, verbal *intend* and *plan* are more frequent than *make plans*. This suggests that predicates that permit LIN are generally more frequent than ones that don't. A scale of markedness could be constructed using an appropriately weighted average of these two factors, and possibly others, providing a measure of markedness among predicates of roughly equal strength in a class, which answers to our intuitions of how prototypical or marked a predicate is.

None of this will have any significance for a predicate that clearly is used to describe an event of which the subject is an agent, or a state experienced by the subject; sentences with such a predicate will have the full interpretation. But for the other predicates, the situation of the language is that some of them, but not all, could be designated as having the minimal interpretation as a bare modalized content condition with respect to the content of the complement relative to the experiencer doxastic and/or bouletic ordering source. If there is a markedness scale among these predicates, it is reasonable to suppose that there is a language-particular threshold at which this cut will be made, so that predicates above the threshold in markedness participate only in the full interpretation, while predicates below the threshold permit the minimal interpretation. The threshold is not a semantic property of any predicate, but rather a way for the language to resolve the middle ground, so the exact specification of the threshold can be reasonably expected to vary from language to language. Two cognates, such as German *hoffen* and English *hope*, could then by chance turn out to be on the opposite sides of the thresholds of their respective languages, even if, as seems to be the case, there is otherwise no significant semantic difference between the two. If, as suggested here, the availability of the minimal interpretation is determined by such a markedness scale, then a pattern of lexical, dialectal, and cross-linguistic variation of exactly the sort observed would be expected.

## 7.4 Other recent accounts of the lower interpretation of negation

An excluded middle assumption for a propositional attitude predicate $P$ or other matrix predicate, with respect to the proposition $p$ expressed by the complement clause, is the condition $P(x_i, p) \vee P(x_i, {\sim}p)$. With regard to Alex and the belief that Bill stole the artifact, it would be the condition that either Alex believes that Bill stole the artifact, or Alex believes that Bill did not steal the artifact. If this holds as background, then the statement that Alex does not believe that Bill stole the artifact automatically entails the interpretation with LIN, that Alex believes that Bill did not steal the artifact. When such a background condition holds, LIN is obtained as a matter of course, even with predicates that don't otherwise permit it. This, of course, is not the phenomenon of LIN traditionally identified as Neg-Raising, which holds only for certain predicates, and which involves a slight weakening of the matrix predicate. Nevertheless, it is open for one to propose, as did Bartsch (1973), that NRPs are distinctive exactly in that they allow for a pragmatic presupposition of an excluded middle condition defined for the proposition expressed by their complement clause. They don't require such a pragmatic excluded middle presupposition, but when it is in force, the negated ascription can be interpreted with LIN.

Epstein (1976) and Horn (1978, 1989) cast doubt on this approach. To begin with, Epstein (1976) reinforced the irreducibly lexical nature of LIN by noting that even the most favorable pragmatic conditions will not produce LIN with a non-NRP, short of imposing an explicit excluded middle condition. To update Epstein's example, suppose that a family reunion has the unharmonious property that some in attendance are admirers of Sarah Palin and strong supporters of her for the vice-presidential nomination, while the rest abhor her views and think she is demonstrably unqualified to hold high political office. In this context, Jill asserts *Alex doesn't fear that Palin will be nominated.* Even in the highly favorable context of this family reunion, Jill's assertion doesn't seem to go so far as to assert that Alex fears that Palin will not be nominated. Furthermore, as Horn (1989: 321) argues, there is nothing in the suggestion that LIN proceeds from a pragmatic excluded middle presupposition that accounts for why LIN would be available quite generally with *suppose*, but with *guess* only in certain dialects, and with *want* but not with *desire*.

Nevertheless, despite the disfavor this approach has endured during the twentieth century, a few recent proposals revive the spirit of the excluded middle approach to LIN. Three of these will be reviewed in this section.

### *7.4.1 Tovena (2000)*

Tovena (2000) borrows from logic programming the idea that, in a closed world database, the failure to establish a proof of a proposition expressed by a phrase marker **S** is grounds to infer **not-S**. In evaluating **Alex believes that S**, the speaker and addressee's assessment of the information state of Alex is modeled as a search for the proposition $p$ expressed by **S** in the information state of Alex. The search is necessarily finite, both in principle and in practice, and when it comes to an end, the database is closed, and if $p$ is not found listed among the beliefs of Alex, then the inference to the presence of $\sim p$ in the database is made automatically. The inference is not due to an excluded middle assumption, but to the processes of belief evaluation that are, in effect, set up to render a verdict.

The database can be closed for an NRP such as *believe*, according to Tovena, precisely because belief is not an exacting condition; it is tentative and subject to revision, so it is evaluated in a temporally and logically local way. This locality is modeled by closing the database, allowing an interpretation to be obtained through a process of "negation as failure." A predicate that is not traditionally Neg-Raising, such as *be certain*, in contrast, is not provisional in this way; to check if one is sure of something requires an exhaustive search; the database remains open, and negation as failure cannot apply.

For Tovena, the distinguishing properties that make a predicate Neg-Raising include not just the property of permitting a LIN, but the property of being nearly self-dual. This seems to hark back to the fact that the most salient way for it to be false that Alex believes **S** is for Alex to believe **not-S,** so that a failure to verify that Alex believes **S** results in closing the database and concluding that Alex believes **not-S**. In contrast, if it is not true that Alex is certain that **S,** this could be for a variety of reasons, and Alex being certain that **not-S** is only one among them, and no more salient than the others. A failure to verify that Alex is certain that **S** does not lead to a parallel conclusion that Alex is certain of **not-S**, given the range of equally salient alternatives, so the database stays open and no conclusion is drawn.

It is tempting to view this as a processing counterpart of the theory of LIN by NID developed in the present work.[13] The whole point of the present approach is to have identical semantic interpretive mechanisms for NID with both classical NRPs and with predicates such as *suspect* and *be sure*, which have strength above or below the midpoint of their scale. The classical NRPs are then distinguished simply by the fact that they are nearly self-dual, so that what is special about them is emergent from the analysis. It could be argued that this is parallel to Tovena's excluded middle approach, in that a single interpretive mechanism of searching a database of beliefs, certainties, desires, and so on, is involved in interpreting the negation of any attitude ascription, and that what distinguishes classical NRPs from non-NRPs is the closure of the search when the predicate is nearly self-dual, as described. The main obstacle to comparing the two approaches in this way, regarding one as potentially a processing counterpart of the other, is that Tovena's account draws its primary distinction between predicates that are nearly self-dual and those that aren't, whereas the possible worlds account developed here draws its primary distinction between predicates that don't permit LIN and ones that do, whether or not they are nearly self-dual.

[13] An anonymous reviewer suggests consideration of this possibility.

### *7.4.2 Gajewski (2007)*

Gajewski (2007) revives the spirit of Bartsch's proposal by offering an account of LIN as a product of the interaction of the assertion of the matrix predicate with a soft (easily canceled, following Abusch [2010]) excluded middle presupposition (EMP) associated with the complement of the NRP as a conventional property of the predicate, an interaction achieved through explicit, semantically realized mechanisms of presupposition projection.[14] For example, *John thinks Mary left* has the EMP in (36), in Gajewski's notation, where $p$ is the proposition expressed by the complement clause: the beliefs of John in $w$ include $p$, or they don't.

(36) $B_{j,w} \subseteq \{v: p(v) = 1\} \vee B_{j,w} \subseteq \{v: p(v) \neq 1\}$

This presupposition is inherited by *John doesn't think Mary left*, and combines with the truth conditions of the negated clause, $\sim(B_{j,w} \subseteq \{v: p(v) = 1\})$, to yield LIN, expressed as $B_{j,w} \subseteq \{v: p(v) \neq 1\}$.

LIN will not be obtained with a verb such as *say* in place of *think* since the corresponding EMP with *say* is not true; *say* does not have an excluded middle condition of the sort in (36) associated with it as a conventional property of the verb.

Gajewski combines this with an account of the licensing of strict NPIs, including *until*-phrases, in terms of their occurrence in the scope of an antiadditive operator, based on the previous work by Zwarts (1998).[15] He notes that negated NRPs are distinguished from negated non-NRPs by being antiadditive, not merely Downward Entailing, as demonstrated by the contrast in (37)–(38) [Gajewski's (64) and (65)].[16]

(37) John doesn't think Mary left and John doesn't think Bill left. $\Rightarrow$
John doesn't think Mary left or Bill left.

(38) John isn't certain that Mary left and John isn't certain that Bill left. $\not\Rightarrow$
John isn't certain that Mary left or Bill left.

[14] Gajewski formulates his account with standard propositional attitude predicates as NRPs. In explicating the account, he mentions in passing (pp.303–304) that the propositional attitude predicate has a modal base, an idea that has evident affinities to the account of the interpretation of propositional attitude ascriptions developed here in Chapters 3 and 4. But Gajewski's truth conditions for a propositional attitude ascription for a predicate with modal base $M$, subject $x$, and propositional argument $p$, are $M(x) \subseteq p$. This is essentially the account of belief ascriptions we considered in (1) of Chapter 3, repeated in (2) of Chapter 4.

[15] An Anti-Additive function is one that satisfies the condition $f(X) \wedge f(Y) \Rightarrow f(X \vee Y)$, in addition to being Downward Entailing (satisfying the conditions $f(X) \vee f(Y) \Rightarrow f(X \wedge Y)$ and $f(X \vee Y) \Rightarrow f(X) \wedge f(Y)$).

[16] Gajewski (2011) and others have since argued that antiadditivity is not the correct licensing condition for strong NPIs. This conclusion is supported by the discussion of (41) and (42).

One difference between Gajewski's account of LIN and that presented here in Section 7.2 is that the representations derived on Gajewski's account do not capture the well-recognized weakening of the matrix predicate. Even if we admit that (36) is in force as a soft, easily canceled presupposition in a context in which the ascription *John doesn't think Mary left* is used, the interpretation obtained from it, according to Gajewski, which is $B_{j,w} \subseteq \{v\text{: } p(v) \neq 1\}$, seems to be too strong to represent the correct interpretation, which, on the notation used here, is, John thinks$^{-}$ ~ (Mary left).

Another difference is that Gajewski's account would not derive LIN for predicates such as *sure* and *imagine*, predicates that do permit LIN in examples such as (39), but that do not fit the profile of traditional NRPs since they are not even approximately self-duals.

(39) a. I'm not sure we can get in there.
b. I don't imagine we can get in there.

For example, applied to *sure* in (39a), Gajewski's account would rely on the implausible EMP in (40), where $S_{a,w}$ is the set of worlds (with $w$ as the world of evaluation) consistent with what the speaker is sure of.

(40) $S_{a,w} \subseteq \{v\text{: } p(v) = 1\} \vee S_{a,w} \subseteq \{v\text{: } p(v) \neq 1\}$

By the mechanisms of Gajewski's account, this would yield the interpretation with LIN given by $S_{a,w} \subseteq \{v\text{: } p(v) \neq 1\}$, which is too strong since the speaker in (39a) is not in a state of certitude about getting in. The account derives LIN only with mid-scalar predicates.

Ascriptions with *be sure* also cast doubt on the role of antiadditivity in licensing strict negative polarity items such as *until*-phrases in the complements to such predicates. In (41), the strict NPI *until Sunday* can be interpreted with the lower clause, where (41) means that the speaker suspects that the referent of the pronoun *he* doesn't want to leave until Sunday.

(41) I'm not sure he wants to leave until Sunday.

The test of antiadditivity for negated *sure* is the entailment in (42).

(42) I'm not sure Mary left and I'm not sure Bill left ⇒
I'm not sure Mary left or Bill left.

Although there is a temptation to say that this entailment fails as obviously as that for negated *certain* in (38), it actually takes some consideration to arrive at this conclusion. Note that, as applied to a predicate that permits LIN, the criterion of antiadditivity works only insofar as it applies to the interpretation with LIN. Consider the entailment with negated *think* in (37), which, on Gajewski's account, certifies *think* as an NRP licensing strict NPIs in its complement. The entailment in (37) does not hold for an interpretation of its component sentences with wide scope negation (e.g., as applied to a case where John has evidence that one of Mary or Bill left, but the evidence does not determine which of them was the one that left). The entailment holds when its component sentences are interpreted with LIN: if John thinks (actually, thinks$^{-}$) that Mary didn't leave, and thinks that Bill didn't leave, then he must think that neither Mary nor Bill left. Similarly, in (42), we must put aside the interpretations of the component sentences with wide scope negation, which indeed would make (42) parallel to (38). Interpreting the component sentences with LIN, the entailment in (42) would be the following: if the speaker *a* thinks it possible (is sure$^{\perp}$) that Mary didn't leave, and likewise that Bill didn't leave, then *a* will think it possible (be sure$^{\perp}$) that ~(*Mary left or Bill left*). But this does not hold. In a situation in which one of Bill and Mary must pick up their child from school, and the other must stay home to receive a delivery, but the speaker is unsure who will do which, the speaker can think it possible that Mary didn't leave, and likewise that Bill didn't leave, while being sure that it is not possible that neither left.

A third difference between Gajewski's account of LIN and that proposed in Section 7.2 concerns the availability of LIN with multiply embedded clauses. Horn (1978: 191) noted that when the predicates of the matrix and intermediate clause of (33b) are reversed, LIN is impossible, as in (43) [Horn's ex. (153′)].

(43) I don't want (him) to believe that she would do it (*until Sunday).

Gajewski concludes that LIN is systematically excluded when a predicate of volition occurs in the matrix clause, and a belief predicate occurs in the intermediate clause. Referencing the example in (44) [his exx. (107a) and (111a)], Gajewski provides an explanation for this based on the interaction between the EMP and mechanisms of presupposition projection.

(44) a. John doesn't think Fred wants Mary to leave.
b. John doesn't want Fred to think Mary left.

For the intermediate clause of (44a), Gajewski specifies the truth conditions in (45), along with the indicated EMP generated by *want*, where *p* is the proposition expressed by the lowest clause, *Mary to leave*.[17] ($D_{f,u}$ is the set of desires of Fred at world *u*.)

(45) TC: $\lambda u.D_{f,u} \subseteq \{v: p(v) = 1\}$
Presupp: $\lambda u.D_{f,u} \subseteq \{w: p(w) = 1\} \vee D_{f,u} \subseteq \{w: p(w) \neq 1\}$

[17] In addition, the intermediate clause inherits presuppositions of the lowest clause, such as the presupposition that Mary is at a certain place and has not yet left it; Gajewski assumes that these are inherited by the intermediate clause as beliefs of Fred.

Without negation, (44a) is interpreted as in (46), where the highest clause inherits the presupposition in (45) as a belief of John, Presupp (i); in addition, the attitude verb *think* in the highest clause has an EMP expressed in terms of a belief operator on the content of the intermediate clause, Presupp (ii).

(46) TC: $B_{j,w} \subseteq \{u: D_{f,u} \subseteq \{v: p(v) = 1\}\}$
Presupp (i): $B_{j,w} \subseteq \{u: D_{f,u} \subseteq \{v: p(v) = 1\} \vee D_{f,u} \subseteq \{v: p(v) \neq 1\}\}$
Presupp (ii): $B_{j,w} \subseteq \{u: D_{f,u} \subseteq \{v: p(v) = 1\}\} \vee B_{j,w} \subseteq \{u: D_{f,u} \nsubseteq \{w: p(w) = 1\}\}$

Negation in (44a) negates the truth conditions in (46), and the negated truth conditions combine with Presupp (ii) to yield $B_{j,w} \subseteq \{u: D_{f,u} \nsubseteq \{w: p(w) = 1\}$, which combines with Presupp (i) to yield $B_{j,w} \subseteq \{u: D_{f,u} \subseteq \{v: p(v) \neq 1\}\}$. This represents LIN in (44a).

In contrast, for the intermediate clause in (44b), Gajewski specifies the truth conditions in (47), along with the EMP generated by *think*.

(47) TC: $\lambda u.B_{f,u} \subseteq \{v: p(v) = 1\}$
Presupp: $\lambda u.B_{f,u} \subseteq \{w: p(w) = 1\} \vee B_{f,u} \subseteq \{w: p(w) \neq 1\}$

Without negation, (44b) has the truth conditions and presuppositions given in (48), where Presupp (ii) is the EMP of *want*, and Presupp (i) is the presupposition of the intermediate clause, inherited by the highest clause as a belief of John.

(48) TC: $D_{j,w} \subseteq \{u: B_{f,u} \subseteq \{v: p(v) = 1\}\}$
Presupp (i): $B_{j,w} \subseteq \{u: B_{f,u} \subseteq \{v: p(v) = 1\} \vee B_{f,u} \subseteq \{v: p(v) \neq 1\}\}$
Presupp (ii): $D_{j,w} \subseteq \{u: B_{f,u} \subseteq \{v: p(v) = 1\}\} \vee D_{j,w} \subseteq \{u: B_{f,u} \nsubseteq \{w: p(w) = 1\}\}$

In this case, the negation of the truth conditions of the highest clause combine with Presupp (ii) to yield $D_{j,w} \subseteq \{u: B_{f,u} \nsubseteq \{w: p(w) = 1\}$, but this condition on John's desires about Fred's beliefs does not interact with Presupp (i), which pertains to John's beliefs about Fred's beliefs. An inference to LIN with the lowest clause, expressed by $D_{j,w} \subseteq \{u: B_{f,u} \subseteq \{v: p(v) \neq 1\}\}$, is not available in (48).

This argument against LIN in (44b) goes through because desires and beliefs are not commensurable. The same result will be obtained whenever the higher predicate is a volitional predicate, represented by $D_{j,w}$ in Presupp (ii) in (48), and the intermediate predicate is any other belief predicate, or a modal or perceptual predicate such as *likely* or *seem*. With such a predicate (and corresponding operator $X_{f,u}$), the derivation of LIN will always be blocked by the failure of John's desire, $D_{j,w} \subseteq \{u: X_{f,u} \nsubseteq \{w: p(w) = 1\}\}$, to interact with an EMP on what John believes, what is likely, or what seems to be the case, in Presupp (i).

The account is elegant, but it may be too strong in excluding all examples along the lines of (44b), with a higher volitional predicate and an intermediate predicate of belief, modality, or perception. A job candidate could well utter (49) before an interview, with negation interpreted with the lowest clause. If rescuers have spotted a missing hiker, lying motionless on the other side of a chasm, out of reach, then, while waiting for the backup equipment and personnel to arrive, one could utter (50) with negation interpreted in the lowest clause.

(49) I don't want to seem to be (too) eager.
(50) I really don't want to think she's dead, just asleep.

To get closer to Horn's and Gajewski's examples, if the speaker is reporting on a situation in which Maria was mortally injured and died, and knows that Maria's parents had rushed to be by her side, but doesn't know if they arrived on the scene before she died, the speaker can use (51) to report a desire to think Maria did not die until her parents arrived.

(51) I really don't want to think she died until her parents got there.

But curiously, even in this very facilitating context for LIN, the presence of a modal in the lowest clause is very disrupting of LIN.

(52) I really don't want to think she might have died (*until her parents got there).

Thus, the order of verbs in (43) doesn't seem to be fully determinative of whether or not LIN is available.

A full assessment of the status of the failure of LIN with multiply embedded clauses, with predicates that ordinarily permit it, is beyond the scope of the present work. But these failures of LIN seem to depend on multiple factors, including the predicates chosen, and the presence of the modal in (52) versus (51). They also seem highly sensitive to context, as we saw comparing (50) and (51) with (44b). It is not clear that the order of predicates should be credited with systematically excluding LIN.

### *7.4.3 Romoli (2013)*

Another form of excluded middle account of LIN was offered by Romoli (2013), who argues that there is an excluded middle semantic presupposition lying within an alternative set defined by a question under discussion in the discourse context. As in Gajewski's account, the interaction of this presupposition with the assertion yields the interpretation with LIN by implication. But for Romoli, the assertion is "exhaustified" – in effect, it affirms the asserted proposition while at the same time negating all relevant alternatives.[18] Relevant alternatives are defined on a scale, so LIN emerges as a scalar implicature of the interaction between the assertion and the alternative set.

Romoli illustrates by taking the alternative set for the determiner quantifier *every* to be that given in (53), containing the denotations of relevant lexical alternatives whose semantics lies on the same scale as *every*.[19] The alternative set for *Every student came* is then given in (54), using functional composition notation for the denotations in the alternative set.

(53) Alt (**every**) = {**every, some**}

(54) Alt (**every (student) (came)**) = {**every (student) (came), some (student) (came)**}

Next, we add negation in (55a) in order to derive the scalar implicature of (55a) given in (55b), that some student came. (This is over and above the entailment that some student didn't come.)

(55) a. Not every student came.
b. Some student came.

The alternative set of (55a) is that given in (56).

(56) Alt (~**every (student) (came)**) =
{~**every (student) (came)**, ~**some (student) (came)**}

In asserting (55a), we combine this assertion with the denial of alternatives to obtain the assertion along with its scalar inference, as in (57), called the exhaustification of (55a) with respect to its alternatives in (56).

(57) EXH (~**every (student) (came)**) =
~**every (student) (came)** ∧ ~~**some (student) (came)** = ~**every (student) (came)** ∧ **some (student) (came)**

Note that the assertion of *Every student came* on this approach, combining the assertion with denial of alternatives in (54), would yield **every (student) (came)** ∧ ~**some (student) (came)**, a contradictory assertion whenever the domain of quantification is nonempty. For this reason, the operative "excludable" alternatives are "all the ones that can consistently be negated without contradicting the assertion" (Romoli 2013: 306).

[18] This builds upon the work by Chierchia et al. (2012), Chemla (2009), Fox and Katzir (2011), Magri (2010), and others.

[19] Curiously, Romoli does not include *no* in the alternative set of *every*. This ultimately makes no difference since the alternative with *no* naturally drops out of all the calculations Romoli considers.

A non-NRP such as *be-certain* has an alternative set with no excluded middle presupposition. If $a$ is the individual who is the experiencer subject, and $p$ is the propositional content of the complement clause, Romoli expresses this as in (58).

(58) Alt ($\mathbf{be\text{-}certain}_a\ p$) = {$\mathbf{be\text{-}certain}_a\ p$, $\mathbf{be\text{-}possible}_a\ p$}

No non-contradictory scalar implicature can be generated from this. Along the lines described, if we add negation, this changes since it can be seen that the assertion that Alex is not certain of proposition $p$ will have the implicature that Alex considers $p$ possible. That seems to be the correct result.

For an NRP like *believe*, on the contrary, the alternatives include the asserted attitude together with an excluded middle presupposition, as in (59).

(59) Alt ($\mathbf{believe}_a\ p$) = {$\mathbf{believe}_a\ p$, ($\mathbf{believe}_a\ p \vee \mathbf{believe}_a\ {\sim}p$)}

Negating this in (60a), Romoli gets the assertion in (60b) and the alternative set in (61).

(60) a. Alex doesn't believe that it is raining.
b. $\sim \mathbf{believe}_a\ p$

(61) Alt (${\sim}\mathbf{believe}_a\ p$) = {${\sim}\mathbf{believe}_a\ p$, $\sim$ ($\mathbf{believe}_a\ p \vee \mathbf{believe}_a\ {\sim}p$)}

Combining the assertion in (60b) with denial of the alternatives in (61), the assertion yields the scalar inference given in (62); clearly, in the presence of the assertion, the excluded middle presupposition reduces to the interpretation with LIN.

(62) EXH ($\sim \mathbf{believe}_a p$) = ${\sim}\mathbf{believe}_a p \wedge {\sim\sim}$ ($\mathbf{believe}_a p \vee \mathbf{believe}_a {\sim}p$)
= ${\sim}\mathbf{believe}_a\ p \wedge$ ($\mathbf{believe}_a\ p \vee \mathbf{believe}_a\ {\sim}p$)
iff ${\sim}\mathbf{believe}_a\ p \wedge \mathbf{believe}_a\ {\sim}p$

The alternative set in (61), according to Romoli, is obtained from the main question at issue in the discourse. For both (60a) and its positive counterpart, that question is the one given in (63).

(63) What does Alex believe about whether it is raining?

The *wh*-complement in this question correlates with the excluded middle presupposition in the alternative set. For comparable examples with the non-NRP *be-certain*, the question at issue would be something like, "What is Alex's degree of conviction that it is raining?" This question leads to the alternative set in (58), which doesn't support LIN.

With this approach, Romoli notes that we don't get presupposition projection out of non-negative contexts such as the protasis of a conditional. This successfully argues against an analysis of the excluded middle assumption as an ordinary presupposition, as opposed to a semantic presupposition arising from the question under discussion in the discourse.

The main question to raise about this approach is whether the excluded middle presupposition, generated from the question under discussion in (63), is a plausible semantic presupposition of (60a) as a general, lexical matter rather than just in special circumstances when there is explicit reason to think that Alex has an opinion one way or the other. This issue may reduce to the appropriateness of (63) as the question under discussion in general or typical uses of (60a), rather than other formulations of the question under discussion, such as "What is the degree of Alex's conviction that it is raining?" or "What is Alex's stance toward the proposition that it is raining?" The latter two would have alternative sets akin to that in (58), except that we could fill in further points on the scale in light of discussions in Chapter 3, as in (64).

(64) Alt (**believe**$_a$ $p$) = {**be-certain**$_a$ $p$, **be-likely**$_a$ $p$, **be-possible**$_a$ $p$}

These would not permit interpretation of (69) with LIN on Romoli's account. One might answer on behalf of Romoli's account that these other questions are possible questions under discussion for (60a), and that they lead to an interpretation with wide scope negation, which is empirically possible.[20] But this response (which isn't Romoli's) would leave it unclear why, if ascriptions with *believe* are so flexible in responding to different questions under discussion, an ascription with *be-certain* is more restricted, appropriate only when the question under discussion is that in (63).

Furthermore, like Gajewski's account, Romoli's admits no change in the strength of predicates under LIN. In particular, it doesn't address the slight weakening of the mid-scalar (just above the middle of the scale) predicates addressed by Horn. According to the account developed in the present work, Horn was on the right track, but the slight weakening is just a special case of dualization with Neg-Incorporation.

Putting aside further consideration of the basic distinction between NRPs and non-NRPs, consider other facts about LIN discussed in the first sections of this chapter. One of these facts was that extra content in the matrix clause blocks LIN. Romoli gives one example of this, involving a stressed auxiliary verb in the matrix clause. His example is in (65b), where the follow-up shows that the first sentence is not interpreted with LIN; this contrasts with (65a), where the first sentence can be interpreted with LIN, in which case the follow-up is infelicitous.

(65) a. Alex doesn't think that it is raining. #He isn't sure.
b. Alex DOESN'T think that it is raining. He isn't sure.

[20] Romoli doesn't take this line, allowing that ascriptions such as (60a) lack LIN only when there is stress or some other element in the matrix clause. Romoli's approach to these are discussed in the following sections.

Romoli's analysis is that the extra content in the matrix clause changes the question under discussion and thereby blocks the Neg-Raising inference, which depends on the question being as in (63). Specifically, for (65b), and its positive counterpart with stressed *DOES*, the question under discussion is *Does Alex think that it is raining?* This generates the binary alternative set in (66) rather than one with an excluded middle presupposition.

(66) Alt ($\sim$**think**$_a$ $p$) = {$\sim$**think**$_a$ $p$, $\sim\sim$ **think**$_a$ $p$}

Romoli doesn't discuss other matrix clause material that would block LIN even in the presence of an NRP, such as manner adverbials or subject-oriented adverbials. Consider such an example in (67).

(67) a. Alex earnestly believes that Fred left.
b. Alex doesn't earnestly believe that Fred left.

What is the question under discussion for these ascriptions? It could be something like, "What is the nature of Alex's belief that Fred left?" And this would clearly be the case when the adverbial is stressed, in which case it is in focus and attracts negation, leading to constituent negation of the adverbial. ("It isn't with earnest that Alex believes that Fred left.") But when the adverbial is not stressed, it is not clear that the question would have to be as adverb-centered as this one. And it's not clear that the question couldn't be, "What does Alex earnestly believe about whether Fred left? This would seem to admit a modified variant of (61) along the lines of (68), for *q* the proposition that Fred left.

(68) Alt ($\sim$**earnestly-believe**$_a$ $q$) = {$\sim$**earnestly-believe**$_a$ $q$,
$\sim$ (**earnestly-believe**$_a$ $q$ $\vee$
**earnestly-believe**$_a$ $\sim q$)}

This would yield an interpretation of (67b) with LIN. To prevent LIN in such examples then, Romoli's account would seem to require that the question under discussion in an utterance of (67a,b) is explicitly adverbial-centered, leading to an alternative set for (67b) other than that in (68).

If the question under discussion in (67a,b) with unstressed *earnestly* is indeed, "What is the nature of Alex's belief that Fred left?" this has implications for unstressed, *really* in the examples in (69), or the example from Section 7.3, *She doesn't really want to see them again.*

(69) Alex doesn't really believe that Fred left.

Remember that *really* has the VERUM reading of Romero and Han (2004), which takes wide scope. But, as discussed in Section 7.3, it can also be interpreted as a degree strengthener in the positive clause, which is weakened through NID to a slight degree strengthener in (69), interpreted with LIN. Romoli's account would give the right result if the use of *really* is a good match to the context where the question under discussion is just, "What does Alex (really) believe about whether Fred left?" But that context is arguably the one where *really* is a VERUM operator. The example in (69) should have a question under discussion pertinent to degrees, rather than either of the other two discussed earlier ("What does Alex believe about whether Fred left?" and "What is the nature of Alex's belief that Fred left?"). This suggests that the question under discussion for (69) is, "What is the degree of Alex's belief that Fred left?" The alternative set generated for this would be akin to (64), and thus would not contain an excluded middle presupposition, predicting that it could not exhibit LIN.

It may be possible to formulate questions under discussion systematically in a way that would retain the *wh*-question for (65a) and for (69), but with a different focus of attention for the example with *earnestly* in (67a,b), explaining why the latter lacks LIN. One candidate for the question underlying (67a,b) would be "What is the nature of the eventuality in which Alex has a belief about Fred leaving?" An alternative set could be generated from this question, containing options with various combinations of alternatives to the adverbial denotation combined with various alternatives to $\textbf{believe}_a$ $q$. But it's not clear that an excluded middle presupposition would be among them, and even if it were, it seems that combination of these possibilities with the negated assertion would leave too many alternatives standing for a simple interpretation with LIN to emerge.

The point is that additional material in the matrix clause blocks LIN precisely where it takes the ascription away from being just a simple comment on a propositional attitude about *p*, of whatever strength. To a large extent, that is coextensive with cases when the question under discussion is more than "What does such-and-such a person *attitude* whether *p*?" But the discussion earlier suggests that the right result is obtained just when it happens to coincide with cases when something more than a minimal ascription is asserted.

Like Gajewski (2007), Romoli sets up interpretive mechanisms whose results comport with the observations by Horn and others that the interpretation of main clause negation is possible into the lower clause of multiply embedded attitude ascriptions when the higher attitude predicate is doxastic and the lower one bouletic, as in (70a), yielding the Neg-Raising inference in (70b), but critically not when the order of predicates is reversed, as in (71a).

(70) a. I don't believe Bill wanted Harry to die.
b. I believe Bill wanted Harry not to die.

(71) a. I don't want Bill to believe Harry died.
b. I want Bill to believe Harry didn't die.

As discussed in the previous section, it isn't clear we want to make this inference generally unavailable. Taking examples introduced in the previous section, LIN seems possible in (72a) with exactly the order of predicates in (71a), but not in (72b), which has the same order of predicates along with an additional modal in the lowest clause.

(72) a. I don't want to think she died until her parents got there.
b. I don't want to think that she might have died (*until her parents got there).

The idea was that other factors than just the order of predicates are at play here, so it is not clear that LIN should be blocked in principle when a doxastic ascription is embedded under a bouletic ascription. Furthermore, Romoli's account relies critically on the assumption that bouletic ascriptions presuppose that the truth of the embedded proposition is an open question. As Romoli notes, a bouletic ascription with *want* "comes with the presuppositions that the complement of *want* is not settled in the attitude holder's mind" (2013: 331–332).Thus, Romoli claims that (73a) would presuppose (73b).

(73) a. Mary wants Fred to come.
b. It's possible for Mary that Fred comes and it's possible that he doesn't.

It was argued previously (Chapter 4, Section 4.3.2) that this assumption is not generally true, and that the interpretation of a bouletic ascription such as (73a) is based on an ordering source reflecting any assumptions that are made about the relation of the bouletic base to the doxastic base. It is not necessary to exclude the extreme cases (or boundary values) as doxastic possibilities, as (73a,b) do, and in fact, it is sometimes necessary not to do so.

### *7.4.4 Conclusion regarding excluded middle accounts of LIN*

The three excluded middle accounts of LIN considered here seek to ground the excluded middle assumption in something more basic: a closed database for deciding the truth of an ascription (Tovena); triggers for soft presuppositions (Gajewski); or a member of the alternative set derived from a question under discussion that is formulated with an embedded yes/no question (Romoli). Like any account, including the one developed in Section 7.2, these will have to come to terms with the vagaries of the distinction between NRPs and non-NRPs. It is even possible they could adopt the same approach to those issues as the present account, assuming that the distinction is a basic one; that every language has to make it; that the distinction largely tracks markedness; and that for predicates in the middle ground, the distinction can become rather arbitrary as each language simply draws a line somewhere to make the distinction, producing instances of dialectal variation in the classification of a predicate and cross-linguistic variation in the classification of cognates. The present section has argued for Neg-Incorporation under Dualization as the mechanism of LIN, primarily on more narrow empirical grounds. But the basic excluded middle proposition still seems to be not very well motivated as the property demarcating the most neutral use of predicates like *think*, *believe*, and *want* from their less neutral uses (e.g., with matrix clause adverbials), as well as from neutral uses of other predicates, such as *be-certain*. And none of the excluded middle accounts naturally includes cases of LIN with predicates that permit LIN but are not mid-scalar, such as *imagine* and *be-sure*.

# 8 Events in epistemic and root modal interpretation

As discussed in Chapter 5, Katz (2000, 2003, 2008) argued that no stative attitude ascriptions need to be interpreted as descriptions of an eventuality corresponding to a Davidsonian event variable introduced by the attitude predicate; in a parsimonious semantic theory it follows that none should be so interpreted. As will be discussed in Section 8.1, Hacquard (2006, 2010) gave an analysis of epistemic modal interpretation that assumes that all attitude ascriptions are based on a Davidsonian event variable, in the context of a broader theory of root and epistemic modal determination and interpretation. The view developed in Chapters 3–6 charts a middle ground between these two proposals, arguing that some stative attitude ascriptions are dependent on a Davidsonian eventuality variable for the attitude predicate, and some are not. In the present chapter, it will be argued that epistemic modals are not strictly event-oriented (as Hacquard acknowledged), but best understood as dependent on an appropriate "cloud of contexts" of the sort described by von Fintel and Gillies (2008b), discussed in Chapter 1. The epistemic modal base and ordering source are retrieved from this cloud of contexts by a process more akin to pronoun resolution, fixing the values of base and ordering source associated with the lexical entry of the modal auxiliary.

## 8.1 Hacquard's event-based theory of modal determination and interpretation

Hacquard (2006, 2010) set forth to explain a variety of facts about modal interpretation, including the relative scopes of root and epistemic modals with respect to aspect, and the availability of actuality entailments for root modals in the perfect in French and Italian, but not for epistemic modals. The approach is to explain these facts in terms of different syntactic positions of root and epistemic modals, following Cinque (1999).

As noted by Picallo (1990) and Iatridou (1990), epistemic modals take scope over tense and aspect, and root modals take scope under tense and aspect; in symbols, epistemic modals > T/Aspect > root modals; see also Abusch (1997), Cinque (1999), Abraham (2001a, 2001b), and Stowell (2004). These relative scopes can be observed in languages having different morphological or syntactic representations of tense and aspect. In English, *could* appears in the perfect as either an epistemic or root modal, as in (1); when *could* is epistemic, (1) means it is possible, as judged by the speaker at the speech time, that Mary took the train at the past time given by the tense and aspect. In contrast, when *could* is a root modal, say, deontic or ability, (1) means that at the time given by tense and aspect, Mary had permission, or respectively, was able, to take the train.

(1) Mary could have taken the train.

This fact is also illustrated by the French example in (2), cited by Hacquard (2010: 90), where the possibility modal *pouvoir*, inflected for simple past and third person singular subject agreement, can be an epistemic or root modal.

(2) Mary put prendre le train.
Mary may/can take the train.

When it's epistemic, *pouvoir* is interpreted with scope over tense, asserting that it is possible there is a past time at which Mary took the train. When it is a root modal (say, ability), it is interpreted below tense, asserting that at some point in the past, Mary was able to take the train. The Italian data in (3), cited by Hacquard (2010: 88), exhibit the same effect, reflected syntactically in the relative position of the modal and the perfect auxiliary, where the modal preceding aspectual *HAVE* in (3a) can be interpreted only as an epistemic modal, not a root modal, and the participial form of the modal following the tensed aspectual *HAVE* in (3b) can be interpreted only as a root modal, not epistemic.

(3) a. Gianni può aver parlato aver Maria. (epistemic only, not deontic/ability)
Gianni can have talked to Maria
"Gianni may have talked to Maria."
b. Gianni ha potuto parlare a Maria. (deontic/ability, not epistemic)
Gianni has could talked to Maria
"Gianni was able to talk to Maria."

As Cinque (1999) emphasizes, this implicates a higher syntactic position for epistemic modals relative to root modals, with aspect in between.

Hacquard finds further support for this, along with the motivation for her event-based account of modal determination, in the fact, originally noted by Bhatt (1999), that French and Italian root modals, but not epistemic modals, exhibit an actuality entailment when they occur in the perfect.[1] The modal in (4) (Hacquard 2010: 90) can be interpreted as an ability modal (that Mary

[1] For illustration, see the data cited in Hacquard (2010: 89–90), her exx.17, 18, and 19.

was able to lift the table), a teleological modal (lifting the table was a way to achieve a goal), or a deontic modal (Mary had permission to lift the table), but with any of these root interpretations, the follow-up clause is infelicitous because the first clause entails that Mary did lift the table.

(4) Mary put soulever cette table, # mais elle ne la souleva pas.
Mary can-PAST-PERF lift this table, but she NE it lifted not
"Mary could lift this table, but she didn't lift it."

Critically, no such entailment is obtained when the modal is interpreted as epistemic.

Following Kratzer, modal interpretation requires a modal base and ordering source. Hacquard proposes that these are determined through uniform lexical entries of the modal auxiliaries. The syntactic positions of different modals then determine the semantic interactions that produce the observed interpretive differences between root and epistemic modals. A root modal, appearing low in clausal phrase structure, just above the VP predicate of the clause, has a circumstantial modal base relative to the abilities, goals, obligations, or preferences of the referent of that predicate's subject, and is evaluated at the time given by the tense/aspect of the clause. In an unembedded clause, an epistemic modal, appearing above tense and aspect, has an epistemic modal base reflecting the knowledge of the speaker of the clause, evaluated at the speech time. And, as noted by Stephenson (2005, 2007), an epistemic modal in the complement clause to an attitude verb is evaluated with respect to the doxastic state of the attitude experiencer, typically given by the subject of the higher clause, evaluated at the time specified by the tense/aspect of the higher clause. In (5), it is the knowledge base of Alex, or what Alex believes he knows, that counts for the evaluation of the embedded epistemic modal, not the speaker (or Bill).[2]

(5) Alex believes that Bill must/should/might be in the next room.

Hacquard's implementation of this idea involves postulating novel lexical entries for modal auxiliaries. To account for the interactions summarized, Hacquard assumes that a modal introduces an event variable. The position of the modal determines not only the modal's scope with respect to tense and aspect, but also how its event variable is bound and identified. The modal base and ordering source are functions of this event variable, so the syntactic position of the modal, and thus the identification of this event, determines whether

[2] For reasons to be divulged shortly, Hacquard assumes that the epistemic modal base is the information state (doxastic context) of the higher subject in (5), or the speaker of an unembedded clause, not the set of possible worlds consistent with what that person knows.

the modal is a root modal, associated with the event described by the local VP, determining a circumstantial base and ordering source, or an epistemic modal, associated with the speech event and its context, determining an epistemic base and ordering source.

More specifically, the event variable introduced by a modal is bound by the most local available binder, depending on the position of the modal in the syntactic structure of the clause, and the modal is thereby anchored to whatever event is associated with that binder. A root modal, occurring in the syntactic scope of aspect, has its event variable bound by aspect and is thereby anchored to the event that aspect places in time, which is the event associated with the local predicate. It is therefore evaluated relative to the abilities, goals or desires of the most prominent argument of that predicate (typically but not invariably given by the subject of the clause), or the laws or social conventions under which that individual falls, using a circumstantial modal base determined by the circumstances of that individual, and relative to the time specified by the tense and aspect of that clause.

A modal in the high position of an unembedded clause has its event variable bound by a default speech operator and is thereby anchored to the speech event; by virtue of the propositional content of the speech event, the high modal is accordingly interpreted as epistemic according to speaker beliefs at the speech time. Likewise, a modal in the high position of the complement to an attitude verb has its event variable bound by the aspect node of the higher clause and so is anchored to the eventuality of the attitude; by virtue of the propositional content of the attitude ascribed, it is interpreted accordingly as epistemic relative to the experiencer of the attitude, the referent of the subject of the higher clause, and the time specified by the tense and aspect of the higher clause. By being anchored to an event of assertion or propositional attitude, the high modal in either case is interpreted as epistemic, relating the propositional content of that event to possible worlds in the information state of the speaker or of the attitude experiencer.

In recognition of issues in the literature surveyed in Chapters 1 and 2, Hacquard (2010: 92) refines this proposal. She notes that Peggy Speas, in a talk in 2004, and Stephenson (2007) "show that epistemics are not always relative to the speaker, but rather to the local knowledge bearer: the *speaker* in matrix contexts, the *attitude holder* in embedded contexts." Hacquard further notes that the speaker in matrix contexts may have to be generalized to the speaker's community, citing DeRose (1991) and von Fintel and Gillies (2008b). Hacquard does not confine the participant of the VP event to the local subject, citing examples of Hackl, including, *It can rain hard here*, in which the local subject is pleonastic, and *A lot of people can jump in this pool*, where the ability of accommodating lots of people is attributed to the pool itself, not to the people who jump in it. In a footnote, she allows that the "participant" in a root modal can be a location, as in, *It can rain hard here*. Hacquard (2010: 92–95) stresses that the modal, in being anchored to an event, is keyed to an individual and the time associated with that event: the speaker or local knowledge bearer for an epistemic, at the speech time, and a participant in the VP event for a root modal, at the time specified by the tense and aspect of the local clause.

To spell out some of the details of implementation, Hacquard modifies the semantics of modals to be relative to events rather than to worlds of evaluation, and follows Percus (2000) in taking possible worlds to correspond to variables (pronouns) introduced in the object language, and bound by modals, attitude verbs, or a default binder at the periphery of the clause. A world variable is introduced by the main predicate of a clause and can be modified or qualified by aspect. Aspect binds the Davidsonian event variable introduced by the main predicate. A low modal binds the world variable introduced by the predicate, and aspect, which raises above the modal to just below T, binds the event variable in the world of evaluation (which we will take to be the actual world, $w_0$). A syntactic structure for (6a), simplified from that assumed by Hacquard (2010: 98), is given in (6b), and compositional semantic interpretation results in the truth conditions in (6c), using Hacquard's representation of the perfect.

(6) a. Il put (enfin) peuvoir.
it could-PERF (finally) rain

b.

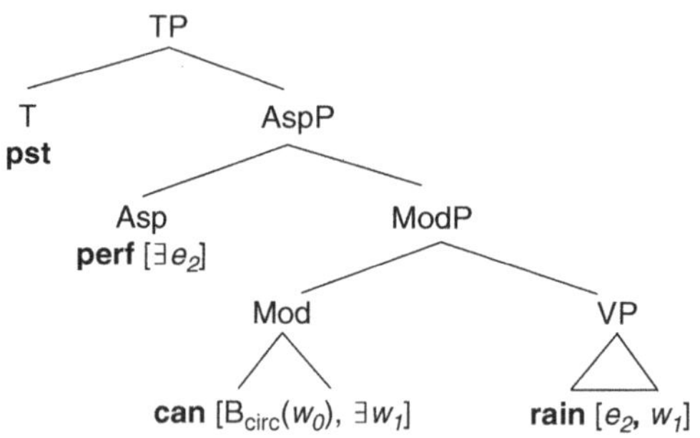


c. $\exists e_2$ [$e_2$ in $w_0$ & $\tau(e_2) \subset$ t & $\exists w_1$ compatible with the circumstances of $e_2$: rain ($e_2$, $w_1$)]

This ensures that the event $e_2$ in the actual world is an event of rain in a world $w_1$ compatible with the circumstances of $e_2$. But Hacquard (2006) outlines a theory of event identification across possible worlds that ensures that, under any feasible notion of compatibility in (6b), if $e_2$ is an event of rain in a world $w_1$, then $e_2$ is an event of rain in $w_0$. This yields the actuality entailment of the perfect modal. The modal then takes the circumstances associated with the predicate as its modal base.

If the modal is merged above T, the main clause event, bound by aspect below T, is not accessible to be bound by the modal auxiliary. In a root clause, the high modal instead qualifies the speech event introduced by an ASSERT operator, which also introduces the doxastic context of the speaker, which the high modal picks up as its modal base. The world variable and the event variable are bound by default binders at the periphery of the clause. In the complement of an attitude predicate, the attitude verb plays essentially the role of the ASSERT operator in the root clause, introducing the doxastic context of the attitude experiencer, which serves as the modal base of a modal verb merged above T in the complement clause to the attitude verb.

This summary leaves out some technical details of implementation of Hacquard's account that are not of concern here.

The facts in (1)–(4) follow from the position of the modal. Furthermore, the root versus epistemic interpretation of a modal is determined by its syntactic position, so a uniform lexical entry can be assumed for root and epistemic modals. Hacquard assumes the uniform lexical entries for *must* and *can* represented in our notation in (7a,b), where *f* and *g* are modal base and ordering source of an as-yet-undetermined nature, and *q* is a propositional variable for the prejacent of the modal. Where the standard syntactic type of a modal is ⟨s, tt⟩, Hacquard gives the modal an event variable, which makes its type ⟨ε, stt⟩. This analysis adopts the Limit Assumption.

(7) a. $[[\textbf{must}]] = \lambda f\, \lambda g\, \lambda q\, \lambda e\text{: } \forall w' \in \max_{g(e)} (\cap f(e))\text{: } q(w') = 1$
b. $[[\textbf{must}]] = \lambda f\, \lambda g\, \lambda q\, \lambda e\text{: } \exists w' \in \max_{g(e)} (\cap f(e))\text{: } q(w') = 1$

When the modal appears in the high position, *f*(*e*) is the set of propositions compatible with what is known in *e*. When it appears in the low position, *f*(*e*) is the set of propositions compatible with the circumstances of *e*.

This compositional semantics outlined by Hacquard will largely serve to derive the similar interpretation proposed in Chapters 3 and 6 of the present work. There are, however, a few qualifications, and one notable sticking point, all of which are taken up in the next section.

## 8.2 On the role of events in modal determination

In Section 8.2.1, evidence will be given that supports Hacquard's proposal for the dependence of root modals on the Davidsonian event of the verb within the VP, although in some cases the interpretation accesses a related event. The implication of these results is that the present account can embrace Hacquard's approach to the derivation of truth conditions for root modals. Of course, the present account does not adopt the Limit Assumption, and expands the quantifiers in Hacquard's (7a,b) into Kratzer's modal quantificational schemas. So the present account cannot adopt (7a,b) from Hacquard, but it can adopt the abstraction over an event variable, along with Hacquard's syntactic structures and compositional semantics.

Hacquard (2010: 92) acknowledges that the dependence of main clause epistemic modals on the event of assertion might not provide an epistemic modal base and ordering source as a direct function of that event, but instead serve to invoke the cloud of contexts of von Fintel and Gillies (2008b), from which the modal base and ordering source are retrieved. This comports with the approach to epistemic modality taken here, so the present account could adopt Hacquard's compositional semantics for epistemic modals as well, with the assertion event serving as an index to the cloud of contexts of von Fintel and Gillies.

To this point, then, the present account is positioned to defer to Hacquard for the syntactic structures and compositional semantic interpretation procedures needed to derive the interpretations argued for in the previous chapters of this work.

The sticking point arises in cases of epistemic modal embedding under an attitude predicate, and concerns the dependence of an embedded epistemic modal on an event associated with the attitude predicate. This will be taken up in Section 8.2.2. This will then leave us with the question of whether the present account can defer to Hacquard's syntactic structures and compositional semantics across the board, with her uniform lexical entry for root and epistemic modal auxiliaries, both of them event-dependent.

### *8.2.1 Root modals*

Hacquard (2006, 2010) proposes that root modals are anchored to the Davidsonian event of the main predicate of the clause, and that they are therefore interpreted relative to the agent of that event, or, if there is no such agent, relative to a prominent participant in that event. To begin this section, we make an observation that adds to the panoply of distinctions between epistemic modal and root modal behavior, one that, moreover, supports Hacquard's theory of the event dependence of root modals. Consider (8a,b), uttered as epistemic modal statements about Bill's behavior during a long-distance race in which contestants are not in sight of one another, to explain why Bill does not seem to be trying hard, despite the fact that speaker and addressee can see that he is losing the race.

(8) a. Bill must not think that he's losing the race.
LIN: He evidently thinks that he's not losing the race.
b. Bill may not think that he's losing the race.
LIN: He may be inclined to think that he's not losing the race.

These can be minimal ascriptions, interpreted as indicated with the LIN, whereby (8a) means that Bill evidently thinks that he's not losing the race, and (8b) means that Bill may be inclined to think that he's not losing the race. This result is, of course, fully compatible with Hacquard's theory of modal interpretation, since the epistemic modal is dependent on the assertion event on her account, and not dependent on a Davidsonian event associated with the predicate of the clause, expressing the prejacent of the modal.

But now consider similar sentences with a deontic modal in (9)–(14). These have the interpretations indicated, which are not interpretations with LIN, produced by Neg-Incorporation with Dualization (NID), as discussed in Chapter 7. For example, (9) does not mean something like, "You are required to rather expect that you cannot get in there," or even, "You are allowed to rather expect that you can get in there."

(9) You must not think you can get in there.
"You are enjoined from thinking you can get in there."
#LIN: You are required to rather expect that you cannot get in there.

(10) You should not think you can get in there.
"You should refrain from thinking you can get in there."

(11) You may not think you can get in there.
"You are forbidden to think you can get in there."

(12) You must not think you eradicated the virus.
"You are enjoined from thinking you eradicated the virus."

(13) You should not think you eradicated the virus.
"You are advised not to think you eradicated it."

(14) You may not think you eradicated the virus.
"You are enjoined from thinking that you eradicated the virus."

There is an interpretation of the ability modal in (15) with a lower negation, but there is reason to think that it is not the result of LIN.

(15) I'm sorry, I cannot believe you eradicated the virus. [nonexclamatory skepticism]
"I'm sorry, I'm quite sure that you did not eradicate the virus."
#LIN (= I am able/inclined to believe you did not eradicate the virus.)

With the nonexclamatory use suggested, (15) means something like, "I'm sorry, I am quite sure that you did not eradicate the virus." This is stronger than the interpretation that would be produced by LIN obtained by NID, which would produce an interpretation that would require paraphrase with a much weaker predicate, as discussed in Chapter 7. So this, too, seems not to exhibit LIN.

In Chapter 7, the explanation for the failure of LIN under NID in cases of clausal complementation was the presence of a Davidsonian eventuality associated with the matrix clause. Therefore, the results in (9)–(15) indicate that the interpretation of the root modal in the main clause is dependent on an event in that clause that the epistemics in (8a,b) don't have. Interpreting (9)–(15) as full interpretations, characterizing an eventuality of some sort across possible worlds, we obtain representations that do not permit NID, thus blocking LIN.

This account seems to be correct for these examples, but on closer examination, it is not clear that root modals are always dependent on a Davidsonian event associated with the main predicate of their clause. Consider the result of combining root modals with stative predicates that would lack a Davidsonian variable according to Katz. We proceed with an examination of deontic, teleological, and ability modals, in turn, and consider for each what is required for interpretation as a root modal. Let S be the prejacent of a root modal statement. Let X($w$) be the modal base and Y($w$) the ordering source of the modal statement. Recall from previous chapters the representations of possibility and necessity modals, which are given in (16) and (17), with phrase structure constituent $\mathbf{S}_1$ for the prejacent.

(16) $[[\text{must-}\mathbf{S}_1]]^{c,w} = 1$ iff
$\forall u \in \cap X(w)$: $(\exists v \in \cap X(w)$: $v \leq_{Y(w)} u)$ $(\forall z \in \cap X(w)$: $z \leq_{Y(w)} v)$
$[[\mathbf{S}_1]]^{c,z} = 1$
$=_{df}$ any sequence of worlds in $\cap X(w)$ weakly ordered by $\leq_{Y(w)}$ has an $\mathbf{S}_1$–uniform tail.

(17) $[[\text{might-}\mathbf{S}_1]]^{c,w} = 1$ iff
$\exists u \in \cap X(w)$: $(\forall v \in \cap X(w)$: $v \leq_{Y(w)} u)$ $(\exists z \in \cap X(w)$: $z \leq_{Y(w)} v)$
$[[\mathbf{S}_1]]^{c,z} = 1$
$=_{df}$ there is a sequence of worlds in $\cap X(w)$ weakly ordered by $\leq_{Y(w)}$, with an $\mathbf{S}_1$–persistent tail.

Since the details of the quantificational structures will not concern us here as we focus on the determination of the modal base, we will paraphrase these as in (18), where X is the modal base and Y the ordering source, both vary with the type, or "flavor," of the modality.

(18) As worlds in X get more and more in accord with a partial order reflecting Y, it is increasingly the case that, in some/all of those worlds, S is true.

For a prototypical deontic modal statement, such as *Sam must show up in court tomorrow morning*, X is a set of laws, rules, conventions, or obligations in force for the referent of the local subject, Sam, pertaining to an event of the type expressed by the predicate of the clause. What happens when the predicate is stative, as in (19)?

(19) a. Alex must resemble Ortcutt.
b. The socks cannot lie on the floor.

Suppose that (19a) is uttered in a context of espionage, where Alex is assigned the job of impersonating Ortcutt. The obligation is then on Alex, but it has force for efforts at disguise and mimicry in which Alex engages. These do not correspond to a Davidsonian event for the predicate *resemble*, and in fact, this is a stative predicate for which Katz very plausibly argues that there is no Davidsonian eventuality. In (19b), the stative predicate *lie* is another that plausibly lacks an event variable, and the obligation is on people retrieved from context (whether addressee or not) in implicit or inferred events of undressing or dispensing with articles of clothing. Here we see that deontic modal interpretation depends on events, but not necessarily a Davidsonian event associated with the main predicate of the clause; it can depend on events that are ancillary to the content of the clause. Deontics in attitude ascriptions, considered earlier, show this quite clearly, as in (20).

(20) a. According to the Ministry of Truth, all citizens must believe that the chairman is a great swimmer.
b. They changed the rules; citizens may now believe that the emperor is mortal.

Intuitively, these are not minimal ascriptions making statements about the doxastic contexts of citizens; they pertain to aspects of the overt behavior of citizens (which is exactly what Katz proposed for the interpretation of attitude ascriptions). This contrasts with minimal ascriptions considered in previous chapters. Stative locatives provide further illustration, in (21).

(21) a. The package must be at the station at 10.
b. Alex must be at the station at 10.

Examples (21a,b) can both be used when the addressee or some third party is responsible for getting the package or Alex to the station. In that use of (21b), the obligation is on that third party, not on Alex. The events referenced are ones that pertain to getting the package or Alex to the station. We summarize the results for the basis of deontic modals in (22).

(22) Deontic:
X = a set of laws, rules, conventions, or obligations in force for a person or group of people retrieved from S, or from the discourse context, and responsible for whether S is true.

The prototypical teleological modal depends on the goals or interests of the overt or implicit agent of the main predicate of the clause, as in (23).

(23) a. Alex must catch this train (to get to the wedding on time).
b. The car must be fixed today (for us to drive to Makinac Island tomorrow).

But they can be used when the goals or interests pertain to a third party involved in some event implicated by the content of the local clause.[3] This is the case when (24a) is used to assert that Alex leaving his office on time is a necessary condition for the realization of the speaker's or Sam's goals, which might not be in Alex's interests; for example, the speaker or Sam might intend to assassinate Alex. Likewise, (24b) might pertain to a bet placed by the speaker or by Sam, and of which the goalie, intent on playing a good, honest game, is unaware.

(24) a. Alex must leave his office on time (for my/Sam's nefarious plans to work out).
b. The goalie must miss this shot (for my/Sam's bet to pay off).

This individual with respect to whose goals and interests the modal is evaluated might be an argument of the local predicate, or it might have to be retrieved from the discourse. These results for teleological modals are summarized in (25).

(25) Teleological:
X = the goals or interests of an individual or set of individuals retrieved from the semantic argument structure of the main predicate of S, or from the discourse.

Hacquard would take an ability modal as pertaining to the latent potential of an individual or object retrieved from the semantic argument structure of the main predicate of the clause S, as in the prototypical example, *Sam can lift 100 pounds*. The examples in (26a,b) are less prototypical examples of this cited by Hacquard, and (26c) is another example of this sort.

(26) a. It can rain hard here. (M. Hackl, cited in Hacquard [2010: 92])
b. A lot of people can jump on this trampoline. (modified from Hacquard [2010])
c. These socks can lie on the radiator for hours – it won't hurt them.

[3] Hacquard and others have noted this, but not pursued it at length.

In (26a), it is the potential of the place denoted by *here*. In (26b), it is the potential of the trampoline to accommodate lots of people jumping at the same time. In (26c), it is the potential of the socks to withstand heat. We can analyze these as involving events, as follows: (26a) says that it is a property of this place that events of rain have the potential to be hard-rains. (26b) says that there can be events of many people jumping on the trampoline together. In both (26a,b), the event retrieved is plausibly that of a Davidsonian event associated with the main predicate. But (26c) says that there can be events in which the socks are left on a hot radiator and do not burn, or generically, that events in which the socks are laid across the radiator are not ones in which they will burn. This has a stative predicate, and retrieves ancillary events in which the socks are placed on the radiator. We also see this in (27), as asserting that Alex has the ability to resemble Ortcutt through events of dressing, disguise, or mimicry implicated by S.

(27) Alex can resemble Ortcutt.

The intended latency can reside in the denotation of the object, not the subject, as in (28), where the latent potential is really in the material, the potential to form a crystal.

(28) Shocks can compress this material into a crystalline pattern.

In (28), there is implicit generic reference to events of the material being compressed by shocks. So the potential can come from the referent of the subject or the direct object, or from other individuals specified in the content of S. But it cannot be wholly implicit, as (29) shows.

(29) a. People can be made rich.
b. People can be made rich by this stock.
c. People can be made rich here. [here = in the situation created by this stock]

Although (29a) can mean that people have the capacity to be made rich, it cannot retrieve the stock from the discourse and assert that making people rich is a latent property of the stock with the meaning of (29b). The pronoun *here* in (29c), though not a thematic argument of the main predicate, suffices to retrieve the context, and thereby the individual with the latent potential. The result for ability modals is summarized in (30).

(30) Ability:
X = the latent potential of an individual or object retrieved from the content of S.

All root modals can be said to involve events related to the content of S, whether generic or a particular event, but the relation is stricter or looser, depending on the type of root modal. The event need not be a Davidsonian event described by S for any type of root modal.

Thus, deontic and teleological root modals seek out a suitable individual, whether retrieved from syntax or the discourse, who is responsible for, or whose goals or interests are related to, the content of S. The ability modal is just slightly different in that the individual or object relative to which the modal is evaluated must be mentioned in S. The event or generic events involved in the ability modal need not be expressed by S, but the individual in which the latent potential resides should be.

In conclusion, root modals are not all event-dependent, and even ones that are do not always depend on a Davidsonian event variable associated with the predicate of the sentence S. They are not, in general, evaluated relative to the local subject or to any individual given in the semantic argument structure of the main predicate of S, although the ability modal is evaluated relative to an individual mentioned in S.

This suggests that the retrieval of the modal base and ordering source in Hacquard's account is a more loosely constrained seeking process – the modal seeks a suitable individual and a suitable set of circumstances associated with that individual, with respect to which it can be evaluated. But the low modal seeks a base and ordering source pertaining to events associated with the circumstances described by the predicate of the clause, so the restriction to circumstantial modal interpretations over an epistemic interpretation is still enforced.

On this view, there is a "cloud of contexts," and in some cases, a "cloud of events," for root modal interpretation, associated with the content of S. The root modal is constrained to scan that set (of contexts, or of related events) for an individual and a suitable circumstantial modal base and ordering source.

### *8.2.2 Epistemic modals*

As discussed earlier, we assume that the high position of a modal in a main clause accesses the assertion event, as detailed by Hacquard, and that this event serves as an index to the cloud of contexts implicated in epistemic modal interpretation, as discussed by von Fintel and Gillies (2008b).

Now consider what happens when a modal statement is embedded as complement to a propositional attitude predicate, as with Sally's reply in (31).

(31) George: What happened to the missing artifact?
Sally: Max thinks that Bill must have stolen it.

The embedded epistemic is clearly evaluated relative to the experiencer subject of the higher clause, as Stephenson (2005) and others have observed. But Sally's response has a plausible minimal interpretation – that Sally is not describing a mental state of Max, but simply stating a condition on Max's doxastic base, which implicates an answer to George's query, depending on the credibility George and Sally accord to Max on the matter of the missing artifact. In this case, Sally is saying that the prospect of Bill having stolen the artifact evaluates as certain relative to what Max thinks he knows.

According to Hacquard, the embedded modal in Sally's response in (31) must be anchored to the eventuality of the higher clause in Sally's utterance in order to type it as epistemic. But with the minimal interpretation of Sally's response, there would be no eventuality variable associated with the matrix clause to which, on Hacquard's account, the modal in the lower clause could be anchored to secure its interpretation as an epistemic modal. Furthermore, in the absence of a Davidsonian event in the higher clause, the epistemic modal is not dependent on the next higher event, the speech event, since this would evaluate the epistemic modal at the speech time, based on the epistemic context of the speaker, Sally. Then Sally's response would assert that Max thinks that Bill must, based on Sally's current evidence, have stolen the artifact, which is not the correct interpretation. Thus, the embedded epistemic modal in Sally's response is not evaluated with respect to any event associated with the higher clause in Sally's response in (31), nor with the speech event or any other event.

To investigate this further, consider (32) and (33).

(32) Sam can believe that Bill is in the room.
(33) Sam believes that Bill must be in the room.

In (32), the ability modal *can* coerces *believe* to be occurrent, so it has a Davidsonian event, as discussed in the previous section. In (33), Hacquard would have the epistemic modal in the lower clause anchored to an event associated with the higher attitude predicate. But on a minimal interpretation of the belief ascription, there is no such event.

On the basis of the discussion in Chapter 7, we can investigate whether the higher attitude has a Davidsonian event associated with it by checking for the LIN. One indication of the LIN is the licensing of a strict Negative Polarity Item in the complement clause. This seems to occur in the last sentence of (34).

(34) We always knew that Sam left the house in the morning or sometime around noon. Then on Tuesday we received evidence that he was home all morning. So, since Tuesday we haven't believed that Sam left the house until noon.

This seems to be a coherent monologue with LIN in the last sentence. To look at this more systematically, based on discussions in Chapter 7, it is relevant to check whether the ascription in (35) can have the interpretation in (36b) as well as that in (36a).

(35) Sam doesn't believe that Bill must be in the room.

(36) a. Sam denies or resists the idea that Bill must be in the room
b. Sam rather suspects that it is not the case that Bill must be in the room = Sam rather suspects that Bill might not be in the room.

The interpretation in (36a) exhibits matrix scope of negation and is obtained, according to Chapter 7, when the belief ascription is interpreted as a full ascription, modifying an eventuality of belief. It is easy to get this interpretation, where Sam is staunchly opposed to the idea that Bill must be in his room, either because Sam is staunchly agnostic on the question of whether it is necessary that Bill is in the room, or because Sam is staunchly convinced that there are no grounds for certainty that Bill is in the room. The interpretation in (36b) is harder to fix upon. First, it must be distinguished from interpretations in which negation associates with focus on a particular subconstituent of the lower clause, as in (37a,b).

(37) a. Sam doesn't believe that Bill MUST be in the room. (stress on the modal)
b. Sam doesn't believe that Bill must be in the ROOM. (stress on the place)

Using the notation from earlier chapters, whereby **believe**$^{-}$ is the dual of **believe** in the metalanguage, the examples in (37) readily allow interpretations such as "Sam believes$^{-}$ that it is not the case that Bill must be in the room," either because some other modality seems preferable to Sam or because Sam can readily imagine Bill being elsewhere. To get this with the most neutral intonation intended for (35) is harder, but can be seen when (35) is seen in consort with similar examples, especially with a LIN-facilitating first person subject.

(38) I don't imagine that Bill must be in the room.
[= I rather think it's possible that Bill is not in the room.]
(39) I don't believe Bill can be in the room.
[= I'm inclined to think that Bill cannot be in the room]
(40) I don't think he must be in the room. I rather think it's possible he isn't.

This poses the question of what aspect acts upon in (41), if there is no Davidsonian event for it to place in time.

(41) Sam has believed that Bill must be in the room.

Here we have evidence that Katz was right about sentences with stative predicates, including stative attitude ascriptions: their perfect does not semantically rely on a Davidsonian eventuality, but quantifies over times.

The perception complements in (42) are also instructive regarding the interpretation of an embedded epistemic modal.

(42) It seems/appears (to Sam) that Bill must/may have Lyme Disease.

The matrix clause in (42) can have a minimal interpretation, which is manifested with the LIN in (43a), as uttered by a doctor reading test results, whereby it is interpreted as paraphrased in (43b), to rule out a conclusive/possible diagnosis of Lyme Disease.

(43) a. It doesn't appear that Bill must/may have Lyme Disease.
b. "It looks like there is no necessity/possibility that Bill has Lyme Disease."

When the experiencer argument of *seem/appear* is overtly expressed in (42), the embedded epistemic is evaluated relative to the doxastic state of this experiencer. When no experiencer is overtly expressed, the speaker, or a group to which the speaker belongs (which can also include the addressee(s)) is the implicit experiencer, and the modal is then evaluated relative to the speaker, or that group. So, as in Sally's response in (31), the embedded epistemic in (43) does not have to target an event expressed by the matrix clause, but effectively targets the cognitive experiencer argument of that higher clause.

The example in (44) is especially instructive.

(44) Sam doubts that Bill must/may have Lyme disease.

As with (42), this expresses a propositional attitude of Sam toward a verdict that Bill must or may have Lyme Disease. But use of the attitude verb *doubt* implicates that the verdict in question, that Bill must or may have Lyme Disease, is already a discourse topic, at issue for Sam and some discourse community to which Sam belongs (which may or may not include the speaker), for purposes of discussing Bill's condition. The embedded epistemic modal in (44) then has a range of interpretive possibilities that is exactly parallel to that outlined by DeRose for main clause epistemics; it can be interpreted based on the knowledge distributed in some way or other over members of this community, possibly deferring to experts, or to the results of tests recorded somewhere, in some repository of information available to Sam or other members of the community. In short, it has the "cloud of contexts" possibilities for epistemic modal interpretation articulated by von Fintel and Gillies (2008b) for main clause epistemics.

The modal in the embedded conditional clause in (45) also poses questions about the event dependence of epistemic modal interpretation.

(45) a. If Bill may have the virus, then we must quarantine him for two weeks.
b. If Bill must have the virus, then he can't have the cancer we suspected.

In a typical context of use, the conditional *if*-clause in these examples would make reference to a prior discussion of the prospects of Bill having the virus, or the results of a test showing a possibility or necessity that he has the virus. The epistemic modal within the *if*-clause is evaluated relative to a group or community involved in that discussion, or experts within that group, or a test outcome recorded in a place accessible to members of the group. But it seems, specifically, evaluated relative to that conditional context, not the speech event. This is clearer when we shift to the perfect, where the epistemic modal actually seems to be evaluable relative to the shifted time, in (46).

(46) a. If Bill may have had the virus, then they should have put him in quarantine.
b. If Bill must have had the virus, then he couldn't have had the cancer they feared.

If (46a,b) are uttered to describe a famous case decades in the past, in which speaker and interlocutors were not involved, it seems possible to evaluate the epistemic modal relative to what doctors involved in that case thought they knew. This is parallel to the judgment of von Fintel and Gillies (2008b: 86) in the Detective Parker example discussed in Chapter 1.

Using Hacquard's interpretive mechanisms, the high modal in the *if*-clauses in (44) and (45) would have to access the event of assertion associated with the matrix clause. This is not absolutely beyond the resources of her account. But, together with results for root modals in the previous section, it suggests that the identification of the event argument of the modal may not be a strict syntactically based relation of binding.

# 9 Attitude ascription in child development and Wernicke's aphasia

This chapter argues that the distinction between two types of attitude ascription developed in Chapter 6 is corroborated by research literature on child language development, specifically, by research on the chronology of the acquisition and use of mentalistic predicates and predicates of perception and assertion in utterances with the form of attitude ascriptions. We argue that this research documents a transition from the use of attitude ascriptions with the minimal interpretation to ones with the full interpretation. The initial stage involved in this transition correlates with a capacity for modal expression; the end stage of the transition correlates with the manifestation of Theory of Mind – the child's understanding that other people have, and act upon, their own representations of the world, which can differ from that of the child. This is discussed in Section 9.2. This chapter also argues that the same distinction is corroborated by research results in the literature on the selectivity of language impairment observed in language production by individuals with Wernicke's aphasia. In such cases, the two types of attitude ascription distinguish between (on the one hand) language-internal operator-variable structures comprehending modal interpretation realized as quantification over possible worlds, anaphor binding, and bound pronoun interpretation, production of which is preserved under this form of aphasia, and (on the other hand) reference, including reference to mental states and events and pronominal coreference, in which language-internal systems interact with language-external systems, the production of which is impaired. This is discussed in Section 9.3. Section 9.1 develops criteria for making this distinction and establishes a baseline of typical adult usage for comparison.

## 9.1 Establishing a baseline

### *9.1.1 Preliminaries*

It will be useful to establish a baseline of typical adult usage of attitude ascriptions, in order to compare it with the production of attitude ascriptions in child language and in the language of nontypical adults with attention to whether ascriptions have the minimal or the full interpretations discussed in Chapters 3, 4, 6, and 7. To do this, it is necessary to establish criteria for identifying an ascription in usage as having, or being intended to have, the minimal versus the full interpretation.

But first, we note the likely preconditions that would be required for expressing an attitude ascription with the minimal and full interpretations, respectively. Choi (1991) argues that incipient modal expression is observable in children as young as two years of age. Her study is particularly telling since she studied the acquisition of sentence-final epistemic and deontic verbal affixes by children learning Korean. Since modality is expressed by verbal affixes in these examples, rather than auxiliary verbs or clause-level adverbials, there was no interference from the development of clausal phrase structure. We take Choi's results to show that incipient epistemic and deontic concepts are acquired early, and can be expressed as soon as the syntactic resources of the language required to express them have developed. Other studies, including Hirst and Weil (1982) and Noveck, Ho, and Sera (1996), show that young children tie both epistemic and deontic modality to a scale or notion of strength or probability, but they are slower to sort out the relative strengths of particular modal operators on this scale.

The development of a child's Theory of Mind has been explored by a number of researchers, primarily through false belief tasks that, through a narrative, cartoon, or acted out scene, give the child grounds to believe that a character has a belief that the child knows to be false, and then test the child on the imputation of this false belief to the character. In a series of such tasks, Wimmer and Perner (1983) obtained evidence that a Theory of Mind first appears around age four, and becomes progressively more reliable and solid, with substantial improvement between the ages of four and six. These results are largely confirmed by other researchers, and are typically taken as the baseline for the study of Theory of Mind on children with cognitive or language deficits.[1]

In the studies presented in Sections 9.2 and 9.3, utterances produced by young children and Wernicke's aphasics are compared to those of nonaphasic adults, focusing on anything with the syntactic form of an attitude ascription. Given an example of an attitude ascription used in a context, it may or may not be intuitively clear whether it was intended with the minimal or the full interpretation. We will develop criteria to determine, as best as possible for our purposes, whether an attitude ascription has the full interpretation based on whether or not the context or neighboring productions indicate reference

[1] Alfandre (2010) found performance on false belief tasks to be still developing in school age children between the ages of six and ten. Notably, she used a stricter administration of the false belief task than those used by Wimmer and Perner and others in the literature.

to a mental state (of believing, doubting, etc.), whether of the speaker, the addressee, or a third party, or reference to an event of cognition, perception, or assertion (such as remembering, concluding, seeing, hearing, saying, or telling something). Given an example of language with the form of an attitude ascription, we will, in collecting evidence for its status, refer to a *candidate* mental state putatively introduced by the ascription, or a *candidate* act of cognition, perception, or assertion characterized by the utterance, before it has been determined whether the utterance actually introduces or describes such an event or state.

With regard to evidence that an example of attitude ascription language is used to characterize an event or a mental state, some factors seem quite decisive in favor of this conclusion, and others less so. Furthermore, the evidence that an attitude ascription reports on a mental state of an experiencer can be more elusive than evidence that an ascription reports on an event of cognition or perception, which in turn can be more elusive than evidence that an ascription reports on an event of assertion. For this reason, we don't look for the same factors in all examples, and we don't seek to formulate a set of criteria that will work equally in all cases.

One straightforward form of evidence demonstrating that an ascription is interpreted as reporting on a mental state of an experiencer, or as describing an event of cognition, perception, or assertion, is when there is subsequent pronominal reference to the imputed state or described event. This criterion is formulated in (1).[2]

(1) Given an attitude ascription, subsequent pronominal reference to a candidate mental state, or an event of cognition, perception, or assertion associated with the ascription shows that it actually describes such an event or mental state.

This criterion can potentially be obtained for the existence of an event or state associated with any type of attitude ascription. The problem is that it occurs fairly rarely in ordinary conversation, when not elicited or prompted.

Third person attitude ascriptions with mentalistic predicates such as *think* or *be sure* can be taken to describe a mental state when they are offered or used to explain the overt behavior of the experiencer subject of the ascription. This criterion is recorded in (2).

[2] An example would be along the following lines, where the demonstrative *that* in the second sentence refers to Max's state of belief: *Max thinks that he is special. That will get him into trouble.* The criterion in (1) is not actually used for any of the examples discussed here. But the criteria were drawn up to serve as the most generally applicable criteria for the full interpretation. The criterion is not used simply because no example along these lines happened to surface in the data examined here.

(2) If an ascription with a mentalistic predicate is offered or used to explain the overt behavior of its experiencer subject, then it has the full interpretation.

If an ascription with a mentalistic predicate imputes what speaker and addressee recognize to be a false belief, then the matrix clause cannot be mere evidential qualification of the content of the complement clause. In this case, too, the ascription can be expected to have the full interpretation, describing the false belief state of the experiencer subject. This criterion is recorded in (3).

(3) If an ascription with a mentalistic predicate imputes a false belief to its experiencer subject (as mutually recognized by speaker and addressee), then it has the full interpretation.

The next criterion references the markedness threshold for the full ascription proposed in Chapter 7, Section 7.3. To illustrate with the class of predicates of belief, it was suggested there that predicates such as *think*, *believe*, *suppose*, and *be sure* are relatively unmarked by virtue of expressing more generic instances of their attitude class, and – what is possibly a related factor – in being more frequent than other members of that class. In contrast, belief predicates such as *remember*, *doubt*, *conclude*, *surmise*, and *be aware* are relatively marked by being less generic and less frequent. It was further suggested that English has a threshold in markedness above which any predicate has only the full interpretation (and thus does not permit LIN); indeed, each language has such a threshold, and although cognate predicates and semantically similar predicates will tend to end up on the same side of the threshold across languages, they don't necessarily do so since the exact placement of the threshold is language-particular. Verbs of assertion are typically used to describe events of assertion, and correspondingly, none of them permit LIN. But *say* and *tell* seem less marked than the rest, and do have a qualifying use in the first person, especially when accompanied by a modal, as in, *I'd say (that) we're stuck here until tomorrow*, or, *I must/will say that I don't like it*, or, *I should tell you that they're not likely to accept your proposal*. This suggests that assertion predicates other than *say* and *tell* are all above the markedness threshold, and thus occur only in ascriptions with the full interpretation, while *say* and *tell* can sometimes have a strictly qualifying interpretation. Finally, among predicates of perception, *seem, see,* and *hear* are below the threshold (and the first two, at least, permit LIN), and therefore have a purely qualifying use in the first person, in which they do not characterize a state of an experiencer subject. But predicates such as *observe* or *note* are above the threshold, and describe only events. The associated criterion is formulated in (4).

(4) An ascription with a predicate above the markedness threshold for its language has the full interpretation, describing a mental

state of an experiencer, or an event of cognition, perception, or assertion.

For verbs of assertion, we recognize that *say* and *tell* can be used with a purely qualifying interpretation, but can also be used to describe events of assertion; we will assume that all other verbs of assertion are above the threshold and can only describe an event.

The final criterion is concerned with informational focus. With a first person subject, a belief ascription can have focus anywhere, and simply be evidential qualification. For example, if we are concerned with having a key to get into a studio, someone can rummage through a purse and assert, ***I think** I have the key here*, interpreted as evidential qualification just as much as, *Wait! I think I have the* ***key*** *here*. But with a third person subject, nuclear stress in the matrix clause with a lower-stressed complement clause does not favor an interpretation as merely evidential qualification, but as mental state ascribing, or false belief ascribing. For example, *Bill* ***thinks*** *he has the key* is making a statement about Bill's belief state – and when the stress is contrastive, it is about a possibly or presumptively false belief state. Likewise, ***Bill*** *thinks he has the key*, and *Bill* ***must*** *think he has the key* would be used to qualify a statement about a mental state, that it is Bill's, or that Bill necessarily has it, not just to report on the location of the key, or the evidential status of the complement clause. The same is true, and less subtly so, when the attitude predicate is one of perception or assertion. Thus, the observation made here for mentalistic attitude predicates is formulated in (5) for all attitude predicates.

(5) If an attitude ascription with a third person subject has its greatest informational prominence in the matrix clause, it has the full interpretation, reporting on a mental state or an act of cognition, perception, or assertion.

A common instance in which the *if*-clause condition in (5) is satisfied is one in which the complement clause expresses previously discussed content, and the informational focus of the utterance is the entire matrix or embedding clause, without contrastive stress; in this case, the attitude clause, including the attitude predicate, is the informational focus of the utterance, not merely qualifying material.

These five conditions should not be regarded as necessary and sufficient conditions for whether an utterance with the form of an attitude ascription has the full interpretation. There may be ascriptions that are clearly intended to characterize an event or state introduced by the attitude predicate, which do not meet any of these five conditions. But these conditions will be used as criteria for demonstrably referential attitude verbs; an ascription accompanied by one or more of these signs will be taken to be demonstrably intended to be interpreted as characterizing an associated event or state.

Finally, derived nominals such as *the belief that we should disinvest now*, *the claim that the market will soon crash*, *the realization that we have lost our investment*, and others, will be taken at face value, as denoting a mental state or an event of cognition, perception, or assertion, characterized in part by the contents of the complement clause.

### *9.1.2 A baseline study*

The ascriptions reproduced in this section are from a transcript of the television talk show *Washington Week* (March 4, 2011), a round-table discussion of weekly news events broadcast on public television stations in the United States. The utterances transcribed here were from a 3,400-word stretch of the 4,126-word transcript excluding an initial segment consisting of the moderator's introduction and the opening lines of the first panelist's remarks, and a final segment consisting of the moderator's closing remarks (all of which are more likely to be scripted). We call this the body of the transcript. The 3,400-word body contained 47 attitude predicates (not counting a few completely parenthetical instances of *you know*). The candidate attitude ascriptions in the examples presented here are all italicized, and include examples with declarative complements of attitude verbs, as well as some with interrogative complements.

The exchange in (6), discussing ongoing uprisings in the Arab world, exhibits some attitude ascriptions of each type, though none satisfying any of the criteria (1)–(5) with compelling clarity.

(6) MS. BISKUPIC: […] But isn't there for the long game an advantage here in terms of all these democratic movements over time?
MR. KITFIELD: Sure. *I think that's right* and *I think you can't help – I mean, we all feel there's sort of this kinship – you can't help to be kind of proud and hopeful for these people trying to get a voice in their own governments*, and they're some pretty rotten governments.
MS. IFILL: Except *we don't know who they are either.*
MR. KITFIELD: We're not sure – you know, each country is different so we don't know – certainly, in Libya, for instance, we don't know the opposition. *I think we have a better handle on it a bit in Egypt.* But, in general, *I think it's pretty well accepted that if these become a more – if it becomes a more normal democratic region, it will quit being quite the breeding ground for terrorism and extremism that we've seen.* [continues]

The four italicized examples with *I think* seem to be minimal ascriptions that meet none of the criteria for denoting a cognitive state of the speaker. Ms. Ifill's *we don't know who they are either*, is different; it uses a borderline mental verb in terms of the markedness threshold (*know* permits LIN in some nonstandard dialects, but not most dialects of English), but one that is frequently used for evidential qualification with a first person singular subject. But Mr. Kittfield's response, fussing over what we do know or don't know about different countries, reinforces the interpretation of Ms. Ifill's negative *know* ascription as describing a state of collective knowledge (or rather, ignorance). This conclusion is reinforced by the fact that it isn't clear what it would mean to have evidential qualification of the content of the interrogative complement "who they are." And although the final instance of *I think* ... is a minimal ascription ("parenthetical" in Urmson's sense), its complement, *it's pretty well accepted that* ..., employs an above-threshold predicate to characterize a state of consensus.

The example in (7) follows shortly upon those in (6).

> (7) MR. KITFIELD: [...] In terms of counterterrorism, there's concern that Libya, if it goes into a civil war, could become a failed state. It looks a lot like Somalia then where there's an al Qaeda affiliate. *We're also worried that Islamists could hijack some of the democratic transitions in places like Egypt.*

The italicized clause with *worry* more plausibly counts as a full ascription, describing a state of worry, since it meets criterion (5): *worry* is quite highly marked in the class of belief verbs.

Another example clearly meeting one of the criteria that it is a full ascription is reproduced in (8), where the topic is the Obama administration's dealing with the Republican-dominated Congress on budget issues.

> (8) MR. HARWOOD: Well, they're playing for time. They're in a somewhat better position than they were in November. The president's certainly in a better position. You know, he cut that bipartisan deal with Republicans in the lame duck session to cut taxes, to additionally – to not only preserve the Bush tax cuts, but to cut payroll taxes. That provided some stimulus for the economy. And so *the administration is trying to make the case that I've proven that I can work with them and I'm being reasonable*. [continues]

In (8), the italicized ascription embeds an attitude ascription of logical inference or assertion, *I've proven that* ..., under another attitude predicate in the same class, *make the case that* ... (with a subject controlled by the higher subject *the administration*). Both attitude predicates are above the threshold for this class, meeting criterion (4).

In (9), the topic was a US Supreme Court decision upholding the free speech rights of antigay protestors at military funerals.

(9) MS. BISKUPIC: Definitely protected. And this is a good example of what the First Amendment is all about – protecting distasteful, offensive speech we hate. And *the Supreme Court itself took pains in the opinion to talk about how the speaker might be viewed as quite offensive*, the Westboro Baptist church in the form of Fred Phelps, the pastor of this small, Kansas-based, family-run church that's known for picketing at military funerals, but *the court took pains to say, we don't like his message necessarily, but he's got every right to do it.*

In the context of the *talk about* ascription, the offensiveness of the speech in the case before the court had already been raised by the previous speaker, the moderator, and Ms. Biskupic reiterates this theme in the sentence before the *talk about* ascription. Thus, the content of the complement clause is informationally old; the informational focus of this utterance is on the material in the matrix clause, referencing the Supreme Court's own acknowledgment of the offensiveness of the speech. This counts as a full ascription under criterion (5). The same is true of the *say*-ascription in the last sentence of (9).

In the example in (10), the topic was the same Supreme Court case.

(10) MS. IFILL: Justice Alito was the one person who –
MS. BISKUPIC: Yes. Justice Samuel Alito, who was the one dissenter. And he also dissented last year on a big free speech case having to do with depictions of animal cruelty that, again, *the Supreme Court said, we might not like it but the statute was too broadly written in that case.* [continues]

The complement clause to *say* is old information: the ascription reports that the Supreme Court decision in the animal cruelty case was essentially the same as in the first amendment case discussed previously, and the complement clause delivers the content of that decision. The focus of the ascription is on the Supreme Court delivering an opinion that is parallel to that in the first amendment case.

In (11), the complements to the verb *announce* and the derived nominal *expectation* both concern Supreme Court justices retiring from the court, a topic brought up just previously; these matrix clauses are thus focal.

(11) MR. HARWOOD: Now, it wasn't reflected in this case, because it was an eight to one case, but in terms of the number of cases that are very evenly balanced in the court, anytime somebody – there's a change in the composition of the court, people are watching closely. This is the time of year when *people do tend to announce that they're going to leave*. Is there any *expectation that any of the sitting justices are going to leave?* And might that affect the contours of the court in future decisions?

Altogether, there were four ascriptions in the 3,400-word body of the transcript that clearly met at least one of the criteria in (1)–(5).

There were a number of other examples in the *Washington Week* transcripts of attitude predicates that seem intuitively to be referring, but that did not satisfy any of the five criteria certifying them as such. The example in (12) occurred in a discussion of the idea of establishing a no-fly zone over Libya.

(12) MR. KITFIELD: Well, you know, John McCain – I was surprised by Senator Kerry. John McCain has been very hawkish on these things as has Senator Lieberman. They just came back from the region, and *even they say we – aren't saying we should do it, but they think we should be contemplating it seriously.* And I think there is serious contemplation. [continues]

The aborted ascription with *say* (meaning, "even the hawkish McCain and Lieberman say we should do it") would have had focal verb and subject with a discourse-old complement, meeting criterion (5) of being referential. In (13), the two ascriptions with *say*, one with *Republicans* as the agent and one with *Democrats* as the agent, are reporting on acts of assertions as political moves, and thus ostensibly referring to them.

(13) MR. HARWOOD: Part of it involves game playing with numbers. Republicans, when they were trying to make their cuts look bigger for the freshmen, because they know how difficult it is to cut the budget, they counted a certain amount of money that *they said had already been cut from the president's initial budget request.* Now *Democrats are counting that money and saying, well, we've cut that money too.*

But none of our five criteria are clearly met by either ascription in this example.

Attitude reports that demonstrably characterize a state or event are a small proportion of the total number of attitude reports in this discussion. We have four clear examples out of forty-seven, or 8.5% in this transcript. Counting the *know* and *pretty well accepted* ascriptions in (6), and the *worried* ascription in (7), and counting each of (12) and (13) as exemplifying one instance, the total is nine out of forty-seven, or 19%. But either way, the proportion is nontrivial, and this example gives us an idea of the frequency and form of ascriptions with the full interpretation in the conversation of unimpaired adults.

The second example of normal adult conversation is a transcript of the discussion "Does NPR have a Liberal Bias?" that aired on March 18, 2011, in the program *On the Media*. This transcript contains 3,171 words, with 40 attitude predicates with clausal complements, excluding parenthetical instances of *you know*. At least thirteen of these, or 32.5%, met at least one of the criteria for having the full interpretation. Some of these involve cognitive verbs that are clearly above the markedness threshold, as in (14a–c), and the ascription with the first italicized predicate in (14d). Others involve predicates of assertion, like the second italicized predicate in (14d), and the examples in (14e–h).

(14) a. we've *decided* to deal with the question both this week and next.
b. As we waded in, we *realized* it's incredibly complicated.
c. Well, you could *convince* people that you were in fact being neutral by sticking to a point in the center [...]
d. You know, in many cases I think it's going to be the right decision for a journalist to say, we're *aware* that the science *says* that there's not a controversy here and [...]
e. and one of the first *observations* that one of the guests *made* was that the situation in Wisconsin would probably galvanize the American labor movement.
f. But we *said* we'd try, so we called Tom Rosenstiel, [...]
g. there's all kinds of academic literature that we've looked at that *suggests* that that wicket is really very sticky.
h. And the research came back and *said*, you do have a bias in your coverage.

None of the instances of predicates of assertion in (14d–h) is the performative use along the lines of *I'd say that* ....

Finally, there are nominals derived from attitude predicates that characterize mental states, reproduced in (15), one of them already noted in (11).

(15) a. Is there any expectation that any of the sitting justices are going to leave?
b. In terms of counterterrorism, there's concern that Libya, if it goes into a civil war, could become a failed state.

These are not surprising in the conversation of normal adults, but it will be useful to keep them in mind when we compare with the speech of young children and Wernicke's aphasics.

## 9.2 Child language development of attitude ascriptions

In discussing "attitude ascriptions" in the present work, we have addressed anything with the standard identified form of an attitude ascription, with a mentalistic predicate, or predicate of perception or assertion, accompanied by a nominal agent or experiencer argument (unless this argument has been passivized away) and a clausal theme argument. We will continue to call these "attitude ascriptions" based on the superficial syntactic form even when, as in the minimal interpretation, the experiencer argument may not be interpreted as actually experiencing a mental state, or when, as in the earliest child utterances with these predicates, it may not be clear that there is true clausal embedding.

Several factors constrain the development of attitude ascriptions in child language. Children must acquire the syntactic structures to express clausal complementation before a clausal theme argument can occur as the direct object of an attitude verb. These structures, which begin to manifest themselves around age two, are typically not robust until age three or four (Limber 1973; Bloom et al. 1989; Radford 1990, 1996). Further complicating matters, Gleitman et al. (2005) argue that the acquisition of the meaning of a relational verb depends in part on the syntax used to express it, which reveals the argument structure of the verb. So a child's acquisition of the meaning of a mentalistic verb may have to await the development of the full structure of finite clause complementation. Finally, in line with concerns in the present work, to grasp the minimal interpretation of an attitude ascription requires one to have command over modal or evidential qualification; to grasp the full ascription requires having Theory of Mind, the understanding that other people have mental states that can represent – and misrepresent – the world.

Of work on child language development of attitude ascriptions, the most relevant for our purposes is that which marks the onset of the minimal and full interpretations, distinguishing the two and detailing differences in their development. Of these, the study by Diessel (2004) is especially relevant since it particularly attends to a variant of this distinction, tracing the details of the development of this language among five children. Examining the use by young children and adults of mentalistic verbs, verbs of assertion, and other predicates taking finite clause complements, Diessel (2004: 80–89) makes a three-way distinction among these examples. First are cases where the matrix clause is formulaic, where Diessel claims there is no syntactic embedding, and the matrix clause is a propositionally empty "clausal operator" that serves as an attention getter, an epistemic marker, or marker of illocutionary force. At the opposite extreme, as it were, are what Diessel calls assertive cases, with full syntactic complementation, where the matrix clause makes an assertion about a mental state of the experiencer referent of the subject, an act of cognition by the experiencer subject, an act of perception that the subject experiences, or an assertion that the subject makes. The content of the complement clause is merely one component of the description of that mental state or act of cognition, perception, or assertion, contributing a specification of what it was about. In between are cases Diessel characterizes as performative, in which there is syntactic complementation and the matrix clause has some propositional content, but the primary purpose of the matrix clause is "to co-ordinate the interaction between the interlocutors" (p.88). For example, the main clauses in (16) are performative when (16a) is interpreted as "This is probably a mistake," and (16b) is interpreted as an evidentially qualified version of "Paul is coming" and (16c) (my example) is interpreted as "You're tired; we should stop now."

(16) a. I believe that this is a mistake.
b. I hear that Paul is coming.
c. I (can) see that you're tired; we should stop.

Note that the function of clause-embedding material might be the same in some of the formulaic and performative cases; for example, both might provide epistemic or evidential qualification. In our own terms (not necessarily acceptable to Diessel), we can nevertheless distinguish them on the grounds that the formulaic cases are frozen collocations, not productively generated or compositionally interpreted, whereas the performative examples are productively generated as such, and interpreted through the semantic and pragmatic interactions that are employed for normal, explicit performatives, as discussed by Austin (1962) and others. Diessel regards the performative cases as analogous to standard performatives, for example, *(hereby) promise*, *(hereby) suggest*, *(hereby) request*. This idea was embraced in Generative Semantics, where it was generalized by positing a performative superordinate structure for most or all utterances; see Ross (1970, 1974). Diessel claims that the subject is invariably first or second person in the performative use.[3]

Diessel (2004: 89) claims that child language development progresses from formulaic examples to performative uses, followed by assertive uses, in sequence. He investigated this through examination of a corpus of 1,812 complex sentences featuring finite complement clauses in the CHILDES transcripts for five children, Adam (Brown 1973), Sarah (Brown 1973), Nina (Suppes 1973), Peter (Bloom 1970), and Naomi (Sachs 1983). We examine his results in order to see how they relate to the minimal and full interpretations of attitude ascriptions proposed in the present work.

Concentrating first on Sarah and Adam, who had more attitude verbs in their earliest utterances, Diessel notes that their initial utterances with *think* were all indicative, present tense with missing subject or first person subject, as the examples for Sarah given in (17) attest, where each utterance is accompanied by the age at which it was made (in the format year;month). The underlines are reproduced from Diessel's presentation of the examples.

[3] There is potential for terminological confusion at this point. What Diessel calls "performative" attitude ascriptions are ones that Hooper (1975) called "assertive" on the grounds that the proposition expressed by the complement clause is asserted, the main clause being "parenthetical" in Urmson's (1952) usage. Simons (2007) points out that the content of the complement clause in the "performative" ascriptions is not necessarily asserted; the material in the main clause serves to evidentially qualify it, and may even cast doubt on it. We will use Diessel's terminology for the exposition of his results, but revert to our own terminology in interpreting his results, and thereafter.

(17) a. I think I'm go in here. 3;1
b. And I think (pause) we need dishes. 3;2
c. Think some toys over here too. 3;3

Starting six months later, interrogative forms, other tenses and other subjects appear, as in (18).

(18) a. You think it does? 3;7
b. [I] Thought it was in the house. 4;0
c. I will sing along with them ... then they think I ... will ... have ... 4;3

Diessel claims (especially pp.111–115) that the attitude verb in (17) is formulaic, where *think* or *I think* is appended to a clause (which, in the adult language, would be the complement clause), where "it functions as an epistemic marker, attention getter, or marker of illocutionary force" (p.111). With the greater productivity in (18), Diessel claims that Sarah has moved on to performative and assertive uses of the ascription form. Diessel notes similar transitions with other attitude verbs, including *see*, *look*, *bet*, *guess*, *wish*, *hope*, and *mean*, with data for all five children, and makes the same point for them. The formulaic collocations depend on the verb, and include *see if* (declarative), *look what* (imperative or interrogative), *guess what* (imperative or interrogative), *know what* and *know where* (interrogative), *you mean* and *what do you mean* (interrogative). The verbs *pretend*, *say*, and *tell* don't appear in formulaic uses; they appear later, and start out in performative or assertive uses.

There is a precedent for taking peripheral, unvarying material early in child language to be formulaic and not syntactically integrated in the adult way; see Hoekstra and Jordens (1994) for such a treatment of early child language auxiliary verbs. The "main clause" material in (17) is plausibly formulaic. But it doesn't follow that it has no semantic content. In recording sessions at age 3;1–17 (3 years, 1 month, and 17 days) and age 3;2–23, Sarah does not produce a single utterance with *think* or *I think*, but produces many declarative utterances, including ones as complex as *I had one yesterday*. So she clearly does not need to embed under formulaic material to make an assertion. The formulaic use of *(I) think* during this period could conceivably be semantically empty, but there is no reason not to credit it with the same meaning it has in its later, more productive, performative uses, as qualification of an assertion. In fact, Sarah has initial expression of epistemic modality during this period, as in *I maybe los(t) it* (3;3–13), and she has regular expression of deontic modality with *can*, as in, *Hey, I can ge(t) it out* (3;2–2). Thus, the question of whether the attitude ascriptions at this point are formulaic or not may not be relevant to their interpretation. Both the initial formulaic attitude ascriptions, and the more productive ascriptions beginning six months later, may be what Diessel calls performative.

Some of the latter could also, imaginably, be full attitude ascriptions, attributing a mental state; Diessel assumes they are performative, and not "assertive" without providing evidence for making this distinction. Although Diessel clearly articulates the distinction between minimal and full ascriptions in his distinction between performative and assertive ascriptions, and compellingly argues that the early formulaic ascriptions are at most performative, he does not actually develop any criteria or grounds to distinguish performative from assertive attitude ascriptions, and instead simply asserts that the child progresses to performative ascriptions, as a middle stage on the way to full "assertive" use. What Diessel actually succeeds in showing is the following: (1) syntactically, the earliest child utterances with the form of attitude ascriptions are formulaic in syntax, with first person present tense indicative collocations (these may be just appended to a clause without syntactic subordination, but Diessel does not prove this); and (2) within about six months, productive ascription forms are in evidence, with other tenses, persons, and moods. That's it. Diessel takes this to establish a sequence from formulaic use (whatever we might take that to be), to performative use (the minimal ascription, evidentially qualifying an assertion), to "assertive" use, the full mental state or event-describing interpretation. Semantically, what we can take Diessel to have shown is that the earliest formulaic utterances with ascription form are probably interpreted with (at most) the minimal interpretation, while later, more productive, utterances with ascription form might be either minimal or full ascriptions ("performative" or "assertive"). Nothing Diessel says shows that the earliest productive utterances have only the minimal interpretation, not the full interpretation.

To decide at what point ascriptions are produced with the full interpretation, we examine CHILDES transcripts (MacWhinney [1995]) of the five children studied by Diessel (2004), employing the criteria developed in Section 9.1. The experimental literature on the development of Theory of Mind leads us to expect that full ascriptions with mentalistic predicates such as *think*, *believe*, *guess*, and *be sure* should begin to appear around age four. Full interpretations of ascriptions of perception and assertion should also appear by this time, where these can be shown to rest on mental states, and possibly earlier if they don't. Typically developing children have acquired the requisite syntax to express finite complements by that time; see Hegarty (2005).

Diessel is correct that productive attitude ascription forms begin to appear just above age 3;5, so we will focus on data from that point onward. Unfortunately, the data for Peter and Nina end before this period, and the data for Naomi have a gap between ages 3;8–19 and 4;7, leaving out most of the developments of interest for the present study. Of the five children examined by Diessel, only Sarah and Adam have adequate data posted on CHILDES for the purposes of this study.

The present study focused on the mentalistic verbs *think*, *believe*, *guess*, *remember*, *know*, because adults in the transcripts use these verbs very frequently long before the children acquire them, and children exhibit early use of them relative to other mentalistic predicates. The use of *say/said* and *tell/told* was also examined, with attention to when they are first used in any form other than present tense with first person singular subject, on the assumption that a present tense first person singular use might well be a performative presentation of the content of the complement clause, but that past tense use, and third person use in any tense is more likely to report on an event of assertion.

Sarah exhibits evidential *think* from 3;1 onward, including early formulaic uses, and later, less formulaic, performative uses. Sarah shows clear reports of assertion at 3:10–1 and 3:10–9, using past tense *said* and *told*, each with third person singular subjects. At 3;8–27, Sarah used *know* in a performative way to express a query, as in her examples in (19a,b).

(19) a. Know where my games are?
b. Know what Easter Bunny gave me?

Sarah's first clear example of reporting on a mental state is at 4;0–14. The conversation concerns a doll that was left out in the rain. Since the context helps make clear that it is mentalistic, some of the conversation is reproduced in (20), with repetitions and hesitations recorded in the CHILDES file rendered here in more ordinary notation.

(20) Sarah: But I didn't know where it … where it is … [unclear] here it was.
Mother: You didn't know where what was?
Sarah: The doll. I didn't know <where>/where it was. Thought it w- was in the house.

Sarah's final thought-ascription in (20) reports on her prior false belief that the doll was in the house, and therefore qualifies as having the full interpretation under criterion (3) of Section 9.1. At 4;1–11 and again at 4;2–1, Sarah uses the above-threshold cognitive verb *remember* repeatedly to report on her own memory of an incident and inquire after her mother's memory of it. These qualify as having the full interpretation under criterion (4).

So, Sarah exhibited a clearly mentalistic use of *thought* at 4;0–14, mentalistic *remember* at 4;1–11. And she uses mentalistic *know* and *remember* at 4;2–1. The utterances using verbs of assertion and above-threshold mentalistic verbs such as *remember* and *know* meet our criteria for being ascriptions with full interpretations. Sarah first displays these within a month of the onset of her fifth year (age 4;0), in line with the average timeline for the development of Theory of Mind.

Adam exhibits evidential *think* (in *I think* …) and performative uses of *think* and *know* to express the interrogative (*do you think…, do you know…*) at 3;4–0 and 3;5–0, and onward. But mentalistic predicates in ascriptions meeting our criteria for the full interpretation don't occur until 3;11–0. The example in (21) at 3;11–0 meets criterion (2), where Adam's final utterance is a mentalist attitude report offered as an explanation for why his little brother Paul is whimpering.

| | | |
|---|---|---|
| (21) | Mother: | It's an opera. |
| | Adam: | How does dat stay on (i)n (th)ere and not turn? |
| | Adam: | What is he doing? |
| | Mother: | [pəl] is for "parasol." |
| | Mother: | No. No. Paul. |
| | Paul: | [whimpers] |
| | Adam: | He thinks he doesn't [?] have nothing. |

By this time, Adam uses *think* freely with third person subjects to talk about other people's mental states. And in the example reproduced in context in (22), Adam uses the ascription, *He's gonna think if dis Mommy*, to ascribe a false belief to Paul; this ascription meets criterion (3) for the full interpretation.

| | | |
|---|---|---|
| (22) | Adam: | No you put de mask on … now. |
| | Mother: | Paul doesn't like for me to put it on. |
| | Adam: | Put it on. |
| | Adam: | It's gonna go … [blows out the party blow-out]. No. No. No. He's gonna think if dis Mommy. |
| | Mother: | He's going to think that that's Mommy. |

At 3;8–0, Adam produced a number of ascriptions of the form, *I thought* … including, *I thought I said a flamingo*, and, *I thought you couldn't do it*. But they seem to be offered as evidential qualification, and none of them meet any of the criteria in (1)–(5). At 3;8–26, Adam utters the intriguing, *I thought you said the animals gonna have space*. But there is no evidence that it expresses anything other than the simple modal, *should*, or *should be the case*. There is some possibility that *think* and *know* are used with full mentalistic interpretation at 3:10–0. But there is no clear evidence of that until the examples at 3:11–0 discussed earlier. Thus, Adam clearly uses mentalistic attitude ascriptions with the full interpretation by 3;11, but there is no clear evidence that he does so before that.

What is remarkable about Adam is that he has ascriptions of assertion very early, by 3;5, six months before he has mentalistic attitude ascriptions at 3;11. The example in (23) at 3;5–0 is reproduced with context to show that Adam's *say*-ascription is referring to an event in which his mother said that he can't have a kitty (or doesn't want to have a kitty), and is not just a performative use.

(23) Adam: Can [?] I have a kitty?
Mother: I don't think so.
Adam: Why not? What might happen?
Mother: Well ... nothing might happen.
Adam: Why you say I don't want to have a kitty?
Mother: I just didn't think you'd like to have a kitty.

The conversation is about a recent visit to a pet shop. Adam might be referring to an event during this visit in which his mother said he doesn't want a kitty, or he might be referring to his mother's denial of permission to have a kitty, at the beginning of this segment, altering the facts a bit. In either case, it refers back to an event in which his mother said something. And the exchange in (24) occurs when Adam is 3;8–0.

(24) Mother: I think perhaps something **is** missing.
Adam: I know.
Adam: Mommy said something is missing.

These indicate that Adam produces ascriptions of assertion with the full interpretation, characterizing an event, months before he produces mentalistic ascriptions with the full interpretation, characterizing a mental state. This is not surprising. Adam's full mentalistic ascriptions at 3;11 are in line with the timeline for Theory of Mind development. The description of an event of assertion in terms of the asserted content is not obviously dependent on Theory of Mind in any way. This shows that there can be an advantage in child development studies in separating the capacity to use propositional content to characterize events of assertion from the use to characterize mental states. Only the latter use seems to require Theory of Mind.

The investigation of Sarah and Adam show that full mentalistic ascriptions, certified as such by meeting any of the criteria in (1)–(5), are not produced until within a month or two of age 4;0, in line with results in the literature on the onset of Theory of Mind. Attitudes of assertion with full event-describing interpretations may occur a couple of months earlier. This completes the developmental arc proposed by Diessel (2004), which we summarize here in our terms. Children's first uses of mentalistic predicates, around age 3;1 or 3;2, are formulaic, typically with present tense predicates and a first person singular subject, and with minimal, evidential qualifying or performative, interpretations. The forms diversify in productivity within six months, exhibiting other persons and tenses, but continue to have minimal interpretations until around age 4;0, when mentalistic ascriptions with full interpretations appear, meeting criteria in (1)–(5) for describing a mental state.

## 9.3 Wernicke's aphasia: selective impairment of full ascriptions

In this section, we review proposals in the literature to explain the dissociation between aspects of language production that are lost with Wernicke's aphasia, and ones that are preserved. The distinction between full and minimal attitude ascriptions empirically fits into this dissociation, and the theoretical account of the distinction between the two types of attitude ascription helps clarify the explanation of the dissociation manifested in Wernicke's aphasia. This explanation will be tested by comparisons between conversational data discussed by Buckingham et al. (1975) of a Wernicke's patient with extensive neolinguistic jargon and the conversational data for typical (unimpaired) adults discussed in Section 9.1. But we begin with a brief survey of studies of pronoun resolution and sentence comprehension in the literature. On the one hand, some of these studies show that people with Wernicke's aphasia have trouble resolving pronoun interpretation when that interpretation involves reference (or coreference) as opposed to interpretation as a bound variable; these studies indicate that Wernicke's aphasics have particular difficulty with processes involving reference, as opposed to language-internal relations. On the other hand, other studies show that sentence comprehension by Wernicke's aphasics is much less impaired overall than their production, and mainly involves a failure to assign the correct thematic roles to nominals in structures such as passives and object clefts in which the linear order of arguments of a verb is reversed from the canonical order. These results in the literature provide the context against which the results in Section 9.3.3 will be interpreted.

### *9.3.1 Referentiality in comprehension deficits of aphasia: results in the literature*

*9.3.1.1 Evidence of a role for referentiality in comprehension deficits in aphasia* Grodzinsky et al. (1993) investigated the interpretation of reflexive and nonreflexive pronouns by people with Broca's (agrammatic) aphasia and Wernicke's (fluent) aphasia, and compared with results of a similar study with typically developing young children conducted by Chien and Wexler (1990). The experiments of Grodzinsky et al. involved presenting research subjects with the questions in (25) and (26) twice, once accompanying a cartoon that accurately depicted an event matching that described in the question, and once accompanying a cartoon that depicted a similar event, but not matching the conditions described in the question. The subjects indicated "yes" or "no" to each question-picture pair by pointing to a smiling face or frowning face set before them (thus removing confounds stemming from any inability of the subject to respond verbally).

(25) a. Is Mama Bear touching herself?
b. Is Mama Bear touching her?

(26) a. Is every bear touching herself?
b. Is every bear touching her?

In their study, Broca's aphasics answered normally (above chance levels) in all picture-question combinations except one; they had trouble answering correctly only with (25b) under a condition of a mismatch, where the cartoon depicted Mama Bear touching herself, and the correct response was "no." They correctly discerned that (25b) accurately describes a cartoon in which Mama Bear is touching another (female) bear. This was the result Chien and Wexler (1990) obtained for young children as well, and it is the result Grodzinsky et al. (1993) focused on, arguing that Reinhart's (2000) theory of intrasentential coreference predicts that exactly this example should be problematic in the case of a mismatch, since processing this example in this case requires computation and comparison of two parallel semantic structures, and thus requires processing power that exceeds that available to young children and aphasics.

But the results of Grodzinsky et al. (1993: 407–408) for Wernicke's aphasics are equally striking, even though the authors admit (p.409) that they basically ignored these results in order to concentrate on the agrammatic individuals. They found that the Wernicke's aphasics perform like unimpaired adults at interpreting anaphor binding by a subject nominal or quantifier, in (25a) and (26a), where the locally A-bound anaphor is interpreted as a bound variable. In contrast, they perform at chance levels in interpreting parallel structures with a pronoun, in (25b) and (26b), both when the condition described in the sentence matches the cartoon, and when it fails to match the cartoon. Thus, they cannot interpret (25b) by matching it to a cartoon in which Mama Bear is touching another female bear, or by denying its match with a cartoon in which Mama Bear is touching herself. Likewise, they cannot interpret (26b) by successfully matching it to a cartoon in which each bear is touching a bear distinct from itself, or by denying its match with a cartoon in which each bear is touching itself. As Grodzinsky et al. themselves note, the Wernicke aphasics thus exhibit a more general breakdown of pronoun reference and coreference.

Binding Theory prevents the nonreflexive pronoun *her* in (25b) and (26b) from being locally bound, and given that it cannot be nonlocally bound in these monoclausal examples, the individual must therefore retrieve a referent through broader pragmatic processes of reference resolution, which include ones for establishing coreference intrasententially (without a bound variable) as well as ones for retrieving a referent from the discourse or intersentential context. Thus, these results show that the Wernicke aphasics have trouble with reference determination (including cases that result in coreference with a nonbinding antecedent), but have no trouble with bound-variable binding. This result doesn't implicate a particular theory of pronominal reference resolution, but it points to a contrast in performance based on grammar-internal binding mechanisms for bound variable interpretations, as opposed to the extra-grammatical processes for the resolution of pronoun reference. In these studies, this contrast shows up in processing and/or comprehension.

Further evidence that Wernicke's aphasics have trouble with reference in comprehension comes from a word disambiguation task first performed on normal adults by Swinney (1979), then performed by Swinney et al. (1989) on agrammatic aphasics and Wernicke's aphasics, and performed on young children by Swinney and Prather (1989). (See Grodzinsky et al. 1993 for further discussion.) The original point of the experiment was to test the role of context-dependent meaning in access of lexical items. Subjects were aurally presented with sentences such as those in (27).

(27) a. The FBI agent searched the room for BUGS.
b. He caught spiders, roaches, and other BUGS.

Immediately after hearing the lexically ambiguous word at the conclusion of either pair, the subjects were tested with a secondary priming task in which they were presented with a sequence of letters on a screen, which may spell a word related to one meaning of the lexically ambiguous word (e.g., "SPY") or a word related to the other meaning (e.g., "ANT"), or which may be a nonsense string that does not spell a word at all. The subject had to signal whether the presented letters spell an English word, and their reaction is timed. With unimpaired adults, interestingly, there is no significant difference in reaction time on the secondary task depending on whether or not the letter sequence constitutes a word related to the contextually appropriate meaning in (27a) or that in (27b); a word related to the contextually appropriate meaning was recognized as quickly as one related to the contextually inappropriate meaning. This shows that both meanings of the lexically ambiguous item are accessed, even though this item is introduced at the end of a sentence that clearly favors one meaning only. In contrast, if the secondary priming task is delayed, flashing the letter sequence a few hundred milliseconds after the lexically ambiguous item is heard, reaction time is much better for recognizing letter sequences that spell a word related to the contextually determined meaning. This effectively shows that context acts after the lexically ambiguous word is first accessed to determine which of its meanings is appropriate.

On this task, children and Broca's aphasics showed an immediate preference for one meaning only, that being the statistically most frequent one, whether or not it was the meaning favored by the context. This is interpreted as showing that they cannot access both meanings and hold them in mind simultaneously immediately after the ambiguous lexical item is accessed, until the context determines which meaning is appropriate. Instead, their lexical access is restricted to the most statistically common meaning of the item.

Grodzinsky et al. (1993) discuss the results of these studies, and point out (p.412, footnote 6) that Wernicke's aphasics, although they do not perform similarly to agrammatics and children, also do not perform similarly to unimpaired adults. The Wernicke's aphasics in the experiments of Swinney and associates are very slow to decide whether the presented sequence of letters is a word or not, and although they exhibited a difference in reaction time in recognizing that the sequence is a word, depending on whether the word was related to the contextually appropriate meaning of the ambiguous item or another meaning, this difference was small compared to the overall length of their reaction time.

This result is consistent with the Wernicke's aphasics having trouble resolving the referent of the ambiguous lexical item at all (as well as other referring forms in the presented sentence). This interpretation of the results of Swinney and associates comports with the interpretation we gave previously for the results of Grodzinsky et al. (1993) for Wernicke aphasics.

Compare this interpretation of results on pronoun resolution and word recognition by Wernicke's aphasics to the proposals of Vasić, Avrutin, and Ruigendijk (2006), based on Reuland's (2001) theory of economy in computing anaphoric relations. Reuland proposed an economy hierarchy in anaphoric relations, with narrow syntactic operations such as feature checking (matching person, number, and gender features of anaphor and antecedent) as the most economical, followed by the semantic operation of variable binding, followed by relations of reference and coreference. Vasić, Avrutin, and Ruigendijk propose that processing demands increase with each step up in the hierarchy. Thus, in pronoun resolution, finding a potential antecedent with matching person, number, and gender features is the least demanding part of pronoun resolution; computing bound variable relations as in *Every boy brought his lunch*, *Sam hurt himself*, or *Every boy hurt himself* is a more demanding process; finding an antecedent without the operator-variable relation is the most demanding, as in *Sam hurt them*, where the referent is recovered from the context, or *The person standing next to him caught Sam as he fell*, where the referent is recovered intrasententially, but not under a variable binding relationship.

Vasić, Avrutin, and Ruigendijk (2006) show that similar results are obtained for agrammatic aphasics in the interpretation of VP ellipsis. As they put it (p.192), "the operations that involve variable binding are less problematic for agrammatic aphasics in contrast to the operations that involve an establishment of coreferential dependencies."

Grodzinsky et al. (1993) and Vasić, Avrutin, and Ruigendijk (2006) argue that their results reflect a lack of processing resources on the part of children and agrammatic aphasics to compute referential and coreferential relations, as opposed to language-internal bound variable relations. We might accept this, but draw the conclusion more broadly for Wernicke's aphasics. As shown by the results of these researchers for Wernicke's aphasics, and as the results of Swinney and associates also show, Wernicke's aphasics have trouble with reference and coreference generally. This suggests that the cross-modular interaction between language-internal systems and language-external systems is itself the problem for Wernicke's aphasics. This may be because cross-modular processes require more processing resources. But it could also be that cross-modular processes are problematic for Wernicke's aphasics in themselves, for structural reasons, for example, because a junction of language-internal and language-external neural systems is damaged. In that case, we would expect cross-modular operations to be more difficult for Wernicke's aphasics, even if they don't require more processing resources, but simply because they require some specific resources knocked out at the junction.

Putting aside the last speculative point, the literature surveyed here does seem to show, in a domain having nothing to do with attitude ascription, that Wernicke's aphasics have more trouble with reference, which involves relating language-internal processes to language-external processes, than with anaphor binding and bound pronoun interpretation, which involve language-internal dependencies. The studies surveyed here establish this result in experiments involving language comprehension.

Other results implicating referentiality have been obtained only for Broca's aphasics, but show a clearly delineated distinction based on the referentiality of nominals, and thus contribute to the view that aphasia can differentially impair referential language over nonreferential language. Saddy (1995) noted that these agrammatic aphasics are better at processing the sentence in (28b), with a quantified subject nominal, than (28a), with a referential subject.

(28) a. The man is pushed t by the boy.
b. Every man is pushed t by the boy.

This can be explained on the basis of the Trace Deletion Hypothesis (TDH) of Grodzinsky (1986, 1990), whereby traces of moved nominals are deleted in the grammar of agrammatic aphasia, preventing the recovery of the thematic role of the nominal, and thus the nominal element's role in the interpretation of the sentence. This was refined by Grodzinsky (1995a) to apply only to referential nominals; traces of quantifiers and heads are not deleted.

The explanation has been developed as follows. The definite nominal *the man* in (28a) is generated as the direct object of the verb *push*, and then moved into subject position, leaving a trace in direct object position. For a typical adult speaker, the derived subject is interpreted as the patient argument of the verb by virtue of its trace in the direct object position where that argument is normally expressed. But in agrammatic aphasia, the trace is deleted, preventing the construal of the derived subject with this thematic role. This impedes the comprehension of the sentence. Grodzinsky (1995b) refines this explanation by supposing that, after its trace is deleted, the referential nominal will be assigned a θ-role by a default rule based on its position in linear order (its surface position relative to the verb), according to which the first nominal in a clause expresses the agent of the verb. Thus, the nominal *the man* in (28a), lacking a θ-role through association with a trace, will be construed as an agent, leading to faulty comprehension of the sentence. The quantified nominal in (28b) is not a referential constituent, and so its trace is not deleted after it moves to the subject position in the passive. Thus, it is associated with the θ-role of Patient through the position of its trace, and the performance of agrammatic aphasics on this sentence is consequently high.

The distinction between referential and nonreferential nominals also has been invoked in the explanation of the data in (29) and (30), observed by Hickok and Avrutin (1995), on comprehension of interrogative sentences; the performance of agrammatic aphasics on each item is indicated in parentheses next to the item.

| | | | |
|---|---|---|---|
| (29) | a. | Which boy pushed the girl? | (above chance) |
| | b. | Which boy did the girl push t? | (chance) |
| (30) | a. | Who pushed the girl? | (above chance) |
| | b. | Who did the girl push t? | (above chance) |

Grodzinsky (1995a) invokes the referentiality of the interrogative nominal *which*-phrases in (29), versus the regular *wh*-expressions in (30), to explain these results. (This explanation is recapitulated by Balogh and Grodzinsky 2000.) In syntactic theory, it has been noted that *which*-phrases such as *which boy* have a referential tie that regular *wh*-phrases such as *who* lack, since they presuppose a discourse-determined set, of which the answer to the question is presumed to be a member; see Chomsky (1973), Pesetsky (1987) and Cinque (1990). This referential tie is often called D(iscourse)-linking, following Pesetsky (1987). On the basis of D-linking, Grodzinsky (1995a) assumes that the *which*-phrase is treated as a moved referential nominal. Its trace is therefore deleted, and it is assigned the Agent θ-role by default. This leads to correct comprehension in (29a). But in (29b), where the unmoved referential nominal *the girl* also expresses the Agent role, the agrammatic aphasic is left to guess at the correct interpretation, leading to faulty comprehension and chance-level performance. The regular *wh*-expressions in (30), not being D-linked, are nonreferential, and thus interpreted in the manner of quantificational NPs. Their traces are retained, and the agrammatic aphasics therefore perform better in interpreting them.

Note in all this that agrammatic aphasics have problems only with reference resolution for displaced nominals when reference interacts with the grammar.

They don't have problems with the reference of nominals that don't move, and they don't seem to have problems with the reference of predicates, nor with the composition of predicates and arguments to make an assertion with a truth value. Thus, their problem is ultimately not with reference per se, but with the grammatical structures and mechanisms required to retrieve referents of moved thematic nominals and to link them in the correct thematic relationship to a predicate. When they can associate a nominal reliably with the correct θ-role, they have no trouble combining the referent of the predicate with the referent(s) of its argument(s) in order to obtain a coherent utterance. Furthermore, these are problems with comprehension. Agrammatic aphasics often have other problems with production (including apraxia), but we would expect the problems with the comprehension of referential nominals noted earlier to carry over to production as well.

*9.3.1.2 Arguments against a role for referentiality in comprehension deficits in aphasia* As discussed by Edwards (2005), Wernicke's aphasics exhibit limitations in working memory and in phoneme and word identification (e.g., in detecting rhymes), both of which would contribute to language comprehension difficulties. Edwards notes (p.160) that Zurif and colleagues found that Broca's subjects had difficulties with processing grammatical (thematic) relations due to Grodzinsky's trace deletion, but they also found that this was not the case for Wernicke's aphasics: the latter didn't have a problem with the relationship of a referential nominal to a trace. In a similar vein, Caramazza and Zurif (1976) established that Broca's aphasics had trouble with parsing, and they cautiously drew a contrast with Wernicke's aphasics, who seem to have less obvious trouble with parsing, although subsequent work has shown some parsing difficulty among some Wernicke's aphasics.

Edwards also notes (p.160) problems with word comprehension in both types of aphasia. Nevertheless, Wernicke's aphasia is distinguished by particular difficulty in associating words with meanings, and even difficulty distinguishing individual phonemes.

Bastiaanse and Edwards (2001, 2004) found that fluent aphasics are similar to agrammatic aphasics (as previously reported by Balogh and Grodzinsky [2000]) in having trouble with noncanonical word order. This result is summarized and discussed by Edwards (2005: 165–170). In tests of comprehension of active sentences, passives, subject clefts, and object clefts, testing sentences that varied from what is depicted in pictures by the reversal of θ-role, lexical substitution of a different verb, or both of these factors, the primary difficulty for Wernicke's aphasics was in the reversal of θ-role. (E.g., the subject signals that *The man rescues the woman* properly describes a scene in which the woman rescues the man.) Two of the nine Wernicke's research subjects in this study performed equally as well on active and subject cleft sentences as on passives and object clefts, but five of the nine did better in both active over passive, and subject cleft over object cleft. Edwards concludes that noncanonical order of thematic arguments is the main culprit (p.168): "Thus, these subjects with fluent aphasia exhibited the same kind of pattern of sentence-comprehension deficit as that reported for agrammatic patients."

Nevertheless, the facts in (25)–(27) still suggest that there are some differences, even in comprehension, between agrammatic and fluent aphasics depending on whether relations of reference are involved (with no reversal of θ-roles with respect to linear order), versus relations of anaphor binding or quantification. We still have reasons to suspect that reference might be a particular difficulty for Wernicke's aphasics, even in comprehension. When we turn to production data in Sections 9.3.2 and 9.3.3, this dissociation between referential and nonreferential nominals in the language of Wernicke's aphasia will become much more pervasive and striking.

### *9.3.2 The dissociation between modalizing language and referential language*

For Wernicke's aphasics, the ability to express core propositional content, what Bally (1942) called the dictum, is noticeably impaired. The simplest form of dictum combines a singular term or other referring nominal with a property-denoting predicate to form a proposition that is asserted (and which it is the main point of the utterance to assert); Jackson (1879) gives the example *Gold is yellow*. Nespoulous et al. (1998) characterize the language in the dictum as referring. In contrast, Wernicke's aphasics tend to be much more capable, and some are quite fluent, at expressing modal or evidential qualification of the dictum, or, in the words of Nespoulous et al. (1998: 317), expressing a "personal attitude to what he or she is saying (about the dictum) or about what the interlocutor is saying." This accompanying commentary is what Bally called the modus, and which Nespoulous et al. (1998) call modalizing language. For many Wernicke's aphasics, this leads to an interesting pattern by which the modalizing material in an utterance is more fully or clearly expressed than what it ostensibly qualifies, which is often missing or incomprehensible.

For some individuals with Wernicke's aphasia, the modalizing parts of utterances are replete with attitude verbs such as *believe*, *say*, *tell*, *admit*, as well as the modal auxiliaries *would* and *should*, with or without negation. First person subjects of these predicates tend to predominate, almost exclusively for some, but others exhibit considerable diversity and productivity of subjects and attitude predicates. But the modified dictum, which should occur in the complement clauses to these verbs, is noticeably impaired in comparison to the modalizing material. Sometimes an utterance will fail to supply a dictum in a complement clause to one of these verbs by trailing off after the complementizer, or it might turn to neologistic jargon after the complementizer, or lose coherence after the complementizer. The examples in (31a,b), from BF, a Wernicke's aphasic discussed by Buckingham et al. (1975) exhibit these tendencies.[4]

(31) a. Clinician: You look well.
BF: Ats right do, because we haven neder ha da fun.
Clinician: You look better than you did before. You feeling better?
BF: Yes. She's in the west. I can say that … She wanted to, I know that beenuh … she wanted to seep with me. And sweep on the peet sing sleeps. It sleep lapes slipe nothing, nothing to slentch cheekut.

b. I mus say well dee I'm sorry the tready's not right because that's what the tris is right.

The extended example in (32) has three instances of *know* with an overt or elided complement clause. In the complement of the first, coherence is lost with a jargon predicate *rent lisdisit* and subject pronoun *it* of obscure reference; in the second, the complement is elided, but with no recoverable content, and the third person subject of *know* is obscure in reference; and the third has a coherent sounding complement, *what to do with that dope*, but one which is nevertheless obscure since the referent of *that dope* is not clear, and the structure is one of subject control by the higher subject *he*, which is obscure in reference.

(32) [In response to the clinician query, "Did you work?"]
Oh, I don't know where it rent lisdisit. It came and teared it an godit. It dar un an it drifted um. It didn't anytheme. Chune saulda. He, he didn't have youk. And, why? Well, he didn't seem to know. He didn't know what to do with that dope. That dumb Tony that I had. He said, you're going on a tempo. What's a jumpuder? As anistrudish, I was still so much. I was always ludin like teh shih. That's all I wanted to try mus tha left alma. These little trefteze. He's love ize. He lovelize. Plemickse. Like my shame buths?

[4] BF was a 55-year-old man who had surgery on a left middle cerebral artery aneurysm, following which he had severe aphasia with copious jargon; he also exhibited right hemiplegia and hemianopsia. His segmental phonology was largely preserved under the aphasia, but exhibited some metathesis of adjacent syllable onsets and other minor effects; his intonation was fully natural under the aphasia. His syntactic structures seemed normal where the syntactic categories were not so obscured by jargon as to be unidentifiable. The language of BF in interviews with researchers and clinicians is discussed further by Buckingham (1981), and the neuroanatomy of BF is discussed by Kertesz (1981).

The Wernicke's aphasic may produce neologisms within or in place of the dictum. Some exhibit awareness that the dictum is not clear or is not being understood; others seem to be unaware, exhibiting anosognosia. First and second person pronouns seem to have their normal referents; third person pronouns are almost always obscure in reference.

In contrast, individuals with Broca's aphasia tend to produce referential speech, but with limited verbal inflection and accompanying functional categories.[5] In extreme cases, there is no verbal inflection or functional material at all, and the lexical content exhibits disruptions in thematic relations so that arguments of a predicate are missing or misplaced, or are replaced with jargon or semantically inappropriate arguments. The result is frequently a form of telegraphic speech consisting of incomplete lexical snippets. See Friedman (2006) for a description and theoretical account of this, and for counterpoints to that view, see Wenzlaff and Clahsen (2004) and Faroqi-Shah and Thompson (2004).

Buckingham, Avakian-Whitaker, and Whitaker (1975) argued that the modalizing material does not reduce to automatisms since (first) it is more lexically varied and productive than automatisms in Broca's aphasia, and (second, in their terms) the modalized material is more transformationally varied, including effects of negative transposition and other transformations. They analyzed the modalizing language along the lines of the syntactic structures posited for implicit performative clauses embedding an utterance, as in the Generative Semantic work of Ross (1970).

Nespoulous et al. argue for the clarity and importance of the distinction between referential and modalizing language, and the differential effects of aphasia on them, and survey various accounts of the distinction. They consider whether the distinction between modalizing and referential language might be linguistically determined or encoded, or determined functionally. By linguistically determined, they mean that different linguistic categories would occur in referential and modalizing speech. They put it in Saussurean terms: that the two types of linguistic production would involve different types of signifiers. However, they observe nothing distinct or special about the nature of the signs used in modalizing speech that sets it apart from referential speech other than its function. They conclude that the dissociation between referential and modalizing use "does not seem to be structurally motivated" (p.320).

Nespoulous et al. also take up the question of whether the dissociation depends on neural representation in different hemispheres, but find the

[5] There are some cross-linguistic differences in this, and some controversy in interpreting those differences. For example, Wenzlaff and Clahsen (2004) found that subject-verb agreement is preserved for German-speaking Wernicke's aphasics, while expression of tense is impaired.

results on this to be quite inconclusive. Left hemisphere lesions are highly implicated in aphasias, and right hemisphere lesions much less so, dispelling the idea that modalizing language might be supported by neural systems primarily located in the right hemisphere, or ones that are more bilaterally symmetrical than those that support referential language. There are cases of total left hemispherectomy in which the patient displays a Wernicke-like asymmetry in the preservation of modalizing over referential language, but this is observed only in cases where prior disease in the left hemisphere may have led to some speech functions migrating to, or developing in, the right hemisphere. These cases suggest that when the left hemisphere is compromised by damage or disease, the right hemisphere may be capable of sustaining more systems involved in modalizing than referential speech, but nothing more.

Nespoulous et al. mention a person whose major language areas were destroyed in both right and left hemispheres, but who could still utter the modalizing, *I don't know whether you have understood what I said*, and they rightly ask where that comes from. They consider the hypothesis that this and other modalizing speech could be subcortical, but decide there isn't enough evidence for this. They note in passing that many automatic behaviors are known to be served by subcortical neural structures, but to the extent that modalizing speech in many fluent aphasics does not seem to be automatic or routinized, this would not fully explain the capacity for modalizing language by many fluent aphasics.

They see more hope in making a neural distinction between referential and modalizing speech using the classification system of Yakovlev (1968). This scheme delineates three types of neural systems or "action systems" in the brain, each type localized in a lobe of the brain. The rhinic systems control automatic bodily processes (e.g., heart function). The limbic systems control emotional affective behavior. And the supralimbic systems control deliberate and manipulative activities such as tool handling. Nespoulous et al. conjecture that modalizing speech is controlled by limbic systems, and referential (propositional) speech by supralimbic systems.

Regarding the vocabulary that can be used equally in referential and modalizing speech, Nespoulous et al. (1998: 329) pose a question about its status, whether it resides in duplicate among limbic systems and supralimbic systems. Insisting on the unity of the sign, they conjecture that such items are stored in one place only, but can be accessed and used by either limbic systems (for modalizing speech) or supralimbic systems (for referential speech). As they put it, "We could hypothesize that it is the access and controlling mechanisms or processes of referencing and modalizing that are represented separately within separate neural networks which interact with the networks where the individual words and phrases are represented" (p.329).

### 9.3.3 *Refining the distinction*

*9.3.3.1 A conjecture about attitude ascriptions in Wernicke's aphasia* As observed by Nespoulous et al. (1998), for aphasics such as BF, even the fullest mentalistic language is mostly modalizing language (evidential qualification such as *I don't know...*, attitudes with modal auxiliaries such as *I must say that...*), rather than referential language. We note that aphasics such as BF have relatively unimpaired expression of aspectual morphemes, tense, agreement, and complementizers, in short, the normal inventory of functional categories. This suggests that what primarily characterizes their language deficit is that referential language is impaired, and most of what would normally be referential language is used nonreferentially, including use of attitude verbs and other mentalistic vocabulary.

We conjecture that the modalizing language of such individuals exhibits attitude ascriptions only with the minimal interpretation distinguished in Chapters 3 and 4, not with the full interpretation as characterizing an eventuality in terms of its relation to propositional content across possible worlds. Use of the full interpretation requires reference to an associated state or event; Wernicke's aphasics, though in full possession of the requisite syntactic structures and the minimal interpretation quantifying over possible worlds, cannot bring this apparatus to bear in order to effect predication to an associated event or state. If correct, this supports the conjecture of Nespoulous et al., that for such a person, what is impaired is the propositional language of true predication, in which a predicate-denoting expression is combined with a singular term to characterize or qualify the referent of the singular term. This would now be generalized to a deficit in ascribing a property to any individual, including an event or state introduced by, or otherwise associated with, an attitude predicate under the full interpretation.

In the aphasic productions examined in this section, there is no evident loss of lexical items involved. Rather, what seems to be lost is the full realization of the capacity to refer, where reference is understood in the general Fregean sense: for singular terms to refer to an object, for predicates to denote a set of objects, for subject and predicate to combine to form a proposition with truth conditions in a context. Thus, modalizing speech involves quantification over possible worlds and occurrent "saying" in the context, uttered as evidential comment, but no reference to mental states of belief, or acts of assertion removed from immediate experience, just as reference to objects removed from immediate experience is impaired.

The split can then be construed as follows: What is needed for modalizing speech, and what BF exhibits, includes quantification over possible worlds, and semantic composition to produce evidentially qualifying meanings. What is needed for referential speech, and thus what BF lacks, is reference to individuals and eventualities, and consequently, truth conditions for acts of predication. Predication seems to be available, yet generally not to characterize an object or eventuality, but primarily to effect evidential qualification. On this view, the mechanisms of predication and quantification are not impaired, but just the mechanisms for reference, whether to singular individuals, to kinds, to properties (sets of individuals), or to eventualities (events and states).

The next two sections present and discuss the results of a closer study of language with the form of attitude ascriptions produced by BF. Before proceeding, we note that there are a few occasions when BF seems to assert a proposition, or to formulate a question with propositional content. The best such example is given in (33).

(33) Clinician: You got new shoes?
BF: Oh, yeah? Oh no dear. Do you know that I been wearin' ths in sen years zingk? I bet these my lass minacertick.
Clinician: You've had those for a long time, have you?
BF: Oh, they're bertiful.
Clinician: Are they?
BF: They are.
Clinician: They're comfortable...
BF: I never tooth my toase mateet thy tig. Take them right out and get them to Use my steff all over. Hef some ... fer um.

This is the most coherent sounding segment of discourse in which BF engages in the entire transcript. The candidate attitude ascription has the form of an interrogative, *Do you know that I been wearin' ths in sen years zingk?* If it is asserting that BF has been wearing his shoes for some time, it is a coherent elaboration on his answer, *Oh no dear*, to the clinician's question whether he got new shoes. But if this is correct, its function is modalizing: the question is a way of asserting that he has been wearing the shoes for some time (for some number of years?). Next, BF seems to assert that his shoes are beautiful, and then to agree with the clinician on this point.[6] This indicates that BF is not completely lacking capabilities for reference and propositional form, and still has them at least for the discussion of objects perceptible in the immediate environment. This capacity seems to be corroborated by the example in (34a), uttered in the middle of a long, uninterpretable soliloquy, and the example in (34b), his final utterance at the end of the transcript.

(34) a. There, that's better. [rearranging himself in the chair]
b. Clinician: Are you a doctor?
BF: No.

[6] This was the impression of one observer of the interview, Hugh Buckingham (personal communication).

Clinician: Good. You're doing better than you did before.
BF: What? Are you through? Are you all through?

In (34a), the demonstrative pronoun *that* seems to refer to BF's own state after he arranges himself more comfortably in his chair. And in the last line of (34b), he seems to refer to his interlocutor and what they are doing at the moment. So he seems to have pronominal reference to himself and his interlocutor, and to states and things in the immediate environment. These examples don't really challenge the present conjecture that BF suffers a primary impairment in reference, since the transcript overall still manifests massive impairment in the capacity for reference, with these examples as the best exceptions. The conjecture is not that BF completely lacks the ability to refer, but that this ability is severely impaired, restricting reference to objects, states, and events that are observable in the immediate environment.

*9.3.3.2 A comparative study* Impressionistically, the difference between the productions of a Wernicke's aphasic such as BF and unimpaired adults is quite striking. Partly this is due to the neologistic jargon with which BF's productions are laced, and the dynamic aphasia involved in the overall lack of coherence of his discourse. We take these to be, at least in part, highly visible consequences of impairment in reference. Against this backdrop, some of the modalizing language of a Wernicke's aphasic such as BF is startlingly normal, even if it trails off before a verbal complement, or embeds incoherent or jargon-filled complements, as in the examples in (35).

(35) a. I don't know how to tell you … [trails off]
b. I couldn't say the talk swingis. But I tell you now I do. I do believe …
I've gotta believe they gotta be tirm clothes haven't they?

Nevertheless, there is an important difference between these and the language of an unimpaired adult, in that an unimpaired adult uses the same attitude ascription forms for referential utterances, as well as for modalizing speech. A qualitative study was conducted for the present work in order to establish something more than an intuitive basis for this difference in the potential referentiality of attitude ascriptions in the language of Wernicke's aphasics such as BF and unimpaired adults.

In this study, the productions of BF were compared to those of nonaphasic adults, focusing on anything with the syntactic form of an attitude ascription. The criteria (1)–(5) of Section 9.1 were employed. When one or more of them were satisfied, it was concluded that a candidate mental state or event of cognition, perception, or assertion associated with an attitude ascription is, in fact, an event or state being characterized by the ascription. Examples that meet none of the five criteria may have the full or minimal ascription, but they fail to meet the conditions for being certifiably full ascriptions.

The study surveyed a transcript of a conversation of Wernicke's aphasic BF with a clinician, during which BF spoke approximately 3,400 words, not counting short responses under 5 words in length. This was compared to the conversations by unimpaired adults, of comparable length, discussed in Section 9.1. In each conversation, candidate attitude ascriptions were identified, and the number of certified full attitude ascriptions according to the five criteria outlined previously was tallied.

*9.3.3.3 Results* In Section 9.1, two transcripts of unimpaired adult conversation were examined, of about the same length as the BF transcript. The first contained forty-seven attitude predicates, of which at least 8.5% were used in ascriptions that demonstrably had the full interpretation characterizing an event or state, counting most strictly according to the criteria presented in Section 9.1. Counting less strictly, as many as 19% were used in this way. The second contained forty attitude predicates, of which at least 32.5% were used in ascriptions that demonstrably had the full interpretation.

In the transcripts of BF's conversation with the clinician, BF utters roughly the same number of words as in the examined section of the *Washington Week* transcript, and 7% more than in the NPR transcript. BF uses eighty attitude verbs, which is 77% more than were used in the *Washington Week* segment, and twice the number used in the NPR transcript. But none of BF's uses of these verbs is accompanied by any of the five signs demonstrating reference to an associated mental state or event of perception or assertion. None of them intuitively describe a state or event. And there is nothing in the speech of BF like the derived nouns *expectation* and *concern*, in (15), to refer to a mental state and to specify its contents.

Some of the best prospective candidate ascriptions from BF are given in (36)–(39).

(36) Clinician: What's happening in this picture? [shows a stick picture of a house]
BF: What do ya want me to do? Tes dar send this to you dear muchineit. Why shea at. I don't know what it's gonna see. What's gonna hide it, but...

There is some question whether the well-formed question with which BF's response opens is really a request for information about how he is to perform the task requested of him. Nothing in his follow-up supports the supposition that it is. If it is, it may involve reference to a mental state of the clinician there and then, at the moment of utterance, which would put it on a par with reference to his shoes in (33).

(37) Clinician: Who was your foremost acoushner? [deliberate jargon]
BF: You know, there, that's what makes me wonder, because that's the kahn Darn thing that so many things have made me ladis, and so many suh fedis settle so sring with that with it set. And then all of the sudden *they say what the hell is the matter with them*. And don't mind me, but ... Why don't you? *He says, why the hell can't you take it the me the week it ist that you like it.*

(38) Clinician: Just "yes" or "no." Is your name F*****?
BF: Well, I don't know wowver ih true or not. I don't know that you can say that you can perkly fit what you don't know.

(39) Clinician: Where do you live, Mr. F*****?
BF: Mine? Mine's right at home in kell shoddy. Good home kant. I know what, unless they kiss me off I owe her something about tunny. I mus say well dee I'm sorry the tready's not right because that's what the tris is right. I know my tresses are all right. I know every monk I drew as I love everyone of them. Ever thet thy won 'em.

The italicized ascriptions in (37) seem eventive, with third person subjects and the modifier *all of the sudden*. But it's hard to say since the pronouns are all completely obscure in reference. In (38), BF might be disavowing knowledge of his own name, but his elaboration does not support this. In (39), the normal looking propositional forms in this response are uninterpretable on account of failure of reference. For example, *I owe her something about tunny* does not seem to describe a situation in which BF owes somebody something. *I know every monk I drew as I love everyone of them* does not situate BF relative to any evident event or events of drawing referred to by the verb *drew*. The failure is a failure of reference, for nouns and predicates alike. Assuming Fregean compositionality, the failure of the parts to refer results in a failure of the propositional forms to refer or to have truth values. In (39), BF apparently mentions his home (*Mine's right at home in kell shoddy*) in response to a query about where he lives, but even this prospective reference to an object not present in the environment is garbled. These, recall, are among the best candidates for referential language by BF in the interview.

The speech of BF is replete with overt quantification. This includes quantification in the modalizing domain, as in (32), *That's all I wanted to try mus tha left alma*, where *all* appears in a modalizing matrix clause and seems to quantify over content that BF intended to convey, but not over anything concrete. There is also implicit quantification in (32) in the comparative *I was still so much*. So BF seems to have everything needed to produce and intend the meaning of an attitude ascription on a minimal, affectively or evidentially qualifying interpretation.

*9.3.3.4 Discussion* Nowhere in all the modalizing language in the transcripts for BF is there a production with the form of an attitude ascription that meets any of the five criteria for being a report of a mental state or an event of cognition, assertion, or perception. Only the candidate ascription in (33), repeated in (40), even plausibly makes reference to such a state or event, and if it is indeed referring, it is referring to an occurrent mental state of BF's addressee, in keeping with his limited reference to objects that are immediately present in the environment of the discourse.

(40) Clinician: You got new shoes?
BF: Oh, yeah? Oh no dear. Do you know that I been wearin' ths in sen years zingk? I bet these my lass minacertick.

The present conjecture is that what is most preserved in Wernicke's aphasics is not modalizing language per se over referential and propositional language, but quantification per se (which includes modality and evidential qualification), as opposed to language that requires reference to things, people, eventualities, and situations. In short, quantificational mechanisms are preserved over referential ones. This reflects the idea that variable binding is a purely grammar-internal operation, while reference (including coreference) is determined at the interface of language-internal semantic and pragmatic systems with language-external perceptual and conceptual systems that regulate interaction with the environment.

In contrast, people with agrammatic aphasia seem to largely understand reference, and have the capacity to refer; their impairment lies more in lack of access to the functional categories that support inflection and that mediate between a predicate and its thematic arguments. As a result, their productions lack inflection and fail to unite predicates with their arguments reliably and fluently.[7]

The conclusion drawn here is only a slight variant of the proposal made by Nespoulous et al. (1998), that modalizing language is spared and referential language impaired in Wernicke's aphasia. Although the difference between the two views is primarily one of emphasis, the two versions do point in different directions regarding the most plausible neural explanation. For Nespoulous et al., modalizing language is a special realm of language, possibly one served by limbic neural systems rather than supralimbic systems. This reflects a view that modalizing language is really a different type of language, at least in its

[7] Of course, many people with agrammatic aphasia also have apraxia, difficulty in producing and pronouncing lexical items.

function and how it is accessed. The present view emphasizes language-internal mechanisms versus mechanisms that effect interaction of the language systems with conceptual and perceptual systems, and thus ultimately, with context and the environment. The latter are involved in general reference, by which we mean reference to third parties, reference to out-of-sight objects and individuals by name or description – in short, reference to anything other than speaker and addressee and possibly pronominal or ostensive reference to immediately present objects, events, and states. Quantification, reflexive binding, and bound variable binding are language-internal, while other cognitive systems play a significant role in general reference.

In saying that reference is impaired, we are not saying that BF necessarily lacks the requisite intentionality toward things and properties, but only that, at minimum, he lacks the referential connection between those things and properties and the language used to talk about them. In fact, there are large tracts of the transcripts with BF where it is not clear that he has a topic "in mind" in a sustained way. But there are other tracts where it seems clear that, for a while at least, he is trying to communicate coherent thoughts. The discussion of his new shoes in (33) is one such example, and it was the belief of one observer present during that discussion (Hugh) that BF was discussing his shoes, which were nice, suede shoes that stood out as distinctly conversation-worthy. Another example, given in (41), is BF's response to a clinician's inquiry about his illness, where, like many Wernicke's aphasics who are not anosognosic, BF seems to express frustration with his inability to convey his thoughts and feelings. (The rest of the response is omitted here.)

(41) Clinician: Tell us about your sickness…
BF: You mean whuh I do ferm?
Clinician: When you were sick…
BF: Well I don't know sir fu so serm. I mean I try do. I do. I know that the rearder things have gotta be my perda. They've gotta be. They've gotta me, by dear. But I don't always a get what I wanted. What I wanna do. My nates un nurse ster 'en with them. But they can sir clair in time. I've get them back gee hice. I don't know what to do. They don't come here an talk to thum. I talk to them, talk to them… What do they woik? [goes on]

Note that the productions of Wernicke's aphasics vary in coherence. Some seem very coherent, even while lacking normal lexical referents. For example, Nespoulous et al. discuss a well-known example of a fluent aphasic with the pervasive use of a single jargonaphasic stereotype, whose responses seem entirely appropriate except that the single stereotype appears in place of almost all lexical content. In dynamic aphasia (Luria 1970), narrative structure is impaired; see Sample 4 of Nespoulous et al. (1998). BF exhibits a considerable amount of dynamic impairment in the interview, apparently shifting topics (so far as they can be identified) several times within a single conversational turn. However, in the passage in (41), BF does seem to have a meaning in mind that he is trying to convey: he seems to be describing his frustrating experiences trying to talk. In this case, the deficit is more specifically linguistic: what is impaired is reference – the connection between words and the world, and in this case, it seems not an internal understanding of the reality he is trying to convey.

It remains to place this result relative to the body of research on Wernicke's aphasia, much of it on comprehension abilities. There is clearly considerable disagreement in the literature about the extent to which Wernicke's aphasics are similar to agrammatic aphasics in language parsing and comprehension. But however similar Wernicke's aphasics might be to Broca's aphasics in sentence comprehension, they are usually very different in production. Some of this can be attributed to apraxia for Broca's aphasics, but that doesn't account for the differences between the referential and nonreferential speech of Wernickes's aphasics examined here.

These observations suggest a need, in talking about referentiality as the word-and-meaning relation, to specify the directionality of the relation. Comprehension involves the word-to-meaning (and more generally, language-to-meaning) direction, and production involves the meaning-to-word (meaning-to-language) direction. In production, fluent aphasics exhibit a difference in capability between the use of certain forms of referential and nonreferential language. In view of the analysis of nonreferential language as involving quantificational mechanisms, this implicates a problem with referentiality in the meaning-to-language direction. Wernicke's aphasics can fluently come up with the language to express quantificational meanings, including attitude ascriptions with the minimal interpretation, and the sort of epistemic qualification material that normally accompanies asserted content. And as discussed earlier, they do seem to have meanings in mind requiring referential language to convey. But they are impaired in producing the language to express referential content. The impairment can be localized to the meaning-to-language mapping on referential meaning – how to say that a given predicate applies truly to a given object.

We could frame this as a problem of lexical access, but only if we make the same distinction: that fluent aphasics have difficulties with lexical access of words used referentially, to ascribe properties to objects (including mental states and events). They evidently have no trouble with lexical access of words used in modalizing language, to express evidential qualification or to place themselves with respect to a putative assertion, albeit an assertion they are unable to express.

# 10 Conclusion

One way to view the present work is that it has been an extended investigation of the difference between quantification and reference within one broad, empirical domain. Modal statements quantify over possible worlds, and some attitude ascriptions, interpreted using the quantificational schemas of modality, do likewise. Other attitude ascriptions describe eventualities, mental states or events, or events of assertion, using such quantificational schemas. The present work has shown that the difference is consequential.

The distinction provides an account of the LIN (Neg-Raising) in terms of classical quantifier-negation interactions, converting intervening predicates to their duals. The resulting interpretations match the facts about the semantic effects of LIN. And the distinction matches the empirical distinction between predicates that occur in ascriptions permitting LIN, and those that don't, such as predicates of attitude assertion. Furthermore, the existence of this distinction as a lexically varying property explains the sort of variability observed in LIN across dialects and cognates in different languages, on the basis that any language has to grammaticize the distinction, drawing a line somewhere through the middle ground of predicates whose basic semantics does not necessitate one interpretation or the other.

The study in Chapter 9 suggests that the distinction is manifest in child language development. Children typically produce incipient modal statements a month or two after their third birthday, when they also begin to produce attitude ascriptions that they use with increasing productivity over the ensuing months to express modal or evidential qualification. But they do not show evidence of having attitude ascriptions with the full interpretation, characterizing an eventuality, until around the time of their fourth birthday, at the onset of the period where Theory of Mind is manifested, according to false belief studies. Patients with Wernicke's aphasia also make the distinction manifest through their extensive, and for some patients quite productive, use of attitude ascriptions for modal and evidential qualification, and a complete absence of descriptions of mental states, or characterization of acts of cognition, perception or assertion, apart from some isolated comments on the occurrent mental state of someone present at the time of the conversation, which is consonant with their limited ability to refer to and talk coherently about objects in their immediate environment.

Underlying all of this is a view of doxastic attitude ascription as quantification over all possibilities, intersecting the doxastic base with the ordering source, which allows us to ascribe beliefs to people even if their doxastic base is inconsistent. Bouletic ascriptions, likewise, quantify over all possibilities, with an ordering source reflecting the desires or preferences of the experiencer subject, and a contextually appropriate degree of realism, attained by intersecting the doxastic base of the experiencer with the bouletic ordering source, trimming one or the other, or both, as appropriate in case of inconsistency, thus yielding a range of expressibility of bouletic ascriptions, from highly realistic to counterfactual.

In order to sustain this approach, Kratzer's quantificational schemas for modality were slightly modified in the face of challenges from Lassiter and others. A place for possible worlds semantics of modality was preserved, but it had to be allowed to exist alongside scalar theories of modality, in a spirit of semantic pluralism.

Finally, the distinction between epistemic and root modality for purposes of modal determination, if that is to be rooted in syntactic structure, had to be slightly rethought if we do not assume that all instances of modal quantification, in modal statements and attitude ascriptions, have an associated Davidsonian event or state. The variability in epistemic modal interpretation that von Fintel and Gillies explain by invoking a cloud of contexts was argued to be matched in root modal interpretation, so that the underlying syntactic structures are the ones that determine an epistemic or circumstantial cloud of contexts.

Thus, we come full circle. Kratzer's possible worlds approach to modality had to be rethought somewhat, and placed in the context of scalar approaches to modal interpretation that don't supplant it, but coexist with it as different interpretive frameworks appropriate to different contexts of use. The revisions to Kratzer's theory, though not extensive, were enough to support an account of the interpretation of doxastic and bouletic attitude ascriptions in the same truth conditional interpretive framework while taking account of Heim's problems with bouletic ascriptions. On the present approach, Davidsonian events are thoroughly reified, but flexibly individuated at different scales (as physical objects are) rather than, as Davidson proposed, according to causal properties (a problematic idea in that it fixed event individuation at a low level). Events are identified across possible worlds in exactly the same way as objects are, permitting them to survive the challenges presented by Situation Semantics. This permits us to formulate and distinguish the two types of interpretation of attitude ascriptions, with the consequences for LIN, child language acquisition, and our understanding of what goes on in fluent aphasia, with which this conclusion began.

One objective of this study was to propose some innovations in semantic analysis, but only a handful of such innovations have been proposed here, and some of them have been partially anticipated in the literature. It should be clear that the objective was every bit as much to show how the various topics addressed here illuminate one another, and thus, how a handful of innovations in different areas are woven into a larger tapestry of semantic theory, with a consistent ontology of possible worlds and eventualities within a framework that coexists with conceptually quite divergent frameworks. On this view, our interpretive apparatus is consistent, in a fragmented way, with different interpretive lenses appropriate to different purposes. Possible worlds semantics is one such lens, and the present work helps to support the idea that it has a quite expansive range.

# References


Abraham, Werner 2001a. Modal verbs: Epistemics in German and English, in S. Barbiers, F. Beukema and W. van der Wurff (eds.), *Modality and Its Interaction with the Verbal System*, pp.19–50. Amsterdam: John Benjamins Publishing Company.

2001b. Modals: Toward explaining the "epistemic non-finiteness gap," in R. Mueller and M. Reis (eds.), *Modalität und Modalverben im Deutschen*, pp.7–36. Hamburg: Helmut Buske Verlag.

2012. Covert patterns of modality in typological comparison, in W. Abraham and E. Leiss (eds.), *Covert Patterns of Modality*, pp.386–439. Cambridge Scholars.

Abusch, Dorit 1997. Sequence of tense and temporal *de re*. *Linguistics and Philosophy* 20: 1–50.

2010. Presupposition triggering from alternatives. *Journal of Semantics* 27: 37–80.

2012. Circumstantial and temporal dependence in counterfactual modals. *Natural Language Semantics* 20: 273–297.

Alfandre, Danielle R. 2010. *The Interdependence of Modality and Theory of Mind*. PhD thesis, Louisiana State University.

Anand, Pranav and Hacquard, Valentine 2008. Epistemics with attitudes, in T. Friedman and S. Ito (eds.), *Proceedings of SALT 18*, pp.37–54. Ithaca, NY. http://hdl.handle.net/1813/13025.

Asher, Nicholas 1986. Belief in discourse representation theory. *Journal of Philosophical Logic* 15: 127–189.

1989. Belief acceptance and belief reports. *Canadian Journal of Philosophy* 19: 327–362.

1993. *Reference to Abstract Objects in Discourse*. Dordrecht: Kluwer Academic Publishers.

2000. Events, facts, propositions, and evolutive anaphora, in J. Higginbotham, F. Pianesi and A. C. Varzi (eds.), *Speaking of Events*, pp.123–150. Oxford University Press.

Asher, Nicholas and Bonevac, Daniel 1985. Situations and events. *Philosophical Studies* 47: 57–77.

Asher, Nicholas and Lascarides, Alex 2003. *Logics of Conversation*. Cambridge University Press.

Austin, John L. 1962. *How to Do Things with Words*. (Second Edition). Cambridge, Mass.: Harvard University Press.

Bally, Charles 1942. Syntaxe de la modalité explicite, in *F. cahiers de Saussure*. Genève: Droz.

Balogh, Jennifer E. and Grodzinsky, Yosef 2000. Levels of linguistic representation in Broca's aphasia: Implicitness and referentiality or arguments, in R. Bastiaanse and Y. Grodzinsky (eds.), *Grammatical Disorders in Aphasia: A Neurolinguistic Perspective*, pp.88–104. London: Whurr Publishers.

Bartsch, Renate 1973. "Negative transportation": Gibt es nicht. *Linguistische Berichte* 27: 1–7.

Barwise, Jon and Perry, John 1981a. Semantic innocence and uncompromising situations, in P. A. French, T. E. Uehling and H. K. Wettstein (eds.), *Midwest Studies in Philosophy VI: The Foundations of Analytic Philosophy*, pp.387–404. University of Minnesota Press.

1981b. Situations and attitudes. *The Journal of Philosophy* 78: 668–691.

1983. *Situations and Attitudes*. Cambridge, MA: MIT Press.

Bastiaanse, Roelien and Edwards, Susan 2001. Word order and finiteness in Dutch and English Broca's and Wernicke's aphasia. *Brain and Language* 79: 72–74.

Bhatt, Rajesh 1999. *Covert Modality in Non-Finite Contexts*. PhD thesis, University of Pennsylvania.

2006. *Covert Modality in Non-Finite Contexts*. Berlin: Walter de Gruyter.

Bloom, Lois 1970. *Language Development: Form and Function in Emerging Grammars*. Cambridge, Mass.: MIT Press.

Bloom, Lois, Rispoli, Matthew, Gartner, Barbara and Hafitz, Jeremi 1989. Acquisition of complementation. *Journal of Child Language* 16: 101–120.

Bolinger, Dwight 1968. Postposed main phrases: An English rule for the romance subjunctive. *Canadian Journal of Linguistics* 14: 3–30.

Borthen, Kaja, Fretheim, Thorstein and Gundel, Jeanette K. 1997. What brings a higher-order entity into focus of attention? in R. Mitkov and B. Boguraev (eds.), *Operational Factors in Practical, Robust Anaphora Resolution*. Proceedings of a Workshop at the 35th Annual Meeting of the Association of Computational Linguistics. Madrid: Universidad Nacional de Educación a Distancia.

Brown, Roger 1973. *A First Language: The Early Stages*. Cambridge, Mass.: Harvard University Press.

Buckingham, Hugh W. Jr. 1981. Where do neologisms come from? in J. W. Brown (ed.), *Jargonaphasia*, pp.39–62. New York: Academic Press.

Buckingham, Hugh W. Jr., Avakian-Whitaker, Haiganoosh and Whitaker, Harry A. 1975. Linguistic structures in stereotyped aphasic speech. *Linguistics* 154/155: 5–13.

Buckingham, Hugh W. Jr. and Kertesz, Andrew 1976. *Neologistic Jargonaphasia*. Amsterdam: Swets and Zeitlinger.

Burge, Tyler 1974. Demonstrative constructions, reference and truth. *The Journal of Philosophy* 71: 205–223.

Cappelli, Gloria 2007. *"I reckon I know how Leonardo da Vinci must have felt ..." Epistemicity, Evidentiality and English Verbs of Cognitive Attitude*. Pari: Pari Publishing.

Caramazza, A. E. and Zurif, E. B. 1976. Dissociation of algorithmic and heuristic processes in language comprehension: Evidence from aphasia. *Brain and Language* 3: 572–582.

Cariani, Fabrizio 2011. *Ought* and resolution semantics. *Noûs* 47.3: 534–558.

Cariani, Fabrizio, Kaufmann, Magdalene and Kaufmann, Stefan 2013. Deliberative modality under epistemic uncertainty. *Linguistics and Philosophy* 36: 225–259.

Carlson, Greg 1998. Thematic roles and the individuation of events, in S. Rothstein (ed.), *Events and Grammar*, pp.35–51. Dordrecht: Kluwer Academic Publishers.

Cattell, Ray 1973. Negative transportation and tag questions. *Language* 49: 612–639.

Charlow, Nate 2013. What we know and what to do. *Synthese* 190: 2291–2323.

Chemla, E. 2009. Presuppositions of quantified sentences: Experimental data. *Natural Language Semantics* 17: 299–340.

Chien, Yu-Chin and Wexler, Kenneth 1990. Children's knowledge of locality conditions in binding as evidence for the modularity of syntax and pragmatics. *Language Acquisition* 1: 225–295.

Chierchia, Gennaro, Fox, Danny and Spector, Benjamin 2012. The grammatical view of scalar implicatures and the relationship between semantics and pragmatics, in C. Maienborn, K. von Heusinger and P. Portner (eds.), *Semantics: An International Handbook of Natural Language Meaning, vol. 3*, pp.2297–2332. Berlin: Mouton de Gruyter.

Choi, Soonja 1991. Early acquisition of epistemic meanings in Korean: A study of sentence-ending suffixes in the spontaneous speech of three children. *First Language* 11: 93–119.

Chomsky, Noam 1973. Conditions on transformations, in S. R. Anderson and P. Kiparsky (eds.), *A Festschrift for Morris Halle*. New York: Holt & Reinhardt.

Cinque, Guglielmo 1990. *Types of A-Bar Dependencies*. Cambridge, Mass.: MIT Press.

1999. *Adverbs and Functional Heads: A Cross-Linguistic Perspective*. Oxford University Press.

Coates, Jennifer 1983. *The Semantics of the Modal Auxiliaries*. London: Croom Helm.

Condoravdi, Cleo 2002. Temporal interpretation of modals: Modals for the present and for the past, in D. Beaver, S. Kaufmann, B. Clark and L. Casillas (eds.), *The Construction of Meaning*, pp.59–88. Stanford: CSLI.

Davidson, Donald 1980[1967]. The logical form of action sentences. In N. Rescher (ed.), *The Logic of Decision and Action*, pp.81–95. Pittsburgh: University of Pittsburgh Press. Reprinted in D. Davidson, *Essays on Actions and Events*, pp.105–122. Oxford University Press.

1984[1968]. On saying *that*, in D. Davidson (ed.), *Inquiries into Truth and Interpretation*, pp.93–108. Oxford University Press.

DeRose, Keith 1991. Epistemic possibilities. *The Philosophical Review* 100: 581–605.

Diessel, Holger 2004. *The Acquisition of Complex Sentences*. Cambridge University Press.

Dölling, Johannes, Heyde-Zybatow, Tatjana and Schäfer, Martin (eds.) 2008. *Event Structures in Linguistic Form and Interpretation*. Berlin: Walter de Gruyter.

Donnellan, Keith S. 1966. Reference and definite descriptions. *Philosophical Review* 77: 281–304.

Eckhardt, Regine 2005. Too poor to mention: Subminimal eventualities and negative polarity items, in Maienborn and Wöllstein (eds.), pp.301–330.

2008. The lower part of event ontology, in Dölling, Heyde-Zybatow, and Schäfer (eds.), pp.477–491.

Edwards, Susan 2005. *Fluent Aphasia*. Cambridge University Press.

Egan, Andy 2005. Epistemic modals, relativism, and assertion, in J. Gajewski, V. Hacquard, B. Nickel and S. Yalcin (eds.), *New Work on Modality, MIT Working Papers in Linguistics, vol. 51,* pp.35–62. Cambridge, Mass.: Department of Linguistics and Philosophy, MIT.

Egan, Andy, Hawthorne, John and Weatherson, Brian 2005. Epistemic modals in context, in G. Preyer and G. Peter (eds.), *Contextualism in Philosophy: Knowledge, Meaning and Truth*, pp.131–170. Oxford University Press.

Epstein, S. 1976. *Investigations in Pragmatic Theory*. PhD thesis, University of California, San Diego.

Faller, Martina 2002. *Semantics and Pragmatics of Evidentials in Cuzco Quechua*. PhD thesis, Stanford University.

Faroqi-Shah, Yasmeen and Thompson, Cynthia K. 2004. Semantic, lexical, and phonological influences on the production of verb inflections in agrammatic aphasia. *Brain and Language* 89: 484–498.

Fillmore, Charles. J. 1963. The position of embedding transformations in a grammar. *Word* 19: 208–231.

von Fintel, Kai 2010. The subjectivity of conditionals in a new light. Talk presented at DGfS Workshop on Subjective Meaning, Humboldt University, Berlin.

von Fintel, Kai and Gillies, Anthony 2005. "Might" made right. Handout for talk presented to the Philosophy Colloquium of the University of Texas, Austin.

2008a. An opinionated guide to epistemic modality, in T. Sz. Gendler and J. Hawthorne (eds.), *Oxford Studies in Epistemology 2*. 2007 draft http://mit.edu/fintel/fintel-gillies-ose2.pdf, accessed February 15, 2010.

2008b. CIA leaks. *Philosophical Review* 117: 77–98.

von Fintel, Kai and Iatridou, Sabine 2003. Epistemic containment. *Linguistic Inquiry* 34: 173–198.

2008. How to say *ought* in foreign: The composition of weak necessity modals, in J. Guéron and J. Lecarme (eds.), *Time and Modality*, pp.115–148. Berlin: Springer Science + Business Media B.V.

Fox, Danny and Katzir, Roni 2011. On the characterization of alternatives. *Natural Language Semantics* 19: 87–107.

Frege, Gottlob 1966[1892]. On sense and reference, in P. Geach and M. Black (eds.), *Translations from the Writing of Gottlob Frege*. Oxford University Press.

Friedmann, Naama 2006. Speech production in Broca's agrammatic aphasia: Syntactic tree pruning, in Y. Grodzinsky and K. Amunts (eds.), *Broca's Region*. Oxford University Press.

Gajewski, Jon. R. 2007. Neg-Raising and polarity. *Linguistics and Philosophy* 30: 289–328.

Gajewski, Jon. 2011. Licensing strong NPIs. *Natural Language Semantics* 19: 109–148.

Gettier, Edmund 1963. Is justified true belief knowledge? *Analysis* 23: 121–123.

Gleitman, Lila, Cassidy, R., Kimberly, Nappa, Rebecca and Papafragou, Anna 2005. *Language Learning and Development* 1: 23–64.

Goble, Lou 1996. Utilitarian deontic logic. *Philosophical Studies* 82.3: 317–357.

Grice, H. P. 1975. Logic and conversation, in P. Cole and J. L. Morgan (eds.), *Syntax and Semantics*, vol. 3, *Speech Acts*, pp.41–58. New York: Academic Press.

Grodzinsky, Yosef 1986. Language deficits and the theory of syntax. *Brain and Language* 27: 135–159.

1990. *Theoretical Perspectives on Language Deficits*. Cambridge, Mass.: MIT Press.

1995a. A restrictive theory of agrammatic comprehension. *Brain and Language* 50: 27–51.

1995b. Trace deletion, theta-roles, and cognitive strategies. *Brain and Language* 51: 469–497.

Grodzinsky, Yosef, Wexler, Kenneth, Chien, Yu-Chin, Marakovitz, Susan and Solomon, Julie 1993. The breakdown of binding relations. *Brain and Language* 45: 396–422.

Groenendijk, Jeroen and Stokhof, Martin 1991. Dynamic predicate logic. *Linguistics and Philosophy* 14: 39–100.

Groenendijk, Jeroen, Stokhof, Martin and Veltman, Fred 1996. Coreference and modality, in S. Lappin (ed.), *The Handbook of Contemporary Semantic Theory*, pp.179–213. Oxford: Blackwell Publishers.

Grosz, Barbara J., Joshi, Aravind K. and Weinstein, Scott 1983. Providing a unified account of definite noun phrases in discourse, in *Proceedings for the 21st Annual Meeting of the Association for Computational Linguistics*, pp.44–50.

Grosz, Barbara. J., Joshi, Aravind K. and Weinstein, Scott 1995. Towards a computational theory of discourse interpretation. *Computational Linguistics* 21: 203–225.

Grosz, Barbara and Sidner, Candace 1986. Attention, intention and the structure of discourse. *Computational Linguistics* 12: 175–204.

Gundel, Jeanette K., Borthen, Kaja and Fretheim, Thorstein 1999. The role of context in pronominal reference to higher order entities in English and Norwegian, in P. Bouquet et al. (eds.), *Modeling and Using Context: Lecture Notes in Artificial Intelligence*. Berlin: Springer Verlag.

Gundel, Jeanette K., Hedberg, Nancy and Zacharski, Ron 1993. Cognitive status and the form of referring expressions in discourse. *Language* 69: 274–307.

Gundel, Jeanette K., Hegarty, Michael and Borthen, Kaja 2003. Cognitive status, information structure, and pronominal reference to clausally introduced entities. *Journal of Logic, Language and Information* 12: 281–299.

Hacking, Ian 1967. Possibility. *The Philosophical Review* 76: 143–168.

Hacquard, Valentine 2006. *Aspects of Modality*. PhD thesis, MIT.

2009. On the interaction of aspect and modal auxiliaries. *Linguistics and Philosophy* 32: 279–315.

2010. On the event relativity of modal auxiliaries. *Natural Language Semantics* 18: 79–114.

Halpern, Joseph Y. 1997. Defining relative likelihood in partially-ordered preferential structures. *Journal of Artificial Intelligence Research* 7: 1–24.

2003. *Reasoning about Uncertainty*. Cambridge, Mass.: MIT Press.

Hawthorne, John 2004. *Knowledge and Lotteries*. Oxford University Press.

Hegarty, Michael 2003. Semantic types of abstract entities. *Lingua* 113: 891–927.

2005. *A Feature-Based Syntax of Functional Categories: The Structure, Acquisition and Specific Impairment of Functional Systems*. Berlin: Mouton de Gruyter.

2006a. Information update and convert modality in the semantics of propositional attitude verbs. *Proceedings of the Workshop on the Lexicon-Discourse Interface, 8th Conference on Natural Language Processing, Deutsche Gesellschaft für Sprachwissenschaft*. University of Konstanz. http://ling.unikonstanz.de/pages/conferences/konvens06/.

2006b. Type shifting of entities in discourse, in K. von Heusinger and K. Turner (eds.), *Where Semantics Meets Pragmatics: Current Research in the Semantics/Pragmatics Interface*, vol. 16, pp.111–128. Amsterdam: Elsevier.

2007. Context and semantic types in the interpretation of clausal arguments, in N. Hedberg and R. Zacharski (eds.), *The Grammar-Pragmatics Interface*, pp.171–188. Amsterdam: John Benjamins.
Hegarty, Michael, Gundel, Jeanette K. and Borthen, Kaja 2001. Information structure and the accessibility of clausally introduced referents. *Theoretical Linguistics* 27: 163–186.
Heim, Irene 1982. *The Semantics of Definite and Indefinite Noun Phrases in English.* PhD thesis, University of Massachusetts.
1992. Presupposition projection and the semantics of attitude verbs. *Journal of Semantics* 9: 183–221.
Heim, Irene and Kratzer, Angelika 1998. *Semantics in Generative Grammar.* Oxford: Blackwell Publishers.
Herburger, Elena 2000. *What Counts: Focus and Quantification.* Cambridge, Mass.: MIT Press.
Hickok, G. and Avrutin, S. 1995. Representation, referentiality, and processing in agrammatic comprehension: Two case studies. *Brain and Language* 50: 10–26.
Higginbotham, James 1983. The logic of perceptual reports: An extensional alternative to situation semantics. *The Journal of Philosophy* 80: 100–127.
1985. On semantics. *Linguistic Inquiry* 16: 547–593.
1989. Elucidations of meaning. *Linguistics and Philosophy* 12: 465–517.
1991. Belief and logical form. *Mind and Language* 6: 344–369.
2000. On events in linguistic semantics, in J. Higginbotham, F. Pianesi and A. C. Varzi (eds.), *Speaking of Events*, pp.49–79. Oxford University Press.
Higginbotham, James, Pianesi, Fabio and Varzi, Achille C. (eds.) 2000. *Speaking of Events.* Oxford University Press.
Hintikka, Jaakko 1962. *Knowledge and Belief.* Cornell University Press.
Hintikka, Jaakko. 1969. Semantics of propositional attitudes, in J. W. Davis et al. (eds.), *Philosophical Logic*, pp.21–45. Dordrecht: D. Reidel.
Hirst, W. and Weil, J. 1982. Acquisition of epistemic and deontic meaning of modals. *Journal of Child Language* 9: 659–666.
Hobbs, Jerry R. 1990. *Literature and Cognition. CSLI Lecture Notes 21.* Stanford: Center for the Study of Language and Information.
Hoekstra, Teun and Jordens, Peter 1994. From adjunct to head, in T. Hoekstra and B. D. Schwartz (eds.), *Language Acquisition and Studies in Generative Grammar*, pp.119–149. Amsterdam: John Benjamins.
Hooper, Joan B. 1975. On assertive predicates, in J. P. Kimball (ed.), *Syntax and Semantics*, vol. 4, pp.91–124. New York: Academic Press.
Horn, Laurence 1971. Negative transportation: Unsafe at any speed? in R. I. Binnick et al. (eds.), *Papers from the Fifth Regional Meeting of the Chicago Linguistics Society*, pp.120–133. University of Chicago Press.
Horn, Laurence R. 1978. Remarks on Neg-Raising, in P. Cole (ed.), *Pragmatics, Syntax and Semantics*, vol. 9, pp.129–220. New York: Academic Press.
1989. *A Natural History of Negation.* University of Chicago Press.
Iatridou, Sabine 1990. The past, the possible, and the evident. *Linguistic Inquiry* 21: 123–129.
Jackson, Frank 1985. On the semantics of logic and obligation. *Mind* 94: 177–195.

Jackson, Frank and Pargetter, Robert 1986. Oughts, options, and actualism. *The Philosophical Review* 95.2: 233–255.

Jackson, John Hughlings 1879. On affections of speech from disease of the brain, in J. Taylor (ed.), *Selected Writings of John Hughlings Jackson*, vol. 2, pp.129–137. London: Staples Press.

Jespersen, Otto 1917. *Negation in English and Other Languages*. Copenhagen: A. F. Høst.

Johannesson, Nils-Lennart 1976. *The English Modal Auxiliaries: A Stratificational Account*. Stockholm Studies in English 36. Stockholm: Almqvist & Wiksell International.

Kadmon, Nirit 2001. *Formal Pragmatics: Semantics, Pragmatics, Presupposition and Focus*. Oxford: Blackwell Publishers.

Kamp, Hans 1981. A theory of truth and semantic representation, in J. A. G. Groenendijk, T. M. V. Janssen and M. B. J. Stokhof (eds.), *Truth, Representation and Information*, pp.1–41. Dordrecht: Foris Publications.

Kamp, Hans and Reyle, Uwe 1993. *From Discourse to Logic: Introduction to Model-Theoretic Semantics of Natural Language, Formal Logic and Discourse Representation Theory*. Dordrecht: Kluwer Academic Publishers.

Katz, Graham 2000. Anti–neo-Davidsonianism, in C. Tenny and J. Pustejovsky (eds.), *Events as Grammatical Objects: The Converging Perspectives of Lexical Semantics and Syntax*, pp.393–414. Stanford: CSLI Publications.

2003. Event arguments, adverb selection, and the Stative Adverb Gap, in E. Lang, C. Maienborn and C. Fabricius-Hansen (eds.), *Modifying Adjuncts (Interface Explorations 4)*, pp.455–474. Dordrecht: Mouton de Gruyter.

2008. Manner modification of state verbs, in L. McNally and C. Kennedy (eds.), *Adjective and Adverbs: Syntax, Semantics, and Discourse*, pp.220–248. Oxford University Press.

Keenan, Edward L. and Faltz, Leonard M. 1985. *Boolean Semantics for Natural Language*. Dordrecht: D. Reidel.

Kehler, Andrew 2002. *Coherence, Reference and the Theory of Grammar*. Stanford: CSLI.

Kennedy, Chris 1999a. Gradable adjectives denote measure functions, not partial functions. *Studies in the Linguistic Sciences* 29: 65–80.

1999b. *Projecting the Adjective: The Syntax and Semantics of Gradability and Comparison*. New York: Garland.

2001. Polar opposition and the ontology of degrees. *Linguistics and Philosophy* 24: 33–70.

2007. Vagueness and grammar: The semantics of relative and absolute gradable adjectives. *Linguistics and Philosophy* 30: 1–45.

Kertesz, Andrew 1981. The anatomy of jargon, in J. W. Brown (ed.), *Jargonaphasia*, pp.63–112. New York: Academic Press.

Kiparsky, Paul and Kiparsky, Carol 1971. Fact, in D. Steinberg and L. Jakobovits (eds,), *Semantics: An Interdisciplinary Reader in Philosophy, Linguistics and Psychology*, pp.345–369. Cambridge University Press.

Kolodny, Nico and MacFarlane, John 2010. Ifs and ought. *The Journal of Philosophy* 107: 115–143.

Krantz, David H., Luce, R. Duncan, Suppes, Patrick and Tversky, Amos 1971. *Foundations of Measurement*. New York: Academic Press.

Kratzer, Angelika 1979. Conditional necessity and possibility, in R. Bäuerle, U. Egli and A. von Stechow (eds.), *Semantics from Different Points of View*, pp.117–147. Berlin: Springer.

1981. The notional category of modality, in H. J. Eikmeyer and H. Rieser (eds.), *Words, Worlds, and Contexts: New Approaches in Word Semantics*, pp.38–74. Berlin: Walter de Gruyter. Reprinted in P. Portner and B. H. Partee (eds.), *Formal Semantics: The Essential Readings*, pp.289–323. Oxford: Blackwell Publishing.

1991. Modality, in A. von Stechow and D. Wunderlich (eds.), *Semantik: Ein internationales Handbuch der zeitgenössischen Forschung [Semantics: An International Handbook of Contemporary Research*], pp.639–650. Berlin: Walter de Gruyter.

2002. Facts: Particulars or information units? *Linguistics and Philosophy* 25: 655–670.

2006. Decomposing attitude verbs. Handout/talk notes from lecture delivered at the Hebrew University of Jerusalem, July 4, 2006.

2012. *Modals and Conditionals*. Oxford University Press.

Lakoff, Geoge 1970. Pronominalization, negation, and the analysis of adverbs, in R. A. Jacobs and P. S. Rosenbaum (eds.), *Readings in English Transformational Grammar*, pp.145–165. Waltham, Mass.: Ginn and Company.

1974. Syntactic amalgams, in M. LaGaly, R. Fox and A. Bruck (eds.), *Papers from the Tenth Regional Meeting of the Chicago Linguistics Society*. University of Chicago Press.

Lakoff, Robin 1969. A syntactic argument for Negative Transportation, in R. I. Binnick et al. (eds.), *Papers from the Fifth Regional Meeting of the Chicago Linguistics Society*. University of Chicago Press.

Landman, Fred 2000. *Events and Plurality: The Jerusalem Lectures*. Dordrecht: Kluwer Academic Publishers.

Lassiter, Peter 2010. Gradable epistemic modals, probability and scale structure. *Proceedings of Semantics and Linguistic Theory* 20: 197–215.

2011. *Measurement and Modality: The Scalar Basis of Modal Semantics*. PhD thesis, New York University.

Levinson, Dmitry 2003. Probabilistic model-theoretic semantics for want. *Semantics and Linguistic Theory* (SALT) 13: 222–239.

Levinson, Stephen C. 1983. *Pragmatics*. Cambridge University Press.

Lewis, David 1986a. Events, in *Philosophical Papers,* vol. 2, pp.241–269. New York: Oxford University Press.

1986b. *On the Plurality of Worlds*. Oxford: Blackwell Publishers.

Limber, John 1973. The genesis of complex sentences, in T. E. Moore (ed.), *Cognitive Development and the Acquisition of Language*, pp.169–185. New York: Academic Press.

Lindholm, J. 1969. Negative raising and sentence pronominalization, in R. I. Binnick et al. (ed.), *Papers from the Fifth Regional Meeting of the Chicago Linguistic Society*. University of Chicago Press.

Lombard, Lawrence 1985. How not to trip the prowler: Transitive verbs of action and the identity of actions, in E. LePore and B. McLaughlin (eds.), *Actions and Events: Perspectives on the Philosophy of Donald Davidson*, pp.268–281. Oxford: Blackwell Publishers.

Longacre, Robert E. 1983. *The Grammar of Discourse*. Plenum Press.

Luria, A. R. 1970. *Traumatic Aphasia*. The Hague: Mouton.

Lyons, John 1977. *Semantics*. Cambridge University Press.

Magri, G. 2010. *A Theory of Individual-Level Predicates Based on Blind Mandatory Scalar Implicatures*. PhD thesis, MIT.

Maienborn, Claudia and Wöllstein, Angelika (eds.) 2005. *Event Arguments: Foundations and Applications*. Tübingen: Max Niemeyer Verlag.

MacFarlane, John 2003. Epistemic modalities and relative truth. http://johnmacfarlane.net/epistmod-2003.pdf, accessed February 15, 2010.

MacWhinney, Brian 1995. *The CHILDES Project: Tools for Analyzing Talk*. (Second Edition). Hillsdale, New Jersey: Lawrence-Erlbaum.

McCawley, James. D. 1988. *The Syntactic Phenomena of English*, vol. 2. University of Chicago Press.

McConnell-Ginet, Sally 1982. Adverbs and logical form. *Language* 58: 144–184.

Mittwoch, Anita 2005. Do states have Davidsonian arguments? Some empirical considerations, in C. M. A. Wöllstein (ed.), *Event Arguments: Foundations and Applications*, pp.69–88. Tübingen: Niemeyer.

Moltmann, Friederike 1997. *Parts and Wholes in Semantics*. New York: Oxford University Press.

2003a. Context, complex sentences, and propositional content. Ms. Stirling University.

2003b. Propositional attitudes without propositions. *Synthese* 135: 77–118.

Mulkern, Ann Elizabeth 2003. *Cognitive Status, Discourse Salience, and Information Structure: Evidence from Irish and Oromo*. PhD thesis, University of Minnesota.

Nespoulous, Jean-Luc, Code, Chris, Virbel, Jacques and Lecours, André Roch 1998. Hypotheses on the dissociation between “referential” and “modalizing” verbal behavior in aphasia. *Applied Psycholinguistics* 19: 311–331.

Noveck, I. A., Ho, S. and Sera, M. 1996. Children’s understanding of epistemic modals. *Journal of Child Language* 23: 621–643.

Nunberg, Geoffrey 1979. The non-uniqueness of semantic solutions: Polysemy. *Linguistics and Philosophy* 3: 143–184.

Nuyts, Jan 2001a. *Epistemic Modality, Language, and Conceptualization: A Cognitive-Pragmatic Perspective*. Amsterdam: John Benjamins Publishing Company.

2001b. Subjectivity as an evidential mention in epistemic modal expressions. *Journal of Pragmatics* 33: 383–400.

Palmer, F. R. 1979. *Modality and the English Modals*. London: Longman.

1986. *Mood and Modality*. Cambridge University Press.

Papafragou, Anna 2006. Epistemic modality and truth conditions. *Lingua* 116: 1688–1702.

Parsons, Terence 1990. *Events in the Semantics of English: A Study in Subatomic Semantics*. Cambridge, Mass.: MIT Press.

Percus, Orin 2000. Constraints on some other variables in syntax. *Natural Language Semantics* 8: 173–229.

Pesetsky, David 1987. Wh-in-situ: Movement and unselective binding, in A. ter Meulen and E. Reuland (eds.), *Representation of (in)Definiteness*. Cambridge, Mass.: MIT Press.

Picallo, M. Carme 1990. Modal verbs in Catalan. *Natural Language and Linguistic Theory* 8: 285–312.

Pietroski, Paul M. 2005. *Events and Semantic Architecture*. Oxford University Press.

Pinkal, Manfred 1981. Some semantic and pragmatic properties of German glauben, in H.-J. Eikmeyer and H. Rieser (eds.), *Words, Worlds, and Contexts: New Approaches in Word Semantics*, pp.469–484. Berlin: Walter de Gruyter.

Portner, Paul 1992. *Situation Theory and the Semantics of Propositional Expressions*. PhD thesis, University of Massachusetts.

2009. *Modality*. Oxford University Press.

Postal, Paul 1974. *On Raising: One Rule of English Grammar and Its Theoretical Implications*. Cambridge, Mass.: MIT Press.

Poutsma, H. 1928. *A Grammar of Late Modern English*. Groningen.

Prince, Ellen. F. 1976. The syntax and semantics of Neg-Raising, with evidence from French. *Language* 52: 404–426.

Quine, Willard van Orman 1960. *Word and Object*. Cambridge, Mass.: MIT Press.

Radford, Andrew 1990. *Syntactic Theory and the Acquisition of English Syntax: The Nature of Early Child Grammar of English*. Oxford University Press.

1996. Towards a structure-building model of acquisition, in H. Clahsen (ed.), *Generative Perspectives on Language Acquisition*, pp.43–89. Amsterdam: John Benjamins.

Reimer, Marga and Bezeuidenhout, Anne (eds.) 2004. *Descriptions and Beyond*. Oxford University Press.

Reinhart, Tanya 2000. Strategies of anaphora resolution, in H. Bennis, M. Everaert and E. Reuland (eds.), *Interface Strategies*, pp.295–324. Amsterdam: Royal Netherlands Academy of Arts and Sciences.

Reuland, Eric. J. 2001. Primitives of binding. *Linguistic Inquiry* 32: 439–492.

Roberts, Craige 1996. Information structure in discourse: Towards an integrated formal theory of pragmatics. http://semanticsarchive.net/Archive/WYzOTRkO/InfoStructure.pdf, accessed July 25, 2013.

Romero, Maribel and Han, Chung-Hye 2004. On negative “Yes/No” questions. *Linguistics and Philosophy* 27: 609–658.

Romoli, Jacopo 2013. A scalar implicature-based approach to Neg-Raising. *Linguistics and Philosophy* 36: 291–353.

Rooryck, Johan 2001. Evidentiality, parts I and II. *Glot International* 5: 4–5.

Ross, Alf 1944. Imperatives and logic. *Philosophy of Science* 11: 30–46.

Ross, John R. 1970. On declarative sentences, in R. A. Jacobs and P. S. Rosenbaum (eds.), *Readings in English Transformational Grammar*, pp.222–272. Waltham, Mass.: Ginn and Company.

1973. Slifting, in M. Gross, M. Halle and M.-P. Schützenberger (eds.), *The Formal Analysis of Natural Languages*, pp.133–169. The Hague: Mouton.

1974. Three batons for cognitive psychology, in W. B. Weimer and D. S. Palermo (eds.), *Cognition and the Symbolic Processes*, pp.63–124. Hillsdale, NJ: Lawrence Erlbaum.

Rothstein, Susan (ed.) 1998. *Events and Grammar*. Dordrecht: Kluwer Academic Publishers.

Roussarie, Laurent 2009. What might be known: Epistemic modality and uncertain contexts, in *Proceedings of SALT* 19.

Rubinstein, Aynat 2012. *Roots of Modality*. PhD thesis, University of Massachusetts.

Sachs, Jacqueline 1983. Talking about the there and then: The emergence of displaced reference in parent-child discourse, in K. E. Nelson (ed.), *Children's Language*, vol. 4. Hillsdale, NJ: Lawrence Erlbaum.

Saddy, Douglas 1995. Variables and events in the syntax of agrammatic speech. *Brain and Language* 50: 135–150.

Simons, Mandy 2007. Observations on embedding verbs, evidentiality, and presupposition. *Lingua* 117: 1034–1056.

Sperber, Dan and Wilson, Deidre 1986. *Relevance Theory*. Oxford: Blackwell Publishers.

Stephenson, Tamina 2005. Assessor sensitivity: Epistemic modals and predicates of personal taste, in J. Gajewski, V. Hacquard, B. Nickel and S. Yalcin (eds.), *New Work on Modality, MIT Working Papers in Linguistics,* vol. 51, pp.179–206. Cambridge, Mass.: Department of Linguistics and Philosophy, MIT.

Stephenson, Tamina C. 2007. *Towards a Theory of Subjective Meaning*. PhD thesis, MIT.

Stowell, Tim 2004. Tense and modals, in J. Guéron and J. Lecarme (eds.), *The Syntax of Time*, pp.621–635. Cambridge, Mass.: MIT Press.

Suppes, Patrick 1973. The semantics of children's language. *American Psychologist* 88: 103–114.

Swinney, D. 1979. Lexical access during sentence comprehension: (Re)consideration of context effects. *Journal of Verbal Learning and Verbal Behavior* 18: 645–659.

Swinney, D., Nicol, J. and Zurif, E. B. 1989. The effects of focal brain damage on sentence processing: An examination of the neurological organization of a mental module. *Journal of Cognitive Neuroscience* 1: 25–37.

Swinney, D. and Prather, P. 1989. On the comprehension of lexical ambiguity by young children: Investigations into the development of mental modularity, in D. S. Gorfein (ed.), *Resolving Semantic Ambiguity*. Berlin: Springer-Verlag.

Tancredi, Chris 2007a. A multi-model modal theory of I-semantics. Part I: Modals. Ms. Tokyo: University of Tokyo.

Tancredi, Chris. 2007b. A multi-model modal theory of I-semantics. Part II: Identity and attitudes. Ms. Tokyo: University of Tokyo.

Thomason, Richard and Stalnaker, Robert 1973. A semantic theory of adverbs. *Linguistic Inquiry* 4: 195–220.

Tovena, Lucia M. 2000. Neg-Raising: Negation as finite failure? in J. Hoeksema, H. Ruhlmann, V. Sanchez-Valencia and T. van der Wouden (eds.), *Perspectives on Negation and Polarity Items*, pp.331–356. Amsterdam: John Benjamins Publishing Company.

Urmson, J. O. 1952. Parenthetical verbs. *Mind* 61: 480–496.

Vasić, Nada, Avrutin, Sergey and Ruigendijk, Esther 2006. Interpretation of pronouns in VP-ellipsis constructions in Dutch Broca's and Wernicke's aphasia. *Brain and Language* 96: 191–206.

Villalta, Elisabeth 2008. Mood and gradability: An investigation of the subjunctive mood in Spanish. *Linguistics and Philosophy* 31: 467–522.

Walker, Marilyn, Joshi, Aravind and Prince, Ellen (eds.) 1998. *Centering Theory in Discourse*. Oxford University Press.

Webber, Bonnie. L. 1988. Discourse deixis and discourse processing. Technical report, University of Pennsylvania.

1991. Structure and ostension in the interpretation of discourse deixis. *Language and Cognitive Processes* 6: 107–135.

Wenzlaff, Michaela and Clahsen, Harald 2004. Tense and agreement in German agrammaticism. *Brain and Language* 89: 57–68.

Wimmer, H. and Perner, J. 1983. Beliefs about beliefs: Representation and constraining function of wrong beliefs in young children's understanding of deception. *Cognition* 13: 103–128.

Yakovlev, P. I. 1968. Telencephalon "impar," "semipar" and "totopar" (morphogenetic, tectogenetic and architectonic definitions). *International Journal of Neurology* 6: 245–265.

Yalcin, Seth 2005. Epistemic modals. In J. Gajewski, V. Hacquard, B. Nickel and S. Yalcin (eds.), *New Work on Modality, MIT Working Papers in Linguistics,* vol. 51, pp.231–272. Cambridge, Mass.: Department of Linguistics and Philosophy, MIT.

2007. Epistemic modals. *Mind* 116: 983–1026.

2010. Probability operators. *Philosophy Compass* 5: 916–937.

Zucchi, Alessandro 1989. The *Language of Propositions and Events*. PhD thesis, University of Massachusetts.

1993. *The Language of Propositions and Events: Issues in the Syntax and Semantics of Nominalization*. Berlin: Springer.

Zwarts, Frans 1998. Three types of polarity, in F. Hamm and E. Hinrichs (eds.), *Plural Quantification*, pp.177–238. Dordrecht: Kluwer.

# Author Index

# Subject Index

Lightning Source UK Ltd.
Milton Keynes UK
UKOW06n2018070216

267913UK00003B/98/P

9 781107 085763